To Dr Alison Kerr

With Compliments of Editors

Y. Nomura

Age-Related Dopamine-Dependent Disorders

Monographs in Neural Sciences

Vol. 14

Series Editor   *A.D. Korczyn,* Tel Aviv

KARGER

Basel · Freiburg · Paris · London · New York ·
New Delhi · Bangkok · Singapore · Tokyo · Sydney

International Symposium on Age-Related Monoamine-Dependent Disorders
and Their Modulation by Gene and Gender, Tokyo, November 12–13, 1993

Age-Related Dopamine-Dependent Disorders

Volume Editors *M. Segawa,* Tokyo
 Y. Nomura, Tokyo

72 figures and 19 tables, 1995

KARGER Basel · Freiburg · Paris · London · New York ·
New Delhi · Bangkok · Singapore · Tokyo · Sydney

Monographs in Neural Sciences

Library of Congress Cataloging-in-Publication Data
International Symposium on Age-Related Monoamine-Dependent Disorders and Their Modulation by Gene and Gender (1993: Tokyo, Japan)
Age-related dopamine-dependent disorders / International Symposium on Age-Related Monoamine-Dependent Disorders and Their Modulation by Gene and Gender;
volume editors, M. Segawa, Y. Nomura.
(Monographs in neural sciences; vol. 14)
Includes bibliographical references and index. (alk. paper)
1. Extrapyramidal disorders in children – Pathophysiology – Congresses.
2. Dopaminergic mechanisms – Congresses.   3. Basal ganglia – Diseases – Congresses.
4. Extrapyramidal disorders in children – Molecular aspects – Congresses.   5. Extrapyramidal disorders – Age factors – Congresses.   6. Extrapyramidal disorders – Sex factors – Congresses.
I. Segawa, Masaya.   II. Nomura, Y. (Yoshiko), 1940–.   III. Title.   IV. Series.
[DNLM: 1. Basal Ganglia Diseases – in infancy & childhood – congresses.
2. Dopa – therapeutic use – congresses.   3. Basal Ganglia Diseases – in adolescence – congresses.
4. Basal Ganglia – physiopathology – congresses.   5. Age Factors – Congresses.
6. Sex Factors – Congresses.
W1 MO56BC v. 14 1995 / WL 307 I6152a 1995]
RJ496.M68I585   1993
618.92′8–dc20
ISBN 3–8055–5960–7

Drug Dosage. The authors and the publisher have exerted every effort to ensure that drug selection and dosage set forth in this text are in accord with current recommendations and practice at the time of publication. However, in view of ongoing research, changes in government regulations, and the constant flow of information relating to drug therapy and drug reactions, the reader is urged to check the package insert for each drug for any change in indications and dosage and for added warnings and precautions. This is particularly important when the recommended agent is a new and/or infrequently employed drug.

All rights reserved. No part of this publication may be translated into other languages, reproduced or utilized in any form or by any means, electronic or mechanical, including photocopying, recording, microcopying, or by any information storage and retrieval system, without permission in writing from the publisher.

© Copyright 1995 by S. Karger AG, P.O. Box, CH–4009 Basel (Switzerland)
Printed in Switzerland on acid-free paper by Thür AG Offsetdruck, Pratteln
ISBN 3–8055–5960–7

Contents

Preface IX
Acknowledgement X

Introduction 1
   Segawa, M. (Tokyo)

Pathophysiology and Molecular Biology of Dopa-Related Disorders in Childhood and Adolescence

Hereditary Progressive Dystonia with Marked Diurnal Fluctuation and
   Dopa-Responsive Dystonia: Pathognomonic Clinical Features 10
   Segawa, M.; Nomura, Y. (Tokyo)
Juvenile Parkinsonism and Other Dopa-Responsive Syndromes 25
   Yokochi, M. (Tokyo)
Phenotypic Polymorphism of Dopa-Responsive Dystonia (in Russia) 36
   Markova, E.; Ivanova-Smolenskaya, I. (Moscow)
The Pathogenesis of Tourette's Syndrome: Role of Biogenic Amines and Sexually
   Dimorphic Systems Active in Early CNS Development 41
   Leckman, J.F. (New Haven, Conn.)
Pathophysiology of Muscle Tone and Involuntary Movement in Young-Onset Basal
   Ganglia Disorders 50
   Yanagisawa, N.; Hashimoto, T. (Matsumoto)
Voluntary Saccades in Normal and in Dopamine-Deficient Subjects 59
   Hikosaka, O. (Tokyo); Fukuda, H. (Tokyo/Kawasaki); Segawa, M.; Nomura, Y.
   (Tokyo)

Phasic Activity during REM Sleep in Movement Disorders 69
    Fukuda, H. (Tokyo/Kawasaki); Segawa, M.; Nomura, Y.; Nishihara, K. (Tokyo);
    Ono, Y. (Tokyo/Tsukuba)

PET Studies of the Dopaminergic and Opioid Function in Dopa-Responsive
    Dystonic Syndromes and in Tourette's Syndrome 77
    Turjanski, N.; Weeks, R.; Sawle, G.V.; Brooks, D.J. (London)

Fluorodopa PET Scans of Juvenile Parkinsonism with Prominent Dystonia in
    Relation to Dopa-Responsive Dystonia 87
    Takahashi, H. (Vancouver/Yokohama); Snow, B. (Vancouver);
    Nygaard, T. (New York, N.Y.); Yokochi, M. (Tokyo); Calne, D. (Vancouver)

Striatal Dopamine D_2 Receptors in Dopa-Responsive Dystonia and Parkinson's
    Disease 95
    Leenders, K.L.; Antonini, A. (Villigen); Meinck, H.-M. (Heidelberg);
    Weindl, A. (Munich)

Striatal Dopamine in Dopa-Responsive Dystonia: Comparison with Idiopathic
    Parkinson's Disease and Other Dopamine-Dependent Disorders 101
    Hornykiewicz, O. (Vienna)

Dopa-Responsive Dystonia: Clinical, Pathological, and Genetic Distinction from
    Juvenile Parkinsonism 109
    Nygaard, T.G. (New York, N.Y.)

The Gene for Hereditary Progressive Dystonia with Marked Diurnal Fluctuation
    Maps to Chromosome 14q 120
    Endo, K.; Tanaka, H.; Saito, M.; Tsuji, S. (Niigata); Nygaard, T.G.
    (New York, N.Y.); Weeks, D.E. (Pittsburgh, Penn.); Nomura, Y.; Segawa, M.
    (Tokyo)

Early-Onset, Generalized Dystonia Caused by DYT1 Gene on Chromosome 9q34 .. 126
    Ozelius, L.J. (Boston, Mass.); Bressman, S.B. (New York, N.Y.);
    Kramer, P.L. (Portland, Oreg.); Risch, N. (New Haven, Conn.);
    de Leon, D.; Fahn, S. (New York, N.Y.); Breakefield, X.O. (Boston, Mass.)

Neuronal Circuits and Compartments of the Basal Ganglia, and Their Clinical Manifestations

Some Aspects of Basal Ganglia-Thalamocortical Circuitry and Descending Outputs
    of the Basal Ganglia 134
    Nakano, K.; Kayahara, T.; Ushiro, H.; Hasegawa, Y. (Tsu)

Dopaminergic Innervation of Primate Cerebral Cortex. An Immunohistochemical
    Study in the Japanese Macaque 147
    Maeda, T.; Ikemoto, K.; Satoh, K. (Otsu); Kitahama, K. (Lyon);
    Geffard, M. (Bordeaux)

Interactions between Glutamate and Dopamine in the Ventral Striatum: Evidence for
    a Dual Glutamatergic Function with Respect to Motor Control 160
    Svensson, A.; Carlsson, M.L.; Carlsson, A. (Göteborg)

Locomotion and Posture: Modulation by Upper Neurons 168
    Mori, S.; Homma, Y. (Okazaki); Nakajima, K. (Asahikawa)
Function of the Indirect Pathway in the Basal Ganglia Oculomotor System:
    Visuo-Oculomotor Activities of External Pallidum Neurons 178
    Kato, M.; Hikosaka, O. (Okazaki)
Role of the Nigrostriatal Dopamine System in Corticostriatal Signal Transmission in
    Alert Animals 188
    Kimura, M.; Aosaki, T. (Osaka)
The Role of the Descending Pallido-Reticular Pathway in Movement Disorders . . . 197
    Shima, F.; Sakata, S.; Sun, S-J; Kato, M.; Fukui, M.; Iacono, R.P.
    (Loma Linda, Calif.)
Experiences from Human Stereotaxic Surgery 208
    Narabayashi, H. (Tokyo)

Monoamine Neurons: Gene and Gender Differentiation

Catecholaminergic Systems and the Sexual Differentiation of the Brain.
    Cellular Mechanisms and Clinical Implications 216
    Reisert, I.; Pilgrim, C. (Ulm)
Embryonic Striatal Grafting: Progress and Future Directions for Therapeutic
    Approaches to Neurodegenerative Diseases of the Basal Ganglia 225
    Liu, F.-C. (Cambridge, Mass.); Dunnett, S.B. (Cambridge);
    Graybiel, A.M. (Cambridge, Mass.)
Basic Fibroblast Growth Factor in the Substantia nigra in Parkinson's Disease and
    Normal Aging 235
    McGeer, P.L. (Vancouver); Tooyama, I.; Kimura, H. (Otsu);
    McGeer, E.G. (Vancouver)

Closing Remarks 245
    N. Yanagisawa (Matsumoto)

Subject Index 249

Preface

The age- and gender-related occurrence of particular symptoms is a characteristic feature of basal ganglia disorders, and the clarification of the basic pathophysiology of this feature has been one of the main scientific interests in the field of movement disorders.

This volume contains the proceedings of the symposium on 'Age-Related Monoamine-Dependent Disorders and Their Modulation by Gene and Gender' held in Tokyo on November 12–13, 1993 in commemoration of the 20th anniversary of the Segawa Neurological Clinic for Children. Recent knowledge on the pathophysiologies and molecular biologies of dopa-related disorders in childhood and adolescence, neuronal circuits and compartments of the basal ganglia and monoamine neurons and their gender differences was discussed to provide a better understanding of the clinical and basic neurosciences on the roles of age and gender in the clinical features of basal ganglia disorders.

We would like to express our sincere thanks to the distinguished physicians and researchers for their participation in this symposium and for their valuable contributions which enabled the monographs to be completed. We also wish to thank our colleagues and the medical and co-medical staffs of our clinic for their endeavor and cooperation.

Masaya Segawa
Yoshiko Nomura

Acknowledgement

The 20th anniversary symposium of the Segawa Neurological Clinic for Children, held in Tokyo, November 12–13, 1993, was sponsored by the Foundation for Advancement of International Science. We also gratefully acknowledge the strong support provided by the following companies:

Shohei Real Estate Co. Ltd.
Eli Lilly Japan K.K.
Kyowa Hakko Kogyo Co. Ltd.
Mitsui Construction Co. Ltd.
Ciba-Geigy (Japan) Ltd.

Sandoz Pharmaceuticals Ltd.
Kyorin Yakuhin Co., Ltd.
NEC San-ei Instruments Ltd.
Suntory Limited

Introduction

Masaya Segawa

Segawa Neurological Clinic for Children, Tokyo, Japan

The clinical symptoms of basal ganglia disorders are known to have age-related characteristics. In both idiopathic and symptomatic disorders, dystonia tends to be the predominant feature in cases with onset at younger ages, while Parkinson's disease (PD) or parkinsonism occurs in those with onset at older ages. Furthermore, in inherited basal ganglia diseases with lesions in the nigrostriatal (NS) dopamine (DA) neuron, the ages of onset are correlated to gender difference depending on the mode of transmission of the DA terminal.

The NS-DA neuron has a marked age variation in its activity in the terminal during the first two decades while it is not observed in the perikaryon [1]. It is known that each component of the basal ganglia has its own particular developmental course [2]. During development, the DA neuron is considered to exhibit morphological and functional differences between sexes, depending on a genetic rather than an epigenetic hormone-dependent process [3].

This evidence suggests that a certain lesion in the NS-DA system shows a particular clinical feature depending on age and gender or alternatively, a certain component of the basal ganglia, impaired by a certain lesion in the NS-DA neuron, develops a particular symptom or symptoms at a certain age period with a specific gender preference. This also implies that those processes are coded by a particular gene.

Thus, it is important to elucidate the basic pathophysiologies of age- and gender-dependent basal ganglia diseases, and notably to clarify the specific component of the basal ganglia involved in the pathognomonic symptoms of each disease.

The symposium on 'Age-Related Monoamine-Dependent Disorders and Their Modulation by Gene amd Gender' aimed to investigate the basic pathophysiologies of basal ganglia disorders occurring in the first four decades, by evaluating the components of the basal ganglia involved, and the mode of lesion of the NS-DA neuron that is responsible for the pathognomonic features of each disorder.

Among disorders with lesions in the NS-DA neuron, with onset in the first four decades, dopa-responsive diseases were selected. Hereditary progressive dystonia with marked diurnal fluctuation (HPD), or strictly defined dopa-responsive dystonia (DRD), starts in the first decade with a female predominance. Although the main symptom is dystonia, dystonic juvenile parkinsonism (JP), a familial JP, occurs in the second decade and has a marked male predominance. On the other hand, in JP with onset age around the late second or the third to fourth decade, the former with features of the dystonia-parkinsonism complex (DYS-P) has a female predominance, while the latter with features of classical PD or young-onset Parkinson disease (YOPD) has no sex preference.

Besides these, Gilles de la Tourette syndrome (GTS) and idiopathic torsion dystonia are also discussed. GTS, a disorder of exaggerated DA transmission, develops symptoms mainly in the first decade with a marked male predominance. Idiopathic torsion dystonia, with lesions in the components of the basal ganglia, has onset before the age of 15 years, but shows no gender preference.

Studies on the molecular biology of these disorders have already shown the linkage of the early-onset form of torsion dystonia to chromosome 9q34 [4, 5]. Recently, strictly defined DRD was shown by Nygaard et al. [6] to be linked to the long arm of chromosome 14 and the same linkage was observed in HPD [7, this volume]. Since Filipino X-linked dystonia with parkinsonian features was shown to have the locus on Xq12 [8, 9], at least three different genes have been revealed to be responsible for hereditary dystonia with different clinical features [10, this volume]. Although no gene locus or linkage has been detected in dystonic JP and DYS-P, they have a high familial occurrence, suggesting an autosomal recessive inheritance [11, 12], whereas GTS is believed to have an autosomal dominant inheritance [13, this volume]. This evidence suggests that the age and gender dependence of these diseases is modulated by a particular gene. On the other hand, JP with features of classical PD (YOPD) has a significantly low incidence of familial occurrence [11]. So processes other than inheritance might be involved in its etiology.

The characteristic and pathognomonic clinical features of HPD, familial JP and GTS were reevaluated and their pathogeneses were discussed by assessing the results of neurophysiological examinations and neuroimaging studies.

By reviewing the clinical studies of HPD, Segawa and Nomura [14, this volume] implicated the pathophysiology of HPD or strictly defined DRD as follows. HPD is a functional disorder of the NS-DA neuron, mainly affecting its terminal. Among the components of the basal ganglia, the direct projection of the striatum is involved rather selectively, with sparing of its indirect projection. Receptor supersensitivity in the terminal of the NS-DA neuron may not develop. Although postural tremor, a symptom related to the pallidofugal thalamic pathways, develops after functional maturation of the thalamus or the thalamo-cortical pathways, the main symptom of HPD is dystonia throughout the course of the illness.

By neurohistochemical and neuro-immuno-histochemical studies on an autopsy case of DRD, Hornykiewics [15, this volume] showed reduction in DA levels below the lower limits of the control range in the pars compacta of the SN as well as in the caudate and putamen. He also showed similarities in the interregional caudate putamen pattern and subregional rostrocaudal pattern of DA loss between DRD and idiopathic PD. However, in contrast to idiopathic PD, the case with DRD exhibited a greater DA loss in the ventral than in the dorsal subdivision of the rostral caudate. Although tyrosine hydroxylase (TH) activity was decreased in the striatum, it was within the normal range in the SN, thus showing a feature different from PD.

From clinical and neuropathological findings Yokochi [16, this volume] showed the existence of at least three types of JP: the first with immature pigmented cells and a few Lewy bodies in the SN, the second with degenerative changes in the SN but without Lewy bodies and the third with neuropathologies identical to PD. The pathology of the first type was that of a particular female case with dystonic JP with onset at 6 years, in whom histochemical studies had shown a reduction in DA and TH only in the terminal of the NS-DA neuron [17]. However, the fluorodopa PET scan of the brother of this case with similar clinical features had shown levels below normal limits, different from DRD and PD [18, this volume]. The second type seems to be identical to recessive DYS-P with diurnal fluctuation reported by Yamamura et al. [12]. The third type is considered as a typical pathology of YOPD and the neurohistochemistry of a case with this type had shown a decrease in DA and TH in both the perikaryon and the terminal of the NS-DA neuron as in PD [17].

GTS and etiologically related forms of obsessive compulsive disorders are thought to be associated with failure to inhibit subsets of the multiple parallel cortico-striato-thalamocortical circuit, and the dopaminergic projection from the mesencephalic center that modulates its activity has received attention in this connection [13, this volume].

To understand the involuntary movements in basal ganglia disorders, it is important to note the characteristics of underlying muscle tone. Yanagisawa and Hashimoto [19, this volume] reviewed their comprehensive EMG studies on various forms of basal ganglia disease and suggested that the high-frequency tremor of HPD was a characteristic feature frequently encountered in young-onset cases. By analyzing voluntary saccades, Hikosaka et al. [20, this volume] showed differences in the mode of involvement among HPD, PD and JP, suggesting a difference in the mode of modification of the cortical input into the basal ganglia in these disorders. Polysomnographic findings revealed a decrease in the activities of the NS-DA neuron in both HPD and GTS, but those in the latter differed from the former with features suggesting the existence of receptor supersensitivity in DA receptors [21, this volume]. This difference between HPD and GTS may be ascribed to genetic and gender-related factors. In all of these clinical neurophysiological studies, the age variations of the features of signs and components are also stressed.

Neuroimaging studies with PET scan revealed that the D_2 receptors were not affected in DRD [22, this volume] and GTS [23, this volume], while in JP they were affected with upward regulation.

The neural pathways composed of a particular connection to each component of the basal ganglia and the differences of descending and ascending output pathways of SNr and GPM were shown by Nakano et al. [24, this volume]. The existence of a strong cortical projection of the DA neuron to the cingulate gyrus was clarified by Maeda et al. [25, this volume]. They showed that most of the areas of the cortex which in GTS demonstrate decreased or increased regional activity in PET scan [26] are innervated by DA neuron activity, either directly or indirectly, via the pallido-nigro-thalamic pathways.

The roles of each neural connection or component of the basal ganglia were discussed based on the evidence of basic and clinical medicine.

Experimental studies with MPTP parkinsonian monkeys have revealed the involvement of the striatal direct projection in symptoms of dystonia, while the striatal indirect projection was found to play important roles in parkinsonism and L-dopa-induced dyskinesia [27]. In the latter disorder, development of receptor supersensitivity was also detected. Targets of stereotaxic neurosurgery have shown the involvement of the pallidofugal descending pathway in postural dystonia [28, this volume] and the involvement of the pallido-thalamic pathway with the ventral lateral nucleus of the thalamus in rigidity and L-dopa induced dyskinesia [29, this volume]. But for the generation of parkinsonian tremor, the pathways via the ventral intermediate nucleus of the thalamus are involved [29, this volume]. Narabayashi stressed the roles of two output pathways from the globus pallidus, one ascending to the thalamic nuclei and the other descending to the pedunculopontine nuclei (PPN), in the generation of the spectrum of parkinsonian symptoms. This evidence was well confirmed by two deoxyglucose studies on MPTP monkeys revealing suppression of PPN as well as the ventral lateral nucleus of the thalamus in parkinsonism [27]. However, most experimental studies do not clarify the role of the pallidofugal pathways in locomotion, though the SN-PPN-reticulospinal tract is known to integrate posture and locomotor-related signals [30, this volume].

Hikosaka et al. [20, this volume] examined the components of voluntary saccades in normal subjects and revealed that some exhibited age variation while others did not, suggesting that in voluntary saccades the corticostriatal afferent innervation is mediated at least by two neuronal systems in the basal ganglia. In monkeys, Kato and Hikosaka [31, this volume] showed that the striatal indirect projection is also involved in voluntary saccades via the subthalamo-nigral (pars reticulata) pathways. However, in contrast to the striatal direct pathway, the cortical input to the indirect pathway suppresses the contralateral voluntary saccades. In MPTP monkeys, Kimura and Aosaki [32, this volume] suggested that corticostriatal transformation of neural information are under the strong control of the NS-DA system.

Two experimental studies revealed a difference in the modulation between D_1 and D_2 receptors. Svensson et al. [33, this volume] showed the involvement of D_2 receptors in rotatory movements, and with transplantation studies, Liu et al. [34, this volume] demonstrated a connection of the fetal nigral tissue to D_2 receptors of the host striatum but not to D_1 receptors.

Sex-specific development of the brain is thought to depend entirely on the action of the gonadal steroids during certain periods of development. Reisert et al. [35, this volume] reviewed the substantial results of her group suggesting a genetic program for the gender-specific development of the catecholamine neuron.

An age-depedent variation of the DA neuron could exist throughout life even in the process of loss of pigmented cells in the SN. If the pathophysiology of PD is due to early occurrence of this aging process, there may be a certain gender difference in the incidence of PD as observed in young-onset NS-DA disorders. McGeer et al. [36, this volume] showed a significant decrease in the basic fibroblast growth factor positive cells in the SN of PD compared to normal controls, suggesting a more active pathological process in the pathophysiology of PD, which may leave no gender preference.

This evidence obtained in clinical and basic medicine suggests that the age-dependent occurrence of familial or symptomatic DA-mediated disorders is the reflection of an age-related variation of the NS-DA neuron observed in a particular region of the neuron. In these disorders, symptoms involving the direct projection of the striatum develop early in the first decade, while those involving the indirect projection appear later in the second and third decades. It has also been observed that symptoms involving the pallidofugal descending pathways develop at younger ages, while those involving the indirect projection or the pallidofugal thalamic pathways develop at older ages. The latter might be due to the functional maturation of the thalamus.

The gender preference observed in familial or inherited dopa-responsive disorders might be modulated by the gender-specific genetic program of the catecholamine neuron. This might appear as a gender difference in inherited or familial

DA-mediated disorders occurring in the first four decades of life. Absence of gender preference in idiopathic torsion dystonia suggests there is no gender-specific susceptibility in the components of the basal ganglia other than NS-DA neuron, though they show age-related susceptibility.

All of these processes are thought to be coded to a certain gene. However, some JP with features of classical PD, with onset in the third and fourth decades show no gender differences or familial occurrence. In this disease a particular active pathological process might take place in addition to the age-dependent decremental process of the pigmented nuclei occurring earlier than normal.

Soon after the symposium, Ichinose and Nagatsu determined the gene locus of GTP cyclohydrolase I within the region containing the HPD/DRD locus. The mutational analysis of genomic DNA of HPD patients revealed mutations in the GTP cyclohydrolase I gene. The activities of this enzyme in the monocuclear blood cells from HPD patients were less than 20% of those in normal individuals, while those of asymptomatic carriers were within 30–40%. However, mutant enzymes expressed in *E. coli* had no detectable GTP cyclohydrolase I activity. These results show that the HPD gene is located in the gene of 14q22.1-q22.2, that is, GTP cyclohydrolase I gene [37].

This is the first discovery of the causative gene for inherited dystonia and may lead to a breakthrough in clarifying the biochemical components of the basal ganglia and their disorders. Furthermore, this GTP cyclohydrolase I in normal subjects has higher activity in males than females, which might play a role in the gender difference in the dopamine neuron.

References

1. McGeer EG, McGeer PL: Some characteristics of brain tyrosine hydroxylase; in Mandel J (ed): New Concepts in Neurotransmitter Regulation. New York, PLenum Press, 1973, pp 53–68.
2. van der Kooy D, Fishell G, Krushel LA: The development of striatal compartments: From proliferation to patches; in Carpenter MB, Jayaraman A (eds): The Basal Ganglia II – Structure and Function – Current Concepts. New York, Plenum Press, 1987, pp 81–98.
3. Reisert I, Pilgrim C: Sexual differentiation of monoaminergic neurons – Genetic epigenetic. Trends Neurosci 1991;14:468–473.
4. Ozelius L, Kramer PL, Moskowitz CB, Kwiatkowski DJ, Brin MF, Bessman SB, Schuback DE, Falk CT, Risch N, de Leon D, Burke RE, Haines J, Gusella JF, Fahn S, Breakefield XO: Human gene for torsion dystonia located on chromosome 9q32q34. Neuron 1989;2:1427–1434.
5. Kramer PL, de Leon D, Ozelius L, Risch M, Brin MF, Bressman SB, Burke RE, Kwiatkowski DJ, Schuback DE, Shale H, Gusella JF, Breakefield XO, Fahn S: Dystonia gene in Ashkenazi Jewish population located on chromosome 9q32-34. Ann Neurol 1990;27:114–120.
6. Nygaard TG, Wilhelmsen KC, Risch NJ, Brown DL, Trugman JM, Gilliam TC, Fahn S, Weeks DE: Linkage mapping of dopa-responsive dystonia (DRD) to chromosome 14q. Nature Genet 1993;5:386–391.

7   Endo K, Tanaka H, Saito M, Tsuji S, Nygaard TG, Weeks DE, Nomura Y, Segawa M: The gene for hereditary progressive dystonia with marked diurnal fluctuation maps to chromosome 14q. Monogr Neural Sci. Basel, Karger, 1995, vol 14, pp 120–125.
8   Kupke K, Graeber M, Muller U: Dystonia-parkinsonism syndrome (XPD) locus: Flanking markers in Xq12-q21.1. Am J Hum Genet 1992;50:808–815.
9   Wilhelmsen KC, Weeks DE, Nygaard TG, Moskowitz CB, Rosales RL, de la Paz DC, Sobrevega EE, Fahn S, Gilliam C: Genetic mapping of 'lubag' (X-linked dystonia-parkinsonism) in a Filipino kindred to the pericentromeric region of the X chromosome. Ann Neural 1991;29:124–131.
10 Ozelius LJ, Bressman SB, Kramer PL, Risch N, de Leon D, Fahn S, Breakefield XO: Early-onset, generalized dystonia caused by DYT1 gene on chromosome 9q34. Monogr Neural Sci. Basel, Karger, 1995, vol 14, pp 126–132.
11 Yokochi M: Juvenile Parkinson's disease. I. Clinical aspects. Adv Neurol Sci (Tokyo) 1979;23:1060–1073.
12 Yamamura Y, Hamaguchi Y, Uchida M, Fujioka H, Watanabe S: Parkinsonism of early onset with diurnal fluctuation; in Segawa M (ed): Hereditary Progressive Dystonia with Marked Diurnal Fluctuation. Carnforth, Parthenon Publishing, 1993, pp 51–59.
13 Leckman JF: THe pathogenesis of Tourette's syndrome: Role of biogenic amines and sexually dimorphic systems active in early CNS development. Monogr Neural Sci. Basel, Karger, 1995, vol 14, pp 41–49.
14 Segawa M, Nomura Y: Hereditary progressive dystonia with marked diurnal fluctuation and dopa-responsive dystonia: Pathognomonic clinical features. Monogr Neural Sci. Basel, Karger, 1995, vol 14, pp 10–24.
15 Hornykiewicz O: Striatal dopamine in dopa-responsive dystonia: Comparison with idiopathic Parkinson's disease and other dopamine-dependent disorders. Monogr Neural Sci. Basel, Karger, 1995, vol 14, pp 101–108.
16 Yokochi M: Juvenile parkinsonism and other dopa-responsive syndromes. Monogr Neural Sci. Basel, Karger, 1995, vol 14, pp 25–35.
17 Yokochi M, Narabayashi H, Iizuka R, Nagatsu T: Juvenile parkinsonism – Some clinical pharmacological and neuropathological aspects. Adv Neurol 1984;40:407–413.
18 Takahashi H, Snow B, Nygaard T, Yokochi M, Calne D: Fluorodopa PET scans of juvenile parkinsonism with prominent dystonia in relation to dopa-responsive dystonia. Monogr Neural Sci. Basel, Karger, 1995, vol 14, pp 87–94.
19 Yanagisawa N, Hashimoto T: Pathophysiology of muscle tone and involuntary movement in young-onset basal ganglia disorders. Monogr Neural Sci. Basel, Karger, 1995, vol 14, pp 50–58.
20 Hikosaka O, Fukuda H, Segawa M, Nomura Y: Voluntary saccades in normal and dopamine-deficient subjects. Monogr Neural Sci. Basel, Karger, 1995, vol 14, pp 59–68.
21 Fukuda H, Segawa M, Nomura Y, Nishihara K, Ono Y: Phasic activity during REM sleep in movement disorders. Monogr Neural Sci. Basel, Karger, 1995, vol 14, pp 69–76.
22 Leenders KL, Antonini A, Meinck HM, Weindl A: Striatal dopamine D_2 receptors in dopa-responsive dystonia and Parkinson's disease. Monogr Neural Sci. Basel, Karger, 1995, vol 14, pp 95–100.
23 Turjanski N, Weeks R, Sawle GV, Brooks DJ: Positron emission tomography studies of the dopaminergic and opioid function in dopa-responsive dystonic syndromes and in Tourette's syndrome. Monogr Neural Sci. Basel, Karger, 1995, vol 14, pp 77–86.
24 Nakano K, Kayahara T, Ushiro H, Hasegawa Y: Some aspects of basal ganglion-thalamo-cortical circuitary and descending outputs of the basal ganglia. Monogr Neural Sci. Basel, Karger, 1995, vol 14, pp 134–146.
25 Maeda T, Ikemoto K, Satch K, Kitahama K, Geffard M: Dopaminergic innervation of primate cerebral cortex. An immunohistochemical study in the Japane macaque. Monogr Neural Sci. Basel, Karger, 1995, vol 14, pp 147–159.
26 Braun AR, Stoetter B, Randolph C, Hsiao JK, Vlakdar K, Germert J, Carson RE, Herscowitch P, Chase TN: The functional neuroanatomy of Tourette's syndrome: An FDG-PET study. I. Regional changes in cerebral glucose metabolism differentiating patients and controls. Neuropsychopharmacology 1993;9:277–291.

27 Mitchell IJ, Luquin R, Boyce S, Clarke CE, Robertson RG, Sambrook MA, Crossman AR: Neural mechanisms of dystonia: Evidence from a 2-deoxyglucose uptake study in a primate model of dopamine agonist-induced dystonia. Mov Disord 1990;5:49–54.

28 Shima F, Sakata S, Sun SJ, Kato M, Fukui M, Iacono RP: The role of the descending pallidoreticular pathway in movement disorders. Basel, Karger, 1995. Monogr Neural Sci, vol 14, pp 197–207.

29 Narabayashi H: Experiences from human stereotaxic surgery. Monogr Neural Sci. Basel, Karger, 1995, vol 14, pp 208–214.

30 Mori S, Homma Y, Nakajima K: Locomotion and posture: Modulation by upper neurons. Monogr Neural Sci. Basel, Karger, 1995, vol 14, pp 168–177.

31 Kato M, Hikosaka O: Function of the indirect pathway in the basal ganglion oculomotor system. Visuo-oculomotor activities of external pallidum neurons. Monogr Neural Sci. Basel, Karger, 1995, vol 14, pp 178–187.

32 Kimura M, Aosaki T: Role of the nigrostriatal dopamine system in corticostriatal signal transmission in alert animals. Monogr Neural Sci. Basel, Karger, 1995, vol 14, pp 188–196.

33 Svensson A, Carlsson ML, Carlsson A: Interactions between glutamate and dopamine in the ventral striatum. Evidence for a dual glutamatergic function with respect to motor control. Monogr Neural Sci. Basel, Karger, 1995, vol 14, pp 160–167.

34 Liu FC, Dunnett SB, Graybiel AM: Embryonic striatal grafting. Progress and future directions for therapeutic approaches to neurodegenerative diseases of the basal ganglia. Monogr Neural Sci. Basel, Karger, 1995, vol 14, pp 225–234.

35 Reisert I, Pilgrim C: Catecholaminergic systems and the sexual differentiation of the brain. Cellular mechanisms and clinical implications. Monogr Neural Sci. Basel, Karger, 1995, vol 14, pp 216–224.

36 McGeer PL, Tooyama I, Kimura H, McGeer EG: Basic fibroblast growth factor in the substantia nigra in Parkinson's disease and normal aging. Monogr Neural Sci. Basel, Karger, 1995, vol 14, pp 235–243.

37 Ichinose H, Ohye T, Takahashi E, Seki N, Hori T, Segawa M, Nomura Y, Endo K, Tanaka H, Tsuji S, Fujita K, Nagatsu T.: Hereditary progressive dystonia with marked diurnal fluctuation caused by mutations in the GTP cyclohydrolase I gene. Nat Genet 1994;8:236–242.

Massaya Segawa, MD, Segawa Neurological Clinic for Children, 2-8 Surugadai Kanda, Chiyoda-ku, Tokyo 101 (Japan)

Pathophysiology and Molecular Biology of Dopa-Related Disorders in Childhood and Adolescence

Hereditary Progressive Dystonia with Marked Diurnal Fluctuation and Dopa-Responsive Dystonia: Pathognomonic Clinical Features

Masaya Segawa, Yoshiko Nomura

Segawa Neurological Clinic for Children, Tokyo, Japan

Hereditary progressive dystonia with marked diurnal fluctuation (HPD), initially reported by one of the authors (MS) in 1971 [1] and 1976 [2], is a postural dystonia with marked diurnal fluctuation; the symptoms worsen toward the evening and alleviate in the morning after sleep. Originally, the symptomatic characteristics of HPD were drawn from 9 cases including 6 familial cases observed in 3 families.

On the other hand, dopa-responsive dystonia (DRD) is dystonia responding to L-dopa to various degrees. This term was first proposed by Nygaard et al. [3] in 1988. However, it is inevitable that a disease defined with pharmacological characteristics includes heterogeneous disorders with different pathophysiologies or etiologies.

Reappraisal of reported cases with DRD has led to a revision of the clinical criteria of DRD whose definition is now quite similar to HPD [4]. Although diurnal fluctuation of symptoms is not considered as an essential feature in DRD [4], HPD and strictly defined DRD are thought to be one and the same disorder, as shown by linkage studies reported elsewhere in this monograph.

On the other hand, some of the cases reported as HPD with onset in the first decade had different features from our criteria of HPD [5] and others later developed the features of juvenile parkinsonism (JP) [6] described by Yokochi [7]. This evidence suggests the existence of a form of DRD different from HPD or strictly defined DRD and JP with onset in childhood.

Table 1. Clinical characteristics of HPD

1	Onset; in the 1st decade with foot dystonia, a few in adulthood with tremor
2	Marked diurnal fluctuation which, however, attenuates with age
3	Apparent progression in the first two decades, attenuates with age and becomes inapparent after the 4th decade
4	The main symptom; postural dystonia with lower extremity predominance. Postural retrocollis may occur. No axial torsion, action dystonia or oculogyric crisis
5	Tremor; postural tremor with frequency of 8–10 Hz, occurring at later ages; no parkinsonian resting tremor of 4–5 Hz
6	Bradykinesia appears later; locomotor activity is preserved until late
7	DTRs exaggerated in all, some with ankle clonus or striatal toe; no Babinski sign
8	Marked and sustained response to *L*-dopa without any unfavorable side effects; the doses of *L*-dopa can be reduced later, though feeling of ineffectiveness of *L*-dopa develops around the early teens
9	There is a side preference for the left
10	Female predominance
11	Shorter body height
12	Autosomal dominant inheritance with low penetrance is suspected

As to the pathophysiology of HPD, a functional abnormality in the terminal of the nigrostriatal dopamine (NS-DA) neuron has been suggested [5, 8, 9].

We review the clinical characteristics of HPD focusing on the pathognomonic features and discuss the possible pathophysiology of each symptom for demarcating HPD or strictly defined DRD from other disorders.

Clinical Characteristics of HPD

The clinical characteristics of HPD are shown in table 1. The initially reported clinical features were modified by further evaluation of 22 personal cases with long-term follow-up [10–12] and of intra- and interfamilial variations among reported Japanese cases [6]. Although there are cases with onset in adulthood or old age, most HPD patients develop their symptoms in the first decade and there is a predominance of females. Thus, to clarify the pathophysiology of this disorder, it is necessary to evaluate each neurological feature with reference to the age- and gender-dependent abnormalities of the nigrostriatal system.

Pathognomonic Clincal and Laboratory Findings and Their Pathophysiologies

(1) Clinical course, marked diurnal fluctuation, and complete and sustained response to L-dopa of HPD suggest that the lesion is restricted to the terminal of the NS-DA neuron.

Complete and sustained response to *L*-dopa and age-related clinical course suggest that the lesion of HPD is restricted to the NS-DA neuron, and that its pathophysiology is functional and related to the age variation of the neuron [5, 8].

Decremental age variation in tyrosine hydroxylase (TH) activity is restricted to the terminal of the NS-DA neuron [13]; it is marked in the first two decades but attenuates with age to attain a steady level after the middle of the 3rd decade. The clinical course of HPD reflects this age variation well if the TH level in the terminal is lower, but follows the normal variation [5, 8–10]. This lesion also explains the diurnal fluctuation [5, 8, 9] as the activity of the NS-DA neuron exhibits a circadian fluctuation in the terminal [14] while no state-dependent variation is present in its perikaryon [15]. So the lesion of HPD is thought to be a decrease in TH activity limited to the terminal of the NS-DA neuron [5, 8, 9].

Polysomnography supports this hypothesis insofar as it revealed abnormalities only in the phasic components of sleep, that is, gross movements (GM), twitch movements (TM) and rapid eye movements (REM), without abnormalities in the tonic components modulated by other brainstem neurons [2, 16, 17]. This shows the selective involvement of the NS-DA neuron and/or the basal ganglia in HPD.

The number of TM in REM stage normally shows age and nocturnal variations; e.g., decrases with age and increases toward the morning. In HPD both of these variations were also observed but the number of TMs was significantly lower than the normal value and reduced below 20% of the level of normal controls in late childhood and in the first REM stage, but increased above the level in the late REM stage [16] (fig. 1, 2). As the number of TMs in REM stage reflects the activities of the NS-DA neuron [17, 18], these features are considered to reflect age [13] and nocturnal [14] variations of the DA content in the terminal of the NS-DA neuron. Thus, the age and nocturnal variations of TMs in HPD correlated well with onset in the first decade and diurnal fluctuation of symptoms observed in this disorder. This verifies the above speculation. Furthermore, the attenuation of diurnal fluctuation or morning recovery with age is explained as an age-related decrease below the critical level in the morning TMs or DA activity (fig. 2).

Hornykiewicz [19, this monograph] showed the decrease in TH content in the terminal but not in the perikaryon in a case with DRD; the DA content decreased both in the terminal and the perikaryon. These features suggest that the

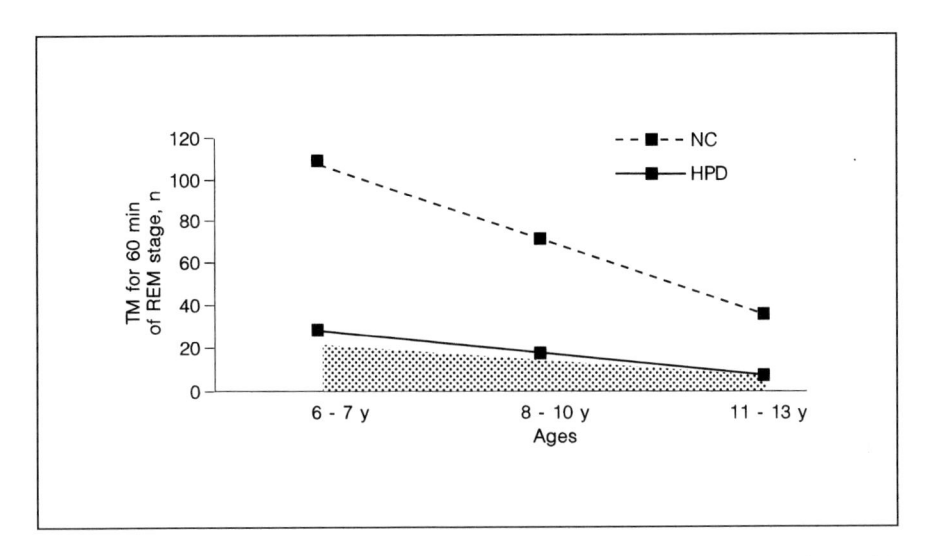

Fig. 1. Age variation of the number of TMs in REM stage observed in patients with HPD and age-matched normal children. NC = Normal children. The top of the hatched area indicates the 20% level of the number of TMs of normal children.

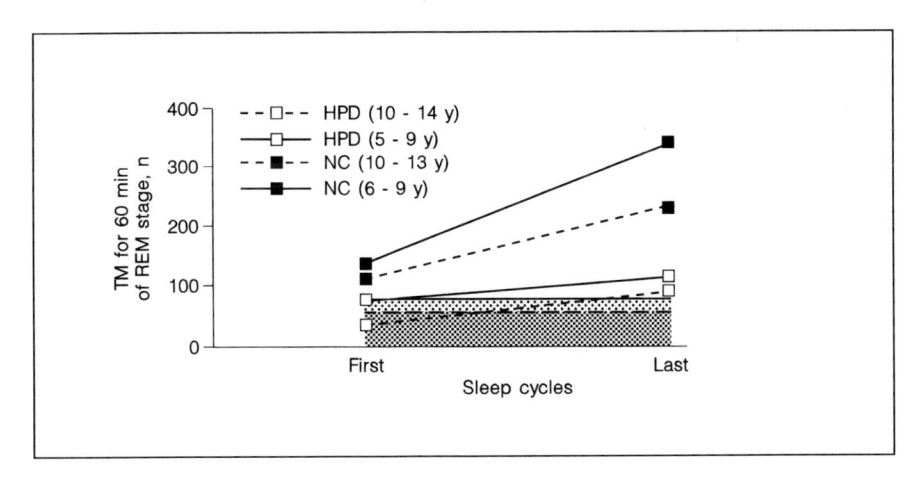

Fig. 2. Nocturnal variation in the number of TMs shown by the difference in values in REM stage between the first and the last sleep cycles. The values were compared between patients of HPD and normal children in two age groups. NC = Normal children; first = the first sleep cycle; last = last sleep cycle. The solid line above the hatched area indicates the 20% level of the number of TMs of the last sleep cycle in normal children of the younger age group (6–9 years). The dotted line above the gray area is that observed in normal children of the older age group (10–13 years).

fundamental function of the NS-DA neuron is preserved in HPD although the TH and DA contents are pathologically low at the terminal. Thus onset in the first decade and diurnal fluctuation in HPD are considered as pathognomonic features.

(2) Absence of torsion implies that the lesion of HPD is localized in the NS-DA neuron and not in the striatum or pallidum.

The pathognomonic lesion of torsion dystonia is considered to be in the striatum or pallidum. This is shown by neuroimaging studies on symptomatic torsion dystonia [20]. Preservation of these basal ganglia nuclei in HPD has been demonstrated by polysomnographic studies [17].

In cases with HPD with left-side predominance, the horizontal REM tend toward the left [18, 21]. This is because the predominantly affected right NS-DA neuron disinhibits the striatal GABAergic neuron, and finally the saccade neuron in the right superior colliculus, resulting in an increase in horizontal REM toward the left. After *L*-dopa, these derangements improve, and horizontal REM show a normal preference for the right.

An identical feature was observed in a case with left hemi-Parkinson's disease (PD) with lesion in the right substantia nigra pars compacta [18, 21] (fig. 3). However, opposite features were observed in a case with left hemidystonia with lesion in the right striatum, who showed a relative increase of horizontal REM toward the nonaffected limbs [18, 21] (fig. 3.). This was due to the lesion in the striatal GABAergic neuron which disinhibited the GABAergic neuron of the right pars reticulata of the substantia nigra and reduced the horizontal REM toward the left by suppressing the saccade neuron of the right superior colliculus.

So the side preference of horizontal REM toward the dominantly affected limb implies that the lesion of HPD is in the NS-DA neuron and not in the striatum or the pallidum. This also explains the absence of torsion dystonia in HPD [21].

(3) Dystonia is a symptom involving the direct projection of the striatum.

Experiments on MPTP parkinsonian monkeys revealed that the direct projection of the striatum was involved in the pathophysiology of the peak dose dystonia [22]. In HPD, the involvement of this projection has been shown by polysomnographic studies.

Although abnormalities of the phasic components of sleep improved completely after *L*-dopa [2, 17, 21], the mode of improvement differed among components, that is, once improved, abnormalities in GM did not reappear, while TMs decreased in number when the patients were in the period of 'feeling of ineffectiveness' in the course of treatment [11].

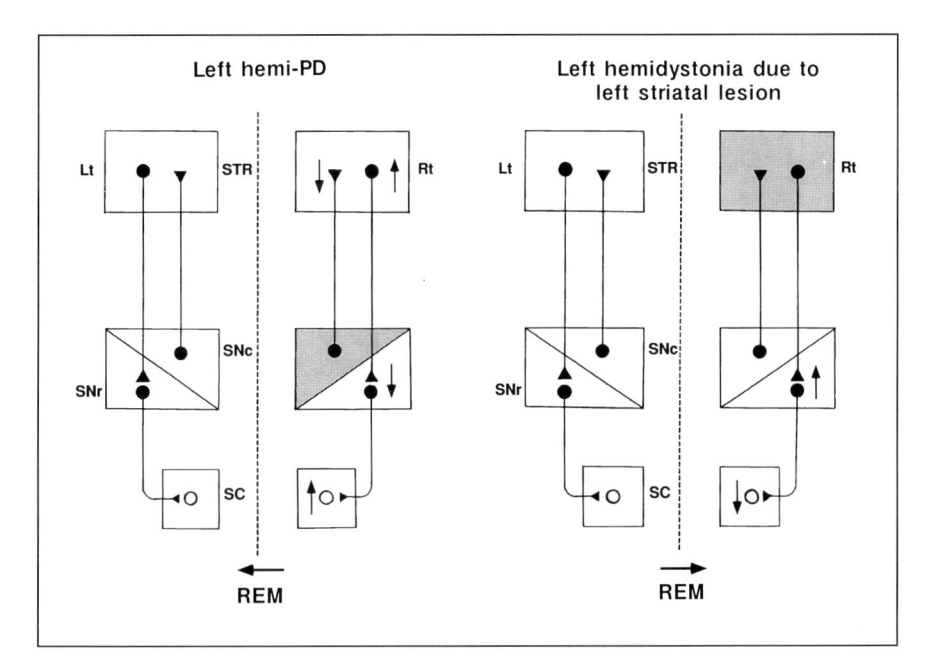

Fig. 3. Schematic diagram of the side preference of the horizontal REMs observed in left hemi-PD and left hemidystonia with axial torsion with right striatal lesion. STR = Striatum; SNc = Substantia nigra pars compacta; SRr = substantia nigra pars reticulata; SC = superior colliculus; the gray area indicates the lesion in the basal ganglia. Horizontal arrows indicate the direction of REMs. Vertical arrows indicate neuronal activity; upward: activation; down: inhibition or suppression; closed circles and triangles indicate inhibitory neurons and then the terminals; open circle shows excitatory neurons. Note: During REM stage, NS-DA neuron to the striatal direct projection has an inhibitory effect [18].

The NS-DA neuron modulates the sleep-stage-dependent occurrence of GMs via the thalamic projection from the medial segment of the globus pallidus [17] and the numbers of TMs and REMs as well as the direction of horizontal REMs via the striatal projection to the substantia nigra [17, 18].

Thus in HPD, hypofunction of the NS-DA neuron becomes clinically apparent by mediating the direct projection of the striatum to the substantia nigra.

(4) Abnormalities in voluntary saccades in HPD indicate hypofunction of the NS-DA neuron and involvement of the striatal direct projection to the substantia nigra.

Voluntary saccades are mediated by the striato-nigro-collicular pathway which is controlled by the NS-DA neuron [23]. Based on the investigation of

MPTP monkeys, the mode of abnormalities in voluntary saccades varied depending on whether MPTP was injected into the caudate nucleus or the putamen; in the former, the memory-guided saccades were affected [24, 25], while in the latter the visually guided saccades became abnormal [25].

Abnormalities of voluntary saccades were observed in HPD and PD and were improved by *L*-dopa. However, the mode of abnormalities differed between them, that is, in HPD both memory- and visually guided saccades were affected, while in PD the memory-guided saccades were mainly affected [25, 26].

Indirect projection of the striatum also involves voluntary saccades [27, this monograph]. However, there is substantial evidence suggesting the preservation of the D_2 receptor in DRD [28, this monograph].

So in HPD, hypofunction of the DA neuron induces abnormality in voluntary saccades via the striato-nigro-collicular pathways. But in contrast to PD, in HPD the abnormality in the NS-DA neuron is mediated by two striatal direct projections to the substantia nigra, one from the caudate nucleus, and the other from the putamen.

(5) The absence of action dystonia implies the absence of receptor supersensitivity and preservation of the pallidofugal thalamic pathway in HPD.

Polysomnography in cases with action dystonia revealed features different from HPD in the sleep-stage-dependent modulation of GM where the rate of occurrence of GMs was markedly decreased in REM stage and did not improve after *L*-dopa [17, 21]. Cases with DRD with action dystonia and/or oculogyric crises had the same features in polysomnographic recordings [17, 21].

This abnormality in the pattern of GM implies the existence of receptor supersensitivity in the DA terminal of the striatum and involvement of the pallidofugal thalamic pathway in the pathophysiology of action dystonia [17, 21]. The latter is supported by evidence on the effect of a stereotaxic surgery. That is, the nucleus ventralis oralis posterior of the thalamus receiving afferents from the medial segment of the globus pallidus is a target for action dystonia [29], but not for postural dystonia [29, 30, this monograph].

Thus, the absence of action dystonia implies the absence of receptor supersensitivity and preservation of the pallidofugal thalamic pathway in HPD.

(6) Absence of L-dopa-induced dyskinesia also implies the absence of receptor supersensitivity and preservation of the indirect pathway of the striatum in HPD.

From studies on MPTP monkeys, the indirect pathway of the striatum has been shown to have important roles in the pathophysiology of *L*-dopa-induced dyskinesia [31]. The marked suppression of the indirect GABAergic pathway by the supersensitized striatal DA receptors finally causes disinhibition of the ventral lateral nucleus of the thalamus and dyskinesia. Involvement of the ventral lateral nucleus in the pathophysiology of dyskinesia of PD is also supported clinically by the effect of ventral lateral thalamotomy [32].

This evidence implies that receptor supersensitivity does not occur in HPD and that the indirect pathway is not involved in the pathophysiology of HPD.

(7) There is substantial evidence for the preservation or minimal involvement of D_2 receptors in HPD.

Although anticholinergic drugs showed sustained and moderate effects [33], bromocriptine, the D_2 receptor agonist showed only minimum effects on clinical and polysomnographic findings of HPD [34]. Leenders et al. [28, in this monograph] estimated striatal dopamine D_2 receptor binding in DRD using [^{11}C]raclopride with positron emission tomography and revealed normal dopamine D_2 receptor binding in DRD compared to age-matched healthy controls, both for putamen and caudate nucleus.

(8) Preservation of locomotor movements in HPD suggests preservation of the pallidofugal pathway to the pedunculopontine nucleus.

The basal ganglia control locomotion via the pallidal projection to the pedunculopontine nucleus (PPN) [35]. MPTP parkinsonian monkeys revealed bradykinesia without interlimb coordination; however, in *L*-dopa induced dyskinesia and peak-dose dystonia, these animals could move with quadripedal locomotion [36]. Investigation with 2-deoxyglucose revelaed hot spot in the PPN and ventral lateral nuclei of the thalamus only in parkinsonism [36], demonstrating suppression of PPN by the pallidal projection. Such uptake was not observed in PPN in dyskinesia or peak-dose dystonia [22, 36]. In parkinsonism, hot spot is also observed in the lateral segment of the globus pallidus. This consequently disinhibits the excitatory neuron of the subthalamic nucleus and finally facilitates the inhibitory output of the medial segment of the globus pallidus. These implicates that in HPD, where locomotive movement is normal, PPN is integrally preserved. It is suggested as well that preservation of the indirect projection also relates to the preservation of locomotion.

(9) Later occurrence of 'parkinsonism' and late-onset cases in HPD families do not make HPD patients candidates for Parkinson's disease. Tremor in HPD appears with age after the functional maturation of the thalamus.

Pathophysiological similarities between HPD and PD have been discussed, mainly because untreated HPD later demonstrates tremor and bradykinesia, which are often called parkinsonism.

In HPD, tremor usually develops after the second decade, in most cases after 30 years of age [5, 6, 8, 10]; furthermore, in contrast to PD, it is postural with a higher cycle of 8–10 Hz and responds well to *L*-dopa [5, 8]. However, tremor does not develop in childhood cases even in the evening, though postural dystonia workens markedly in the evening [10, 16].

In some families with HPD there are cases with onset at adult age, even in the fifties and sixties [6]. They start with tremor and gait disturbance and are often diagnosed as PD. However, the tremor in these cases is postural and their clinical features are milder with minimal progression than their index cases. In these patients, all symptoms, including tremor, respond markedly to smaller doses of L-dopa and the effects are sustained without any side effects or increase in dosage.

Snow et al. [37] examined 6 [^{18}F]dopa PET scans which revealed a normal uptake rate in DRD and HPD, whereas the rate was significantly lowered in PD and young onset PD (YOPD). Snow [38] and Takahashi et al. [39, this monograph] found a normal value in a Japanese adult-onset case in a HPD family. Another Japanese late-onset case also showed a low normal level [40].

These results revealed that HPD or strictly defined DRD is different from PD, JP or YOPD, is not a candidate for PD [38], and can occur in adulthood as a late-onset HPD [6].

Tremor in older HPD patients might be modulated by the pallido- or nigrofugal thalamic pathway which appears with age after the functional maturation of the thalamus.

Clinical Identity of HPD: Differentiation from Other Dopa-Responsive Disorders

(1) DRD with action dystonia and/or oculogyric crises has a pathophysiology different not only from HPD but also from PD.

HPD with preservation of both the striatum and pallidum differs from torsion dystonia, and, with the absence of receptor supersensitivity and preservation of the pallidofugal thalamic pathway, it is different from dystonia with action dystonia and with oculogyric crises [6, 17, 21]. Among cases reported as DRD, those who do not show diurnal fluctuation in childhood may have a disorder different from HPD [5, 6, 8].

Cases with action dystonia and/or oculogyric crises are also different from JP, not only as far as the clinical features are concerned, but also the voluntary saccades of these cases show abnormality both in memory- and visually guided saccades [41]. Involvement of both voluntary saccades is observed in cases of dystonia of various etiologies, but in PD only the memory-guided saccades are affected [26]. These observations suggest that the main pathophysiology of DRD with action dystonia and/or oculogyric crises is dystonia, disorders belonging to DRD but not to HPD.

(2) Differences in age of onset and gender preference among HPD and various types of JP have pathognomonic importance.

HPD can be differentiated from JP by its onset in the first decade, predominance of females, higher frequency of familial occurrence, marked and sustained response to *L*-dopa without any need to increase its dosage and without occurrence of dyskinesia [5, 8, 42]. Furthermore, the clinical course with a stationary period at a later age is also one of the characteristics of HPD in contrast to JP [10, 16].

When classifying JP into dystonic and tremor types [5, 8], the ages of onset of the dystonic type are plotted in the later half period, when the TH activity in the terminal of the NS-DA neuron shows a marked decremental change, and JP of the tremor type in the period when the variation in TH activity subsides [5, 8, 11, 16]. The ages of onset of the familial dystonia-parkinsonism complex (DYS-P) [43] and those of YOPD [44] are similar to those of the tremor type [45]. While dystonia parkinsonism of early onset with diurnal fluctuation and autosomal recessive inheritance, reported by Yamamura et al. [43], starts in the late second decade to early third decade. This implies that the main lesion of dystonic type JP is in the terminal of the NS-DA neuron as in HPD, while the lesions in the tremor type JP, DYS-P and YOPD are in the perikaryon as in classical PD. The cases of Yamamura et al. [43] might have both lesions. The speculation on the pathophysiologies of dystonic and tremor-type JP has been verified by histochemical studies on two JP patients, one dystonic and the other tremor type [45]. Differences in the pathology of the substantia nigra among dystonic-type JP, DSY-P and tremor-type JP are shown by Yokochi [46, this monograph].

Early occurrene of *L*-dopa-induced dyskinesia and the on-off phenomenon in JP suggest the existence of receptor supersensitivity at the striatal DA receptors and involvement of the indirect pathway of the striatum. Leenders et al. [28, this monograph] showed an increase in the D_2 receptor binding in the putamen in PD patients, including cases with onset in the early 5th decade. This suggests upward regulation or receptor supersensitivity in PD of early-onset.

By estimating striatal [^{18}F]-dopa uptake in patients with adult-onset DYS-P, Turjanski et al. [47, this monograph] showed severely reduced putamen [^{18}F]dopa uptake, comparable to that in PD, but the reduction was much greater than in DRD patients. They suggest that DYS-P is a phenotype of PD and not of DRD.

In contrast to HPD, dystonic-type JP shows marked male predominance. Among JP with onset in the second to fourth decade, recessive DYS-P with diurnal fluctuation [43] shows female predominance, while tremor-type JP has no sex preference [5, 8].

The marked difference in gender predominance between HPD and dystonic-type JP might relate to the age-dependent difference in the modulation of the DA system between females and males [5, 8]. This is shown by the reversed sex preference observed in disorders with exaggerated DA transmission occurring in each age period, that is, male predominance in Gilles de la Tourette syndrome (GTS) and female predominance in Sydenham chorea, respectively.

However, considering the apparent familial occurrence in HPD, dystonic type JP and DYS-P as well as GTS, the sex preference in these disorders might be due to a genetically determined sex difference of the DA neuron [48].

Tremor-type JP has a neuropathology similar to PD. So in this disorder a certain degenerative process might be involved which is independent of gender preference as in idiopathic PD [49, this monograph].

These characteristic ages of onset might be due to differences in the components or the internuclear connection of the basal ganglia specifically involved in each disorder.

HPD Dominant Inheritance with Low Penetrance

HPD has been thought to have dominant inheritance with low penetrance [2, 5, 8]. This is supported by studies on Japanese familial cases reported as HPD revealing that all cases with transgeneration occurrence showed typical features of HPD [6]. This was confirmed in the linkage studies on strictly defined DRD [50] and on HPD [51, this monograph]. On the other hand, among those with sibling occurrence without transgeneration involvement, there were cases who later showed features of JP, though they started with features of HPD in the first decade [6]. Furthermore, some of their siblings developed symptoms later, in the second or third decade, with characteristics of JP [6].

Summary and Conclusion

HPD or strictly defined DRD is a functional disorder of the NS-DA neuron. Pathophysiologically, a decrease in TH and DA levels in the terminals connecting to the direct projections of the striatum or the D_1 receptors are suspected. However, at the terminal, normal age and nocturnal variations of TH activities are preserved, without any morphological changes. Among striatal direct projections, those projecting to the substantia nigra might be predominantly affected. On the other hand, in HPD the NS-DA neuron connected to the D_2 receptors, that is, to the indirect projection, is preserved. The involvement of the direct projection in peak-dose dystonia is also shown by MPTP monkeys [22] and preservation of the D_2 receptors in DRD [28, this monograph].

In the first decade, following the marked physiological decremental age variation of the TH activities, but with low levels, DA activity in HPD decreases below the critical levels. This manifests clinically as postural dystonia, mediated by the descending output projection of the basal ganglia via the direct projection, which is already mature in the first decade. This symptom shows diurnal fluctuation reflecting physiological circadian variation of DA activity in the terminal in which DA activity shows decremental variation during the day time [14]. The clinical progression is marked in the first two decades, but it subsides later with attenuation of the physiological age variation of the terminal TH activities. Diurnal fluctuation also becomes inapparent. On the other hand, with the functional maturation of the thalamus with age, symptoms may develop as postural tremor after the second or third decade via the thalamic projection.

As there are no abnormalities in the basal ganglia except in the terminal of the NS-DA neuron, torsion dystonia does not appear. Moreover, with the preservation of the striatal indirect projection, which functionally matures later, HPD does not turn out to be PD even in later life. With the preservation of the descending projection of the medial segment of the globus pallidus to PPN which is regulated by the indirect pathway, locomotion is preserved up to an advanced stage.

Without receptor supersensitivity, action dystonia or oculogyric crises do not appear in HPD. Without receptor supersensitivity and involvement of the indirect striatal projection, unfavorable side effects of *L*-dopa such as dyskinesia or the on-off phenomenon are not observed.

With preservation of the fundamental function of the neuron and without structural abnormalities, *L*-dopa shows marked and sustained effects without need to increase the dosage.

As for the pathogenesis, reduction of TH activity in the terminal, inherited by an autosomal dominant trait with low penetrance, is most probable. The decrease in TH activity in the terminal is supported by findings made in postmortem studies [19, this monograph; 52] where TH was decreased only at the terminal. However, the genetic study on TH was negative [53]. The linkage analyses have mapped the strictly defined DRD [50] and HPD gene to the long arm of chromosome 14 [51, this monograph]. Gender-dependent morphologic and functional characteristics of the DA neuron determined by the genetic process [48] might play a role in the clinical manifestation of female predominance.

For further speculation regarding the pathogenesis of the decrease in TH activity, abnormal pteridin metabolism is suggested. Furukawa et al. [54] examined the levels of biopterin and neopterin in CSF in HPD and various types of JP and showed a decrease in both substances only in HPD, while biopterin was decreased in JP and DYS-P. This evidence suggests abnormalities in GTP cyclohydrolase I in HPD. Further studies, particularly in molecular genetics, are necessary in relation to TH and factors involved in its synthesis such as biopterin and also on the concept of D_1 receptor disorder.

References

1. Segawa M, Ohmi K, Itoh S, Aoyama M, Hayakawa H: Childhood basal ganglia disease with remarkable response to L-dopa, 'hereditary basal ganglia disease with marked diurnal fluctuation.' Chiryo (Tokyo) 1971;24:667–672.
2. Segawa M, Hosaka A, Miyagawa F, Nomura Y, Imai H: Hereditary progressive dystonia with marked diurnal fluctuation. Adv Neurol 1976;14:215–233.
3. Nygaard TG, Marsden CD, Duvoisin RC: Dopa responsive dystonia. Adv Neurol 1988;50:377–384.
4. Nygaard TG, Snow BJ, Fahn S, Calne DB: Dopa-responsive dystonia: Clinical characteristics and definition; in Segawa M (ed): Hereditary Progressive Dystonia with Marked Diurnal Fluctuation. Carnforth, Parthenon, 1993, pp 21–35.
5. Segawa M, Nomura Y, Kase M: Diurnally fluctuating hereditary progressive dystonia; in Vinken PJ, Bruyn GW (eds): Handbook of Clinical Neurology: Extrapyramidal disorders. Amsterdam, Elsevier, 1986, vol 5, pp 529–539.
6. Nomura Y, Segawa M: Intrafamilial and interfamilial variations of symptoms of Japanese hereditary progressive dystonia with marked diurnal fluctuation; in Segawa M (ed): Hereditary Progressive Dystonia with Marked Diurnal Fluctuation. Carnforth, Parthenon, 1993, pp 73–96.
7. Yokochi M: Juvenile Parkinson's disease. I. Clinical aspects. Adv Neurol Sci (Tokyo) 1979;23:1060–1073.
8. Segawa M: Hereditary progressive dystonia with marked diurnal fluctuation (HPD). Adv Neurol Sci (Tokyo) 1981;25:73–81.
9. Segawa M: Catecholamine metabolism in neurological diseases in childhood; in Wise G, Glaw ME, Procopis PG (eds): Topics in Child Neurology. New York, Spectrum Publications, 1982, vol 2, pp 135–150.
10. Segawa M, Nomura Y: Hereditary progressive dystonia with marked diurnal fluctuation; in Segawa M (ed): Hereditary Progressive Dystonia with Marked Diurnal Fluctuation. Carnforth, Parthenon, 1993, pp 3–19.
11. Segawa M, Nomura Y, Yamashita S, Kase M, Nishiyama N, Yukishita S, Ohta H, Nagata K, Hosaka A: Long-term effects of L-dopa on hereditary progressive dystonia with marked diurnal fluctuation; in Berardelli, A, Benecke R, Manfredi M, Marsden CD (eds): Motor Disturbance, Part II. London, Academic Press, 1990, pp 305–318.
12. Segawa M, Nomura Y: Pathophysiology of human locomotion: Studies on pathological cases; in Shimamura M, Grillner S, Edgerton VR (eds): Neurobiological Basis of Human Locomotion. Tokyo, Japan Scientific Societies Press, 1991, pp 317–327.
13. McGeer EG, McGeer PL: Some characteristics of brain tyrosine hydroxylase; in Mandel J (ed): New Concepts in Neurotransmitter Regulation. New York, Plenum Press, 1973, pp 53–68.
14. Phillips AG: Presented at the 3rd International Basal Ganglia Society Meeting, Gagliari, Italy, 1989.
15. Steinfels GF, Heym J, Strecker RE, Jacobs BL: Behavioural correlates of dopaminergic unit activity in freely moving cats. Brain Res 1983;258:217–228.
16. Segawa M, Nomura Y: Hereditary progressive dystonia with marked diurnal fluctuation: Pathophysiological importance of the age of onset. Adv Neurol 1993;60:568–576.
17. Segawa M, Nomura Y, Hikosaka O, Soda M, Usui S, Kase M: Roles of the basal ganglia and related structures in symptoms of dystonia; in Carpenter MB, Jayaraman A (eds): The Basal Ganglia II. New York, Plenum Press, 1987, pp 489–504.
18. Segawa M, Nomura Y: Rapid eye movements during stage REM are modulated by nigrostriatal dopamine (NS-DA) neuron? in Bernardi G, Carpenter MB, Di Chiara G, Morelli M, Stanzione P (eds): The Basal Ganglia III. New York, Plenum Press, 1991, pp 663–671.
19. Hornykiewicz O: Striatal dopamine in dopa-responsive dystonia: Comparison with idiopathic Parkinson's disease and other dopamine-dependent disorders. Monogr Neural Sci. Basel, Karger, 1995, vol 14, pp 101–108.
20. Rothwell JC, Obeso JA: The anatomical and physiological basis of torsion dystonia; in Marsden CD, Fahn S (eds): Movement Disorders, Neurology 2. London, Butterworth, 1988, pp 367–376.

21 Segawa M, Nomura Y, Tanaka S, Hakamada S, Nagata E, Soda M, Kase M: Hereditary progressive dystonia with marked diurnal fluctuation: Consideration on its pathophysiology based on the characteristics of clinical and polysomnographical findings. Adv Neurol 1988, 50, pp 367–376.

22 Mitchell IJ, Luquin R, Boyce S, Carke CE, Robertson RG, Sambrook MA, Crossman AR: Neural mechanisms of dystonia: Evidence from a 2-deoxyglucose uptake study in a primate model of dopamine agonist-induced dystonia. Mov Dis 1990;5:49–54.

23 Hikosaka O, Sakamoto M: Cell activity in monkey caudate nucleus preceding saccadic eye movements. Exp Brain Res 1986;63:659–662.

24 Miyashita N, Matsumura M, Usui S, Kato M, Kori A, Gardiner TW, Hikosaka O: Deficits in task-related eye movements induced by unilateral infusion of MPTP in monkey caudate nucleus. Soc Neurosci Abstr 1990;16:235.

25 Hikosaka O, Fukuda H, Kato M, Uetake K, Nomura Y, Segawa M: Deficits in saccadic eye movements in hereditary progressive dystonia with marked diurnal fluctuation; in Segawa M (ed): Hereditary Progressive Dystonia with Marked Diurnal Fluctuation. Carnforth, Parthenon, 1993, pp 159–177.

26 Segawa M, Hikosaka O, Fukuda H, Uetake K, Nomura Y: Deficits in saccadic eye movements in basal ganglia disorders; in Percheron G, et al (eds): The Basal Ganglia IV. New York, Plenum Press, 1994, pp 525–531.

27 Kato M, Hikosaka O: Function of the indirect pathway in the basal ganglion oculomotor system. Visuo-oculomotor activities of external pallidum neurons. Monogr Neural Sci. Basel, Karger, 1995, vol 14, pp 178–187.

28 Leenders KL, Antonini A, Meinck HM, Weindl A: Striatal dopamine D_2 receptors in dopa-responsive dystonia and Parkinson's disease. Monogr Neural Sci. Basel, Karger, 1995, vol 14, pp 95–100.

29 Lentz FA, Seike MS, Jaeger CJ, Lin YC, Delong MR, Tasker RR, Vitek J: Single unit analysis of thalamus in patients with dystonia. Mov Dis 1992;7(suppl 1):126.

30 Shima F, Sakata S, Sun SJ, Kato M, Fukui M, Iacono RP: The role of the descending pallido-reticular pathway in movement disorders. Monogr Neural Sci. Basel, Karger, 1995, vol 14, pp 208–214.

31 Crossman AR: A hypothesis on the pathophysiological mechanisms that underlie levodopa or dopamine agonist induced dyskinesia in Parkinson's disease: Implications for future strategies in treatment. Mov Dis 1990;5:100–108.

32 Narabayashi H, Yokochi F, Nakajima Y: Levodopa-induced dyskinesia and thalamotomy. J Neurol Neurosurg Psychiatr 1984;46:831–839.

33 Kase M: Pitfalls in neurological disorders. Jpn Med J 1978;2850:3–11.

34 Nomura K, Negoro T, Tagesu E, Aso K, Furune S, Takahashi I, Yamamoto N, Watanabe K: Bromocriptine therapy in a case of hereditary progressive dystonia with marked diurnal fluctuation. Brain Dev 1987;9:199.

35 Garcia-Rill E: The basal ganglia and the locomotor regions. Brain Res Rev 1986;11:47–63.

36 Crossman AR: Animal models of movement disorders. First International Congress on Movement Disorders. Washington, DC, April 1990.

37 Snow BJ, Okada A, Martin WRW, Duvoisin RC, Calne DB: Positron-emission tomography scanning in dopa-responsive dystonia, parkinsonism-dystonia, and young-onset parkinsonism; in Segawa M (ed): Hereditary Progressive Dystonia with Marked Diurnal Fluctuation. Carnforth, Parthenon, 1993, pp 181–186.

38 Snow BJ: Presented at the 10th International Symposium on Parkinson's Disease, Tokyo, Japan, 1991.

39 Takahashi H, Snow B, Nygaard T, Yokochi M, Calne D: Fluorodopa PET scans of juvenile parkinsonism with prominent dystonia in relation to dopa-responsive dystonia. Monogr Neural Sci. Basel, Karger, 1995, vol 14, pp 87–94.

40 Mori M: Presented at the 33rd Annual Congress of the Japanese Society of Neurology, Tokyo, Japan, 1992.

41 Nomura Y, Segawa M: Dopa responsive dystonia (DRD) other than hereditary progressive dystonia with marked diurnal fluctuation (HPD). Mov Dis 1992;7(suppl 1):123.

42 Segawa M, Nomura Y, Kase M: Hereditary progressive dystonia with marked diurnal fluctuation: Clinicopathophysiological identification in reference to juvenile Parkinson's disease. Adv Neurol 1986;45:227–234.
43 Yamamura Y, Hamaguchi Y, Uchida M, Fujioka H, Watanabe S: Parkinsonism of early onset with diurnal fluctuation; in Segawa M (ed): Hereditary Progressive Dystonia with Marked Diurnal Fluctuation. Carnforth, Parthenon, 1993, pp 51–59.
44 Quinn N, Critchley P, Marsden CD: Young onset Parkinson's disease. Mov Dis 1987;2:73–91.
45 Yokochi M, Narabayashi H, Izuka R, Nagatsu T: Juvenile parkinsonism – Some clinical, pharmacological and neuropathological aspects. Adv Neurol 1984;40:407–413.
46 Yokochi M: Juvenile parkinsonism and other dopa-responsive syndromes. Monogr Neural Sci. Basel, Karger, 1995, vol 14, pp 25–35.
47 Turjanski N, Weeks R, Sawle GV, Brooks DJ: PET studies of the dopaminergic and opioid function in dopa-responsive dystonic syndromes and in Tourette's syndrome. Monogr Neural Sci. Basel, Karger, 1995, vol 14, pp 77–86.
48 Reisert I, Pilgrim C: Sexual differentiation of monoaminergic neuron genetic epigenetic. Trends Neurosci 1991;14:468–473.
49 McGeer PL, Tooyama I, Kumura H, McGeer EG: Basic fibroblast growth factor in the substantia nigra in Parkinson's disease and normal aging. Monogr Neural Sci. Basel, Karger, 1995, vol 14, pp 235–243.
50 Nygaard TG, Wilhelmsen KC, Risch NJ, Brown DL, Trugman JM, Gilliam TC, Fahn S, Weeks DE: Linkage mapping of dopa-responsive dystonia (DRD) to chromosome 14q. Nat Genet 1993;5:386–391.
51 Endo K, Tanaka H, Tsuji S, Nygaard TG, Weeks DE, Nomura Y, Segawa M: The gene for hereditary progressive dystonia with marked diurnal fluctuation maps to chromosome 14q. Monogr Neural Sci. Basel, Karger, 1995, vol 14, pp 120–125.
52 Rajput AH, Gibb W, Kish S, Hornykiewicz O: Dopa-responsive dystonia – Clinical, pathological and biochemical studies in one case. Mov Dis 1992;7(suppl 1):124.
53 Tsuji S, Tanaka H, Miyatake T, Hinns Y, Nomura Y, Segawa M: Linkage analysis of hereditary progressive dystonia to the tyrosine hydroxylase gene locus; in Segawa M (ed): Hereditary Progressive Dystonia with Marked Diurnal Fluctuation. Carnforth, Parthenon, 1993, pp 107–114.
54 Furukawa Y, Nishi K, Kondo T, Mizuno Y, Narabayashi H: CSF biopterin levels and clinical features of patients with juvenile parkinsonism. Adv Neurol 1993;60:562–567.

Masaya Segawa, MD, Segawa Neurological Clinic for Children, 2-8 Surugadai Kanda, Chiyoda-ku, Tokyo 101 (Japan)

Juvenile Parkinsonism and Other Dopa-Responsive Syndromes

Masayuki Yokochi

Department of Neurology, Tokyo Metropolitan Institute for Neuroscience, Tokyo, Japan

In 1979, cases of juvenile parkinsonism (JP) were described by the author as a nosological entity separate from Parkinson's disease (PD) [1]. The nosological concept was derived from the study of subjects with an onset age below 40, showing major symptoms of parkinsonism which were improved definitely and markedly by *L*-dopa treatment.

In the past, nosology of extrapyramidal disorders was mainly based on their symptomatological analysis, leading to classifications such as 'parkinsonism', 'dystonia' and 'chorea'. However, the effects of *L*-dopa therapy on patients with extrapyramidal disorders led to their reconstruction. Namely, it became possible to separate a clinical entity of dopa-responsive syndromes from other dopa-non-responsive syndromes of extrapyramidal disorders after *L*-dopa therapy [2]. This entity includes hereditary progressive dystonia (HPD) [3] or dopa-responsive dystonia (DRD) [4], which is characterized by dystonia as well as JP and PD, both of which are characterized by parkinsonism. The common pathophysiology of these diseases is their persistent dopamine-deficient state in the brain.

It is interesting to note that in these disease groups (HPD or DRD, JP and PD), the clinical characteristics are manifested according to age at disease onset, which is in line with the topic 'age-related monoamine-dependent disorders' of this symposium. JP in monoamine-dependent disorders may be a unique clinical model to understand the pathophysiology and pathogenesis of parkinsonism. Furthermore, JP acts as a bridge between HPD or DRD and PD.

In this paper, clinical characteristics of HPD or DRD, and JP and PD according to the age at disease onset are described. Moreover, the pathological

Age at onset of illness

| 10 | 20 | 30 | 40 | 50 | 60 | 70 | yrs |

HPD   JP   PD

Persistent Dopamine-Deficient State in the Striatum

Familial Occurrence

F ≫ M   F ≒ M

Progression of Illness

Diurnal Fluctuation of Motor Symptoms

Dystonia

Parkinsonism

Rigidity

Tremor

Akinesia

Efficacy of Levodopa

Wearing-off Phenomenon

DOPA-Induced Dyskinesia

Extremities   Bucco-lingua

Hallucination

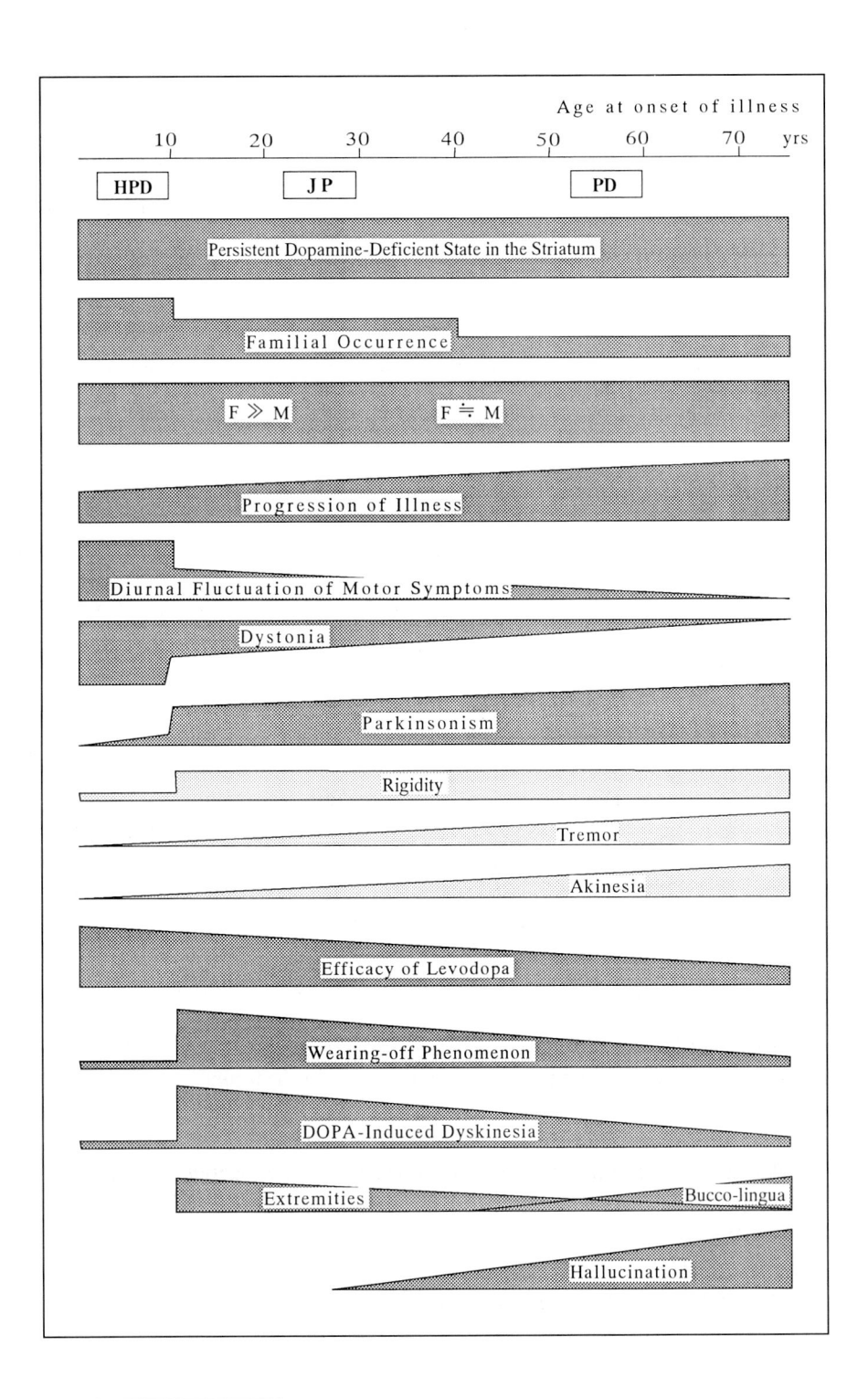

Yokochi

findings from autopsied cases of JP will be presented in a summarized form. Finally, the nosology of these disorders will be proposed, taking into account the results of clinical characteristics and pathological findings.

The clinical characteristics of JP as a case group compared to other groups are given briefly. The symptoms at onset and the degree or the distribution of rigidity, tremor and autonomic dysfunction are different from those of PD. Also, the degree of complication of dystonic feature is different from PD. In this respect, JP has several clinical features of HPD or DRD. The response to L-dopa therapy of JP patients is quantitatively different from that of PD patients and the L-dopa-induced adverse effects in JP patients are qualitatively different from those in HPD or DRD patients.

Comparison of Clinical Findings among HPD, JP and PD

Several clinical characteristics with respect to the age at disease onset are illustrated in figure 1. Ranges of onset age of below 10 years for HPD or DRD, 10–40 years for JP and over 40 years for PD were provisionally given. It was assumed that a persistent dopamine-deficient state in the striatum commonly exists in all groups.

Familial occurrence is much higher in the younger-onset cases. Familial cases are often due to an autosomal-dominant trait in HPD and an autosomal-recessive trait in JP. HPD cases are predominant among females. Progression of illness is slower in JP than in PD, and HPD follows a much more benign clinical course. Diurnal fluctuation of symptoms is characteristic of HPD, and also occurs to a limited extent in some cases of JP.

Symptoms of HPD are characterized not by parkinsonism but by dystonia. JP is characterized by parkinsonism, but patients with younger ages of onset often exhibit dystonic features, especially in the lower extremities. Dystonia-parkinsonism syndrome, often used as a nosological sub-entity [5, 6], seems to focus on this point.

Slight rigidity which appears to depend on the posture and movements is also noted in HPD and is constantly observed in the advanced stage; however, it is not plastic [7]. Typical parkinsonism symptoms such as tremor and akinesia are manifested in increasing severity from JP to PD.

All cases are shown to respond to L-dopa because of a common dopamine-deficient state in the brain. Among them, the efficacy for HPD is marked. In

Fig. 1. Conceptual illustration of clinical characteristics of the disease groups of HPD, JP and PD with respect to their ages at onset.

contrast, *L*-dopa therapy for advanced cases of PD has limited effects. It has a persistent effect on rigidity and tremor, but has reduced efficacy on akinesia, especially on poverty of movement or hypokinesia. Bradykinesia is mainly mediated by the dysfunction of the projection from the substantia nigra (A9) to the dorsal striatum (putamen and dorsal caudate). In contrast, the author proposes that poverty of movement is mainly mediated by the functional disturbance of dopamine projection from the ventral tegmental area (A10) to the ventral striatum, especially to the nucleus accumbens and to the fronal cortex. Furthermore, *L*-dopa might be effective only for the symptoms caused by nigro-dorsal striatal dysfunction. Typical cases of JP actually suffer from very slight hypokinesia, namely, poverty of movement, which is characterized in PD.

Adverse effects of *L*-dopa therapy appear in different forms in all groups. The wearing-off phenomenon occurs more often and earlier after start of *L*-dopa therapy in JP than in PD. Dopa-induced dyskinesia also appears more often and earlier in JP. The distribution of dyskinesia differs between JP and PD: it occurs mostly in the extremities in JP and in the bucco-lingual area in PD [8].

Dopa-induced psychic problems, such as hallucinations and/or delusions, occur often in PD but rarely in JP. HPD patients suffer almost no adverse effects from *L*-dopa therapy [9].

Now, the author would like to elaborate on the following points of JP with an overview from HPD to PD as an age-related monoamine-dependent disorder.

(1) JP is one disease entity due to persistent cerebral dopamine deficiency.

(2) JP has several distinct clinical characteristics such as higher familial incidence and characteristic symptoms mentioned above. The efficacy of *L*-dopa is higher for JP than for PD and lower than for HPD. Adverse effects are rare in HPD, but much more frequent and severe in JP. Alleviation of symptoms by *L*-dopa therapy can be continued for a longer period of time in JP patients than in PD patients. However, most JP is characterized by a slow but sure progressive course of the illness; i.e., JP is a degenerative disease.

(3) Persistent dopamine-deficient state at least in the striatum is common in HPD, JP and PD. However, we must look into whether their mechanism and processes of pathological changes are in common with or independent of one another.

(4) For semiological and pathological analyses, it is possible to exclude physiological aging factors in JP. Thus JP will be an appropriate clinical model for understanding parkinsonism.

(5) From autopsied cases, the pathology of JP was localized in the substantia nigra and did not extend to other regions as compared with that of PD.

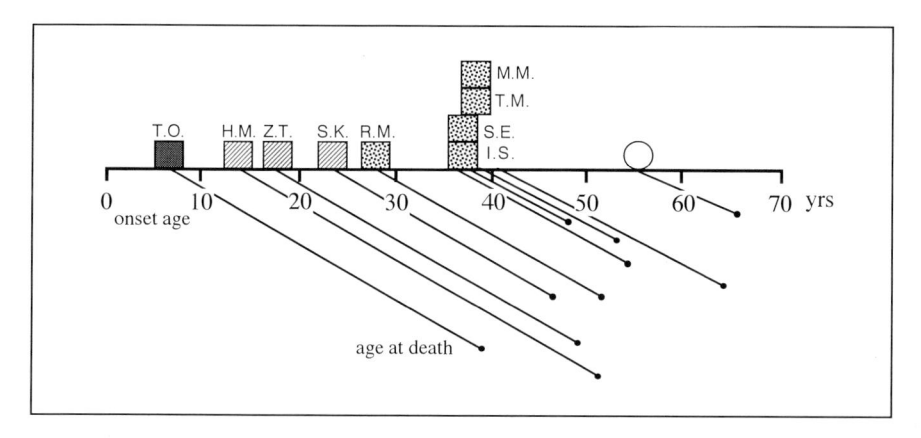

Fig. 2. Autopsied cases of JP arranged according to age at onset. The oblique lines indicate their ages at death. Open circle indicates the average age at onset of PD in patients in Japan. Stippled, hatched and shaded squares indicate the cases represented by compatible pathological findings with PD, by the pathology of lack of Lewy body in the brain and by a unique pathology in the substantia nigra, respectively.

The Pathology of JP

Since 1980, 9 cases of JP were autopsied and their histological examination performed. Details of histopathological findings were published for one case [10] and those for the other cases will be published in the near future. This paper outlines these findings for further understanding of nosology in age-related monoamine-dependent disorders. These 9 cases are arranged according to age at disease onset in figure 2. The single open circle and its corresponding oblique line in the figure indicate the average age at onset and death of PD patients in Japan, respectively. Among the 9 cases of JP indicated by squares, 5 were from the 40 follow-up cases in the original articles by Yokochi et al. [1, 11], 1 was from a sibling and 3 others were those recently investigated. All cases of JP are consistent with PD in terms of the pathological distribution in the brain. Tyrosine hydroxylase activity was significantly decreased in the substantia nigra, putamen and caudate nucleus of 5 cases in which chemical analysis was performed by T. Nagatsu and T. Kondo. These findings indicate that the pathophysiology of JP mediated by dopamine deficiency is associated with that of PD.

However, the nigral histopathological findings varied among the nine cases. Thus we tentatively divided them into three subtypes as indicated by shaded, hatched, or stippled squares in figure 2 and as described in table 1. Type Aa consists of 5 cases (RM, IS, SE, TM and MM) whose findings were compatible with

Table 1. Classification of histological findings in the substantia nigra of JP

Type A: assumed cases of degenerative disease (8/9 cases)
    Aa: cases with findings compatible with PD (5/9 cases)
    Ab: cases with lack of Lewy bodies (3/9 cases)

Type B: assumed case of defect in neuronal development of the substantia nigra (1/9 cases)

PD, and are indicated by stippled squares in the figure. Type Ab consists of 3 cases (SM, ZT and SK) showing lack of Lewy bodies as assessed by a colleague, Y. Mizutani (Tokyo Metropolitan Matsuzawa Hospital), in serial slices of the substantia nigra and the locus ceruleus, and are indicated by hatched squares in the figure. These type A cases are assumed to represent a chronic degenerative disease mediated by cell loss and gliosis in the substantia nigra.

Type B consists of 1 case (TO) indicated by the single shaded square in the figure. Her clinical and pathological findings were described by us [10]. Much more peculiar pathological findings were seen in this case with onset age of six years. The lesion was restricted to the substantia nigra. The cell population was scattered and the number of neurons was abnormally low as shown by cytometric studies. The population of melanin-containing cells is reduced as revealed by the Fontana-Masson stain. In these cells, pigmented granules are sparse and do not fill the cytoplasm. Moreover, immature neurons with round cell bodies among spindle-shaped cells appeared in the zona compacta of the substantia nigra. Lewy bodies existed in the substantia nigra. This case is assumed to be due to a defect in the neuronal development of the substantia nigra.

Relationship between Pathology and Clinical Findings

The relationship between the type of pathology and clinical findings is shown in table 2. The ages at onset of the 5 cases classified as type Aa with pathologies compatible with PD were 38, 38, 39, 39 and 28 years. Three of them belonged to clinical subgroup II which had symptoms and response to *L*-dopa treatment similar to those of PD. The other 2 cases belonged to subgroup I which showed clinical characteristics most typical of JP.

The next 3 cases classified as type Ab had ages of onset of 14, 18 and 24 years. These three cases belonged to clinical subgroup I. Two of these cases were accompanied by dystonia in the extremities.

The last case belonged to clinical subgroup III. This subgroup represented cases with average age of onset of 12 years in the original follow-up cases [1, 11;

Table 2. Relationship between pathological and clinical findings in JP

	Cases								
	I.S.	S.E.	T.M.	M.M.	R.M.	S.M.	Z.T.	S.K.	T.O.
Sex	F	M	M	M	M	F	M	M	F
Age at onset	38	38	39	39	28	14	18	24	6
Age at death	54	53	49	64	52	52	48	44	39
Pathology[1]	Aa	Aa	Aa	Aa	Aa	Ab	Ab	Ab	B
Presence of dystonia	(–)	(–)	(–)	(–)	(–)	(+)	(+)	(–)	(+)
Clinical subgroup[2]	II	II	II	I	I	I	I	I	III
Case No.[3]			(32)	(10)	(3)	(4 Sib.)	(1)		(34)

[1] For definitions see table 1.
[2] Subdivided by Yokochi [1].
[3] As numbered in the original article [1].

the clinical subgroups I, II and III in ref. 1 were revised to Ia, Ib and II, respectively, in ref. 11]. Clinical findings were rather varied in each case. As a group, however, the patients exhibited features of dystonia as well as parkinsonism. Without treatment, they became almost immobile and bedridden. Relatively small doses of *L*-dopa show marked effects, but usually produce adverse effects. We assumed that the conditions exhibited by the cases of clinical subgroup III were different from those of HPD.

Discussion

Arranging the cases on a line representing the age of onset as shown in figure 2 is interesting for the further understanding of the pathogenesis of JP as an age-related monoamine-dependent disorder. Cases indicated by stippled squares (type Aa) with onset ages approaching 40 showed pathologies compatible with those of PD as described in textbooks. Pathological changes in the substantia nigra in PD starts incidentally at an earlier age in these cases, although the reason for this is unknown.

Cases indicated by the hatched squares (type Ab), with onset ages in the adolescent years or in the early twenties, showed degenerative changes with diffuse severe gliosis in the substantia nigra and no Lewy body formation in any part of the brain. Some cases classified under this group have recently been reported by other institutes in Japan (some cases of juvenile onset [12, 13] and some cases of adult onset [14]) and abroad [15]. For instance, Yamamura et al. [12] stressed clinical characteristics such as diurnal fluctuation of symptoms and slight foot dystonia in two cases. However, it seems that definite clinical evidence for the pathological diagnosis of this type does not exist at least in earlier stages of the illness, although a unique clinical feature for identifying cases of this type seems to be the progressive disturbance of postural reflex such as loss of protective reaction for pulsion test in advanced stages of the illness. This symptom is assumed to relate to the pathology in the pars reticulata of the substantia nigra. Either way, the absence of Lewy bodies does not allow its pathological classification under the classical definition of PD pathology.

The case with the youngest age of onset, as indicated by the shaded square (type B), showed unique histological findings which were published by us in 1991 [10]. Her pathology does not seem to be degenerative changes but a developmental disorder of dopamine neurons in the substantia nigra. We then sought to clarify whether her disease is related to the pathophysiological process of HPD or not. The presence of her younger brother who is also affected by the disease is a significant point for this discussion. He was affected by symptoms of prominent dystonia accompanied with parkinsonism manifested by rigidospasticity, high-frequency tremors in posture and the disturbance of pronation-supination alternating hand movement at the age of 6, which was the same age of onset as his sister. His more detailed clinical manifestations prior to *L*-dopa treatment have been described as 'juvenile parkinsonism with pallidal posture and spastic paraplegia' by Yanagisawa in 1974 [16] and 1993 [17]. He is 46 years old, in good condition and holds a job while undergoing treatment with a medium dose (300 mg) of dopa/DCI. Adverse effects such as dopa-induced dyskinesia and wearing-off phenomenon are few, if any. Recently, he was subjected to fluorodopa PET scan. In the paper of Takahashi et al. [18, in this volume] the striatal fluorodopa uptake rate constant was calculated to lie within the range for JP or PD and was lower than that of DRD or normal subjects.

An overview of HPD, JP and PD was presented as 'dopa-responsive syndromes' or 'age-related monoamine-dependent disorders'. The nosological concept of these disorders will be discussed below.

In figure 3, the author would like to show a possible classification based on clinical and pathological findings. Dopa-responsive syndrome, which is caused by persistent dopamine deficiency in the striatum, is assumed to include HPD or DRD, JP and PD. Meanwhile, the concept of JP is rather unclear. Apart from the question of the nosological meaning of 'juvenile', in Japan, JP includes parkinsonian cases with onset age below 40. This was recently divided into 'young-onset Parkinson's disease' with onset age from 20 to 40 years and 'juvenile parkinsonism' with onset age below 20 years [19, 20]. In this paper, JP has been used in a comprehensive manner and is tentatively divided into two categories on a clinical

```
                    Dopa-Responsive Syndrome
         ┌──────────────────┬──────────────────────┐
      HPD or DRD            JP                     PD
                   ┌────────┴────────┐
              Parkinsonism      Early-Onset
              Accompanied with  Parkinson's
              Dystonia          Disease
              (P w D)           (EOPD)
```

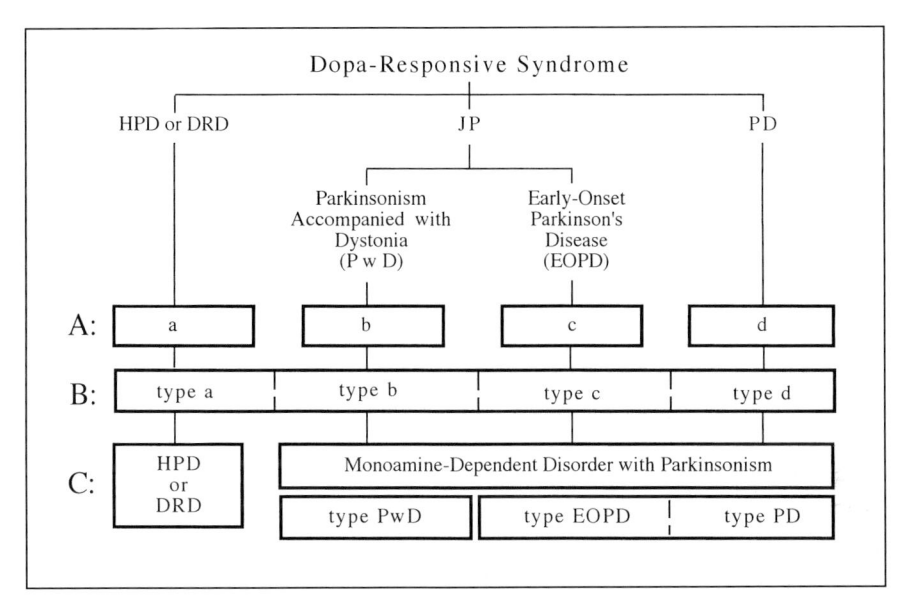

Fig. 3. Concept of nosological classifications of dopa-responsive syndromes referring to clinical and pathological findings. A: Each disease consists of independent pathogenesis; B: all subtypes based on a common pathogenesis; C: modest classification at the present state.

basis, i.e., parkinsonism accompanied with dystonia (PwD) and early-onset Parkinson's disease (EOPD). According to classification C in figure 3, JP and PD are included under one disease entity as 'monoamine-dependent disorders with parkinsonism'. Taking into account the recent study on autopsied cases, both diseases have been categorized based on the main symptoms of parkinsonism and the degenerative condition in the pathological lesions.

On the other hand, some cases of JP might be mediated by independent pathophysiological changes. The group tentatively subdivided as type Ab showing a lack of Lewy bodies gives us cardinal subjects for the discussion of the pathogenesis of parkinsonism. It has been established that the appearance of Lewy bodies is a definite characteristic of PD [21, 22], and that they occur as a result of the disturbance of catecholamine metabolism [23]. If these concepts hold true, cases showing lack of Lewy bodies should not be included under the classical pathogenesis of PD. It seems that this pathology resembles that of an MPTP-induced parkinsonism patient [24, 25] or experimental models [23] in which the typical Lewy body formation is not found. This group may form a new nosological entity among dopa-responsive syndromes with a different dysfunction of catecholamine metabolism manifested by lack of Lewy body formation. However, to date, individually distinct clinical manifestations for the diagnosis of this group remain to be proposed. The cases of PwD are expected to belong to this group; however, an actual case of JP without dystonia was also found in this group. Moreover, from past and present investigations, it seems that cases belonging to this group with a unique pathology are not numerous in JP. The definite conclusion of the nosology based on clinical and pathological findings for 'monoamine-dependent disorders with parkinsonism' should be arrived at by future investigations.

If a difference between EOPD and PD indeed exists, then PD extends its pathology to the regions related with mental abnormality including bradyphrenia and cognitive dysfunction and to dopa-nonresponsive hypokinesia-related structures and systems such as mesolimbicocortical projections, while JP remains mostly within the realm of nigrostriatal dopaminergic dysfunction.

Without autopsied cases showing definite HPD or DRD, a conclusive classification cannot be established. (A case dealing with dopa-responsive dystonia was recently described [26], but it did not seem to be a definite case of HPD or DRD.) However, HPD or DRD can be classified as a distinct disease entity on the clinical basis because it is a childhood disease characterized by dystonia rather than parkinsonism.

Finally, the author defers the following issues for future discussion.

The 9 autopsied cases indicate that nigrostriatal dopamine deficiency is not mediated by a single condition but by various pathological conditions. It should then be asked why there is selective and restricted damage of nigrostriatal neurons.

The evidence that the clinical and pathological characteristics are determined by the age at disease onset is too significant to completely discard classification B (various clinical types based on one etiology). Evidence contradicting the idea that the dopamine deficiency may be based on a certain vulnerability of nigrostriatal neurons is still lacking.

References

1 Yokochi M: Juvenile Parkinson's disease. I. Clinical Aspects. Adv Neurol Sci (Tokyo) 1979;23: 1048–1059.
2 Yokochi M: Nosological concept of juvenile parkinsonism with reference to the dopa-responsive syndrome. Adv Neurol 1993;60:548–552.
3 Segawa M, Nomura Y, Kase M: Diurnally fluctuating hereditary progressive dystonia; in Vinken PJ, Bruyn GW, Klawans HL (eds): Handbook of Clinical Neurology. New York, Elsevier Science Publishers, 1986, vol 5, pp 529–539.
4 Nygaard TG, Marsden CD, Duvoisin RC: Dopa-responsive dystonia. Adv Neurol 1988;50:377–384.
5 Sunohara N, Mano Y, Ando K, Satoyoshi E: Idiopathic dystonia-parkinsonism with marked diurnal fluctuation. Ann Neurol 1985;17:39–45.
6 Nygaard TG, Duvoisin RC: Hereditary dystonia-parkinsonism syndrome of juvenile onset. Neurology 1986;36:1424–1428.

7 Segawa M, Nomura Y: Hereditary progressive dystonia with marked diurnal fluctuation; in Segawa M (ed): Hereditary Progressive Dystonia with Marked Diurnal Fluctuation. Carnforth, Parthenon Publishing, 1993, pp 3–19.
8 Yokochi M: Juvenile parkinsonism as a crucial clinical model for dopamine deficiency syndrome; in Rinne UK, Nagatsu T, Horowski R (eds): International Workshop Berlin Parkinson's Disease from Basic Research and Early Diagnosis to Long-Term Treatment. Bussum, Medicom Europe, 1991, pp 72–83.
9 Segawa M, Nomura Y: Hereditary progressive dystonia with marked diurnal fluctuation. Pathophysiological importance of age of onset. Adv Neurol 1993;60:568–576.
10 Mizutani Y, Yokochi M, Oyanagi S: Juvenile parkinsonism: A case with first clinical manifestation at the age of six years and with neuropathological findings suggesting a new pathogenesis. Clin Neuropathol 1991;10:91–97.
11 Narabayashi H, Yokochi M, Iizuka R, Nagatsu T: Juvenile parkinsonism; in Vinken PJ, Bruyn GW, Klawans HL (eds): Handbook of Clinical Neurology. New York, Elsevier, 1986, vol 49, pp 153–165.
12 Yamamura Y, Arihiro K, Kohriyama T, Nakamura S: Early-onset parkinsonism with diurnal fluctuation – clinical and pathological study. Clin Neurol (Tokyo) 1993;33:491–496.
13 Ikuta F, Takahashi H, Ohama E, Suzuki S, Horikawa Y, Ishikawa A: Juvenile parkinsonism with autosomal recessive inheritance: Report of a family with one patient examined by complete autopsy; in Mannen T (ed): Annual Report of the Research Committee of CNS Degenerative Disease. Ministry of Health and Welfare of Japan, 1992, pp 188–192.
14 Kowa H, Hasegawa K, Ryou T, Yagishita S: Two autopsy cases with autosomal dominant Parkinson's disease; in Mannen T (ed): Annual Report of the Research Committee of CNS Degenerative Disease. Ministry of Health and Welfare of Japan, 1993, pp 204–207.
15 Dwork AJ, Balmaceda C, Fazzini EA, MacCollin M, Cote L, Fahn S: Dominantly inherited, early-onset parkinsonism: Neuropathology of a new form. Neurology 1993;43:69–74.
16 Yanagisawa N: Equilibrium, postural mechanism and the extrapyramidal system. Adv Neurol Sci (Tokyo) 1974;18:767–778.
17 Yanagisawa N: Juvenile parkinsonism with pallidal posture and spastic paraplegia; in Segawa M (ed): Hereditary Progressive Dystonia with Marked Diurnal Fluctuation. Carnforth, Parthenon Publishing, 1993, pp 205–214.
18 Takahashi H, Snow B, Nygaard T, Yokochi M, Calne D: Fluorodopa PET scans of juvenile parkinsonism with prominent dystonia in relation to dopa-responsive dystonia; in Segawa M, Nomura Y (eds): Age-Related Dopamine-Dependent Disorders. Monogr Neural Sci. Basel, Karger, 1995, vol 14, pp 87–94.
19 Quinn N, Critchley P, Marsden CD: Young-onset Parkinson's disease. Mov Disord 1987;2:73–91.
20 Golbe LI: Young-onset Parkinson's disease: A clinical review. Neurology 1991;41:168–173.
21 Gibb WRG, Lees AJ: The relevance of the Lewy body to the pathogenesis of idiopathic Parkinson's disease. J Neurol Neurosurg Psychiat 1988;51:745–752.
22 Gibb WRG, Lees AJ: The significance of the Lewy body in the diagnosis of idiopathic Parkinson's disease. Neuropathol Appl Neurobiology 1989;15:27–44.
23 Forno LS, DeLanney LE, Irwin I, Langston JW: Similarities and differences between MPTP-induced parkinsonism and Parkinson's disease: Neuropathological considerations. Adv Neurol 1993;60:600–608.
24 Davis GC, Williams AC, Markey SP, Ebert MH, Caine ED, Reichert CM, Kopin IJ: Chronic parkinsonism secondary to intravenous injection of meperidine analogues. Psychiatry Res 1979;1:249–254.
25 Langston JW, Ballard P, Tetrud JW, Irwin I: Chronic parkinsonism in humans due to a product of meperidine-analog synthesis. Science 1983;219:979–980.
26 Olsson JE, Brunk U, Lindvall B, Eeg-Olofsson O: Dopa-responsive dystonia with depigmentation of the substantia nigra and formation of Lewy bodies. J Neurol Sci 1992;112:90–95.

Masayuki Yokochi, MD, Department of Neurology,
Tokyo Metropolitan Institute for Neuroscience, 2-6 Musashidai, Fuchu-shi, Tokyo 183 (Japan)

Phenotypic Polymorphism of Dopa-Responsive Dystonia in Russia

Elena Markova, Irina Ivanova-Smolenskaya

Institute of Neurology, Russian Academy of Medical Science, Moscow, Russia

Torsion dystonia (TD), one of the hereditary diseases, is characterized by involuntary movements, pathological postures and changes of muscle tone. Problems of diagnosis, clinical polymorphism, genetics, pathogenesis and treatment remain unsolved in spite of some recent achievements. These problems have been investigated for many years at our institute, particular emphasis being placed on genetics and phenotypic polymorphism.

The purpose of our investigation is a research of the phenotypic and genotypic polymorphism of TD, especially dopa-responsive dystonia.

Materials and Methods

Various methods of investigation have been used: clinical, genealogical, biochemical (especially the study of neurotransmitters and microelements), electrophysiological, CT scan, MRI and other methods which are necessary to establish the diagnosis, differential diagnosis and pathogenesis.

185 patients with TD (92 males, 93 females), aged between 3 and 82 years, from 141 families, were observed for 5–20 years. 80 patients (43.2%) from 15 families suffer from the dopa-responsive form (17 males, 63 females; female-to-male ratio: 3.7).

Results

Our patients with TD were characterized by marked clinical polymorphism concerning age of onset, course of the disease, degree and generalization of dystonia and reaction to pharmacologic agents. Marked interfamilial and intrafamilial polymorphism was observed.

In 1970, we started dividing TD according to clinical signs into two forms which we called: 'rigid' and 'dystonic-hyperkinetic'. The first form corresponds to dopa-responsive dystonia (DRD), named later and reported in the literature as 'Segawa's syndrome' [1, 2]. This form is characterized by the presence of increased muscle tone and fixed pathologic postures, more often in the feet, but also in the hands, neck and trunk. In the second form, local or generalized torsion hyperkinesis is observed [3].

Investigation of neurotransmitters in our DRD patients revealed a decrease in the activity of the dopaminergic and serotoninergic systems, and an increase in the activity of the cholinergic systems. Corresponding changes are observed in cyclic nucleotide metabolism – a decrease of cAMF activity, which depends on the dopamine system, and an increase of cGMF activity, which depends on the acetylcholine system. Opposite changes were observed in hyperkinetic forms of TD [4]. Similar but more marked dopamine deficiency was noted formerly in Parkinson's disease which led to successful use of *L*-dopa treatment of this disease.

With a view to the discovered disorders of the neurotransmitter metabolism, patients with DRD were treated with *L*-dopa-containing drugs (*L*-dopa, madopar, sinemet, nacome).

Analysis of the treatment with dopa-containing drugs demonstrated a very positive dramatic effect in all of our patients, leading to a decrease in muscle tone (sometimes to a normal level), pathologically fixed postures, improvement or normalization of gait, frequent return to work [5, 6].

It is important to stress that the results of this therapy in DRD patients greatly exceeded the effect in parkinsonian patients, and our patients needed lower drug doses (usually 1.0–1.5 g of *L*-dopa, and 0.125–0.250 g of sinemet daily).

During treatment of DRD with *L*-dopa, certain complications can occur, very rarely through, and much less marked than in parkinsonism.

We have used *L*-dopa with a highly positive effect for 15–20 years since 1970 [4, 5].

Analysis of the data demonstrated a correlation between age of onset of the disease and severity, generalization and progress. Age of onset in our DRD patients ranged from 3 to 27 years (mean 12 ± 0.5). We observed a slightly later age of onset in the Russian population. A predominance of familial cases was observed in DRD: 11 of 15 families with 2–12 patients over 3–5 generations. We observed 13 Russian DRD families (fig. 1), 1 Moldavian family and 1 large Jewish family with typical DRD symptoms (fig. 2).

As a rule, patients with childhood onset have a more severe clinical picture of DRD; in patients with adult onset, there are predominantly focal forms with more moderate progression. Initial symptoms in DRD usually were postural dystonia of feet with pes equinovarus, less often in hands (only in 2 families). Features of

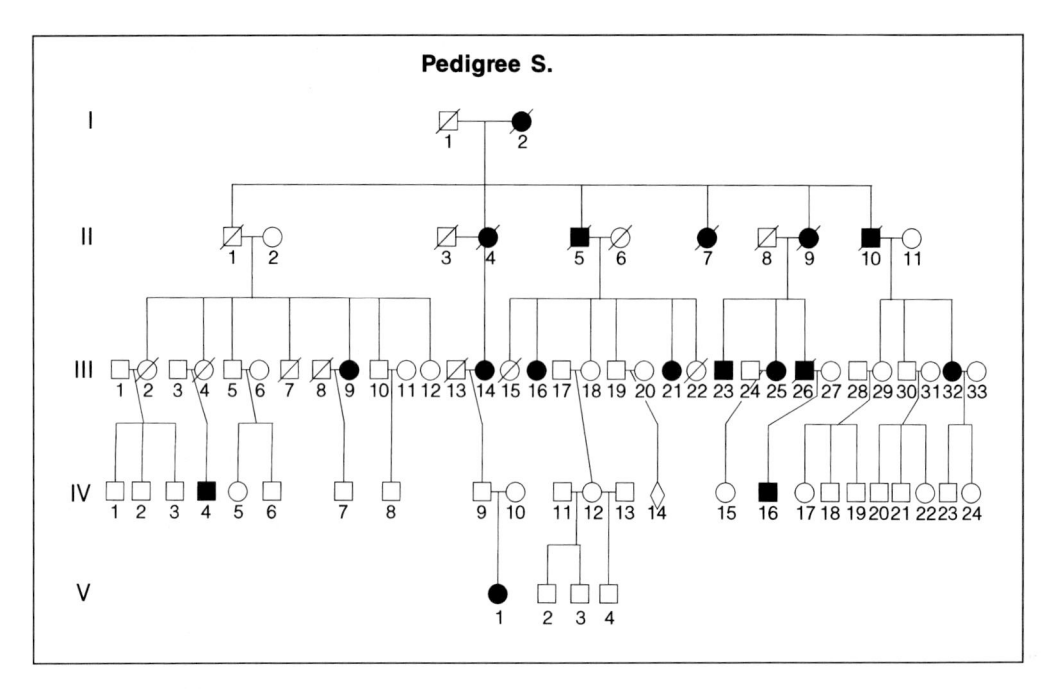

Fig. 1. Pedigree of the Russian DRD family S.

parkinsonism (tremor, slight muscle rigidity) sometimes appeared later, after the age of 25–30 years, in 7 of our families, but the dystonic symptoms were constant. Diurnal fluctuations were observed in 10 families. In our families, clinical-genetic analysis reveals a predominantly autosomal-dominant mode of inheritance; only 2 Russian families had autosomal-recessive type disease, but this needs special study including molecular-genetic analysis [6].

Discussion and Summary

The long-term study of 80 patients with DRD from 15 families demonstrated a wide spectrum of phenotypic polymorphism, both interfamilial and intrafamilial. The following main clinical features of DRD were observed in our patients: dystonia with pathological postures, moderate progression, and a very high positive response to small doses of *L*-dopa during many years.

It is important to note that a high percentage (43.2%) of the Russian patients suffers from a dopa-responsive form. These patients are characterized by a slightly later age of onset and predominance of familial cases.

Pedigree F.-P.

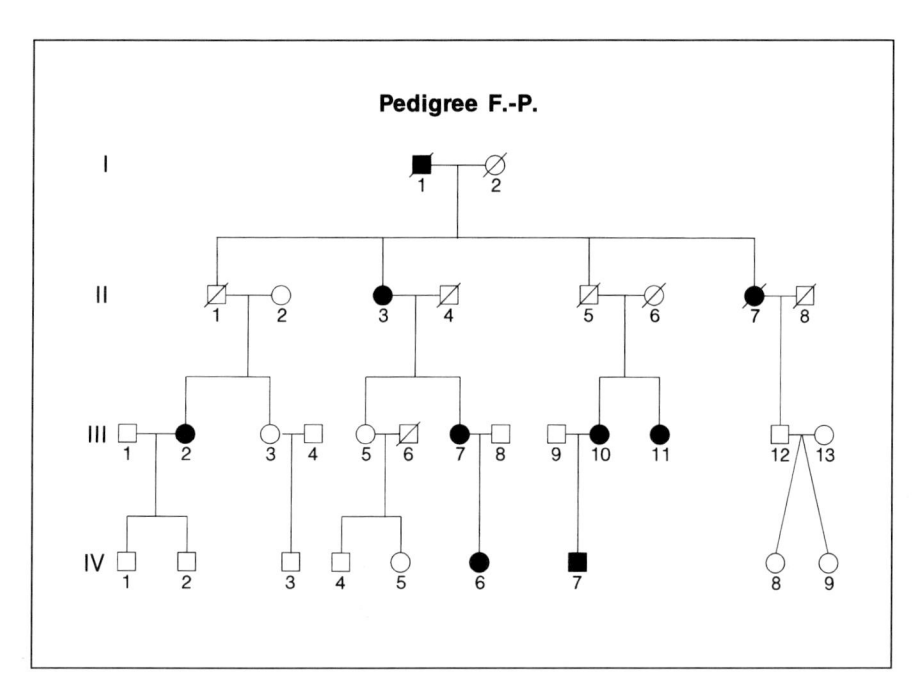

Fig. 2. Pedigree of the Jewish DRD family F-P (fragment).

DRD needs a demarcation from similar syndromes: the rigid form of Huntington's disease, Hallervorden-Spatz disease, some forms of Wilson's disease and Parkinson's disease, especially juvenile paralysis agitans.

The criteria which allowed us to distinguish two forms of TD (DRD and NDRD) not only reflect phenotypic polymorphism but true heterogeneity. This is confirmed by the newest molecular genetic investigations. Only molecular genetic methods can clarify the heterogeneity of TD.

Recent mapping of the TD gene at 9q32 only in patients with NDRD (not DRD) confirms the heterogeneity of TD and the importance of distinguishing this form of the disease [7, 8].

References

1   Segawa M, Ohmi K, Itoh S, Aoyama M, Hayakawa H: Childhood basal ganglia disease with remarkable response to L-dopa, 'hereditary progressive basal ganglia disease with marked diurnal variation'. Shinryo 1971;24:667–672.
2   Nygaard TG, Snow BJ, Fahn S, Calne DB: Dopa-responsive dystonia: Clinical characteristics and definition; in Segawa M (ed): Hereditary Progressive Dystonia with marked Diurnal Fluctuation. New York, Parthenon, 1993, pp 21–36.

3 Barkhatova V, Markova E: Pathogenesis and treatment of torsion dystonia. Zh Nevropat Psikhiatr 1977;78:1121–1128.
4 Markova E: Clinic, pathogenesis and treatment of torsion dystonia in childhood. Zh Nevropat Psikhiatr 1989;8:32–35.
5 Ivanova-Smolenskaya I, Markova E: Parkinsonian hereditary syndromes and their phenotypes. Neurology (suppl July) 1992;7:12–14.
6 Markova E: Torsion dystonia (phenotypic polymorphism, pathogenesis, treatment) (abstract). Mov Disord 1990;5(suppl 1):63.
7 Kramer P, Ozelius L, Brin M, Bressman S, Fahn S, Kidd K, Gusella J, Breakefield X: The molecular genetics of an autosomal dominant form of torsion dystonia. Adv Neurol 1988;50:57–66.
8 Kwiatkowski D, Nygaard T, Schuback D, Perman S, Bressman S, Burke R, Brin M, Trugman J, Ozelius L, Breakefield X, Fahn S, Kramer P: Highly polymorphic microsatellite VNTR within the argininosuccinate synthetase locus excludes the DYT1 dystonia gene as a cause of dopa-responsive dystonia in a large kindred. Am J Hum Genet 1991;48:121–128.

Dr. E.D. Markova, Department of Neurogenetics, Institute of Neurology RAMS,
Volokolamskoe shose 80, Moscow 123367 (Russia)

The Pathogenesis of Tourette's Syndrome: Role of Biogenic Amines and Sexually Dimorphic Systems Active in Early CNS Development

J.F. Leckman

Child Study Center, Yale University, New Haven, Conn., USA

Tourette's syndrome (TS) is a chronic neuropsychiatric disorder of childhood onset that is characterized by tics that wax and wane in severity and an array of behavioral problems including attention deficit hyperactivity disorder and some forms of obsessive compulsive disorder (OCD) [1]. Many of the features of the syndrome including the age of onset, course, anatomic distribution of tics, the premonitory sensory phenomena, the sex ratio, the comorbidity with OCD and attention deficit hyperactivity disorder, and the marked diminution of tics during sleep provide promising clues concerning the underlying neurobiology of this syndrome.

During the course of the past decade, TS and related conditions have emerged as model disorders for researchers interested in the interaction of genetic, epigenetic (environmental), and neurobiological factors which shape clinical outcomes from health to chronic disability over the life span [2]. Although their etiological role has yet to be established, biogenic amines acting in concert with sex-specific hormonal factors are likely to be important influences in shaping the course of TS and related conditions. This chapter provides an overview of an emerging model of the pathogenesis of TS with special attention to the role of these mediating factors.

A Model of Pathogenesis

The proposed model consists of four interrelated areas: phenomenology and natural history, neurobiological substrates, genetic factors, and epigenetic (or environmental factors) [2]. In our view, this model provides a heuristic framework for understanding much of the ongoing research in this area and as well as a basis for clinical care.

Neuroanatomic Substrates

Although there have been relatively few neuropathological studies of TS brains, there is a substantial body of data that implicate the basal ganglia and related cortical and thalamic structures in the pathobiology of TS and OCD [see ref. 3 for a review]. Recent magnetic resonance imaging studies have revealed abnormalities in the structural lateralization of basal ganglia structures in both children [4] and adults with TS [5]. Data have also been presented that indicate that adult TS subjects on average have smaller corpus callosi as measured in the mid-sagittal plane [6]. These structural alterations are likely to have functional consequences as well as provide clues concerning the location and timing of events in the developing CNS that directly influence the emergence and course of TS and related symptomatology such as attention deficit hyperactivity disorder and the premonitory urges.

Functional brain imaging studies also suggest that the abnormalities associated with TS may not be limited to the basal ganglia. Positron-emission tomography (PET) studies have demonstrated both *decreased* regional metabolic activity in frontal, cingulate and insular cortices as well as the striatum, and *increased* metabolic rates in the superior cortical convexities including the supplementary motor area [3].

Functionally, the basal ganglia are composed of fiber tracks that contribute to the multiple parallel cortico-striato-thalamocortical circuits (CSTC) that concurrently subserve a wide variety of sensorimotor, motor, oculomotor, cognitive and 'limbic' processes [7]. We have hypothesized that TS and etiologically related forms of OCD are associated with a failure to inhibit subsets of these CSTC minicircuits [8]. Based on this hypothesis, it is anticipated that the frequently encountered tics involving the face would be associated with a failure of inhibition of those mini-circuits that include the ventromedian areas of the caudate and putamen that receive topographic projections from the orofacial regions of the primary motor and premotor cortex. The potential role of the corpus callosum in mediating this disinhibition has not been explored.

While the neurophysiologic defect that underlies TS and etiologically related conditions remains unknown, a complete understanding of these disorders will likely illuminate mechanisms that regulate the activity of the multiple parallel CSTC circuits that subserve much of our normal cognitive, behavioral, and emotive repertoire. Advances in this area will also lead to the identification of specific neuroanatomical sites that may be crucially involved in the genesis of TS and OCD symptoms which in turn may be of value in isolating candidate genes that are uniquely expressed in these regions.

The Role of Biogenic Amines

Among the most remarkable developments in the neurobiology of the basal ganglia and related structures over the past decade are the extensive immunohistochemical studies which have demonstrated the presence of a wide spectrum of differently distributed classic neurotransmitters, neuromodulators, and neuropeptides. The functional status of a number of these systems has been evaluated in TS [see ref. 9 for a review]. Thus far, the dopaminergic projections from mesencephalic centers that modulate the activity of the CSTC circuits have received the greatest amount of attention.

Dopaminergic Systems
Based largely on parallels between the tics, vocalizations and obsessive-compulsive behaviors seen in some patients with encephalitis lethargica, Devinsky [10] has suggested that TS is the result of altered dopaminergic function in the midbrain. Other data implicating central dopaminergic mechanisms are reviewed elsewhere [9] and include: clinical trials in which haloperidol and other neuroleptics which preferentially block dopaminergic D_2 receptors have been found to be effective in the partial suppression of tics in a majority of TS patients. Tic suppression has also been reported following administration of agents that reduce dopamine synthesis. Alternatively, increased tics have been reported following withdrawal of neuroleptics and following exposure to agents which increase central dopaminergic activity such as *L*-dopa and CNS stimulants, including cocaine. The exacerbation of tics with CNS stimulants remains a controversial area, with most investigators reporting no change in tic symptoms following short-term treatment. In addition, there have been reports of altered mean levels of homovanillic acid, a major metabolite of brain dopamine in CSF.

Recent preliminary PET studies of brain dopamine D_2 receptors, however, do not support the view that there are increased numbers of these receptors in the few TS patients that have been studied [11]. Additional imaging studies are needed to address fully the potential abnormalities of receptor number, affinity and distribution across the growing family of dopamine receptors. Indeed, the recent molecular characterization of a large family of dopaminergic receptors holds considerable promise both in terms of potential treatments as well as concerning the development of safe and effective pharmacological probes. Available postmortem-data also suggest that there are pronounced changes in the number of D_1 and D_2 receptors in human brain over the first two decades of life [12].

More recently, Singer and co-workers have reported increased levels of the dopamine transporter sites in the post mortem striatum of a small number of TS subjects [13]. Studies are presently underway to evaluate this finding using radioligands and in vivo neuroimaging techniques.

In sum, dopaminergic systems have been repeatedly implicated in the pathophysiology of TS and related disorders. The next decade should bring notable advances both in terms of the availability of highly selective pharmacological agents that may have therapeutic value as well as novel ligands that will permit state-of-the-art neuroimaging studies.

Other Biogenic Amines

Evidence of noradrenergic involvement in the pathophysiology of TS is based in part on the reported beneficial effects of clonidine [14]. Although the effectiveness of clonidine remains controversial [15], we have recently completed a large double-blind placebo-controlled trial in TS subjects that again demonstrated a beneficial effect of this agent on motor tics [16]. Additional support has been based on the rebound exacerbations of tics in patients abruptly withdrawn from clonidine [17] and from the finding of blunted growth hormone response to clonidine challenge in children with TS [18]. Most recently, a series of adult TS patients were found to have elevated levels of CSF norepinephrine (NE) and to have excreted high levels of urinary NE in response to the stress of the lumbar puncture [19].

Ascending serotonergic projections from the dorsal raphe have also been invoked as playing a role in the pathophysiology of both TS and OCD [see ref. 9 for a full list of citations]. The most compelling evidence relates to OCD and is based largely on the well-established efficacy of potent serotonin reuptake inhibitors such as clomipramine and fluvoxamine in the treatment of OCD. However, some investigators have reported that the serotonin reuptake inhibitors are less effective in treating tic-related OCD compared to other forms of OCD [20]. Preliminary postmortem brain studies in TS recently have also shown that serotonin and the related compounds, tryptophan and 5-hydroxy-indoleacetic acid (5-HIAA), may be globally decreased in the basal ganglia and other areas receiving projections from the dorsal raphe. Although preliminary, these postmortem findings are consistent with previous observations of significantly lower levels of CSF 5-HIAA, plasma tryptophan, whole blood serotonin, and 24-hour urinary serotonin in TS patients compared to normal controls [see ref. 9 for a review].

Other Neurotransmitter Systems

Despite their central role in the functioning of the CSTC circuits, the data concerning the roles of various excitatory and inhibitory amino acid neurotransmitter systems are sparse and inconclusive [see ref. 9 for a review]. Other evidence has focused attention on the endogenous opioid projections from the striatum to the pallidum and substantia nigra that form one portion of the CSTC circuits [21]. In addition, some clinical data implicate local cholinergic interneurons as well.

Gender-Specific Hormonal Factors

The results of genetic studies that have shown that TS is likely to be transmitted within families as an autosomal dominant trait suggest that males and females should be at equal risk. However, males are several-fold times more frequently affected than females. This observation has led us and others to hypothesize that androgenic steroids acting at key developmental periods may be involved in determining the natural history of TS and related disorders [see ref. 22 for a review]. This may be a direct effect of androgenic steroids or may be due in part to the action of estradiol (formed in key brain regions by the aromatization of testosterone).

Normal surges in testosterone and other androgenic steroids during critical periods in male fetal development are known to be involved in the production of long-term functional augmentation of subsequent hormonal challenges (as in adrenarche and during puberty) and in the formation of structural CNS dimorphisms. In recent years, several sexually dimorphic brain regions have been described, including portions of the amygdala and related limbic areas. These regions contain high levels of androgen and estrogen receptors and are known to influence activity in the basal ganglia both directly and indirectly. It is also of note that some of the neurochemical and neuropeptidergic systems implicated in TS and related disorders, such as dopamine, serotonin, and the opioids, are involved with these regions and appear to be regulated by sex-specific factors.

Further support for a role for androgens comes from case reports of tic exacerbation following anabolic steroid abuse [23] and pilot study data from open trials of antiandrogens in patients with severe OCD and TS [22].

Genetic Factors

The etiology of TS remains unknown. Twin studies indicate that genetic factors are likely to play an important role in the transmission and expression of TS and related phenotypes [24, 25].

Using mathematical models of genetic transmission and either a broad or narrow definition of the affected phenotype, the distribution of affected individuals within families follows a pattern consistent with an autosomal dominant form of transmission [26]. This finding, coupled with recent advances in human genetics, has led directly to the initiation of genetic linkage studies in an effort to identify the chromosomal location of the putative TS gene.

The potential benefits of detecting linkage to a chromosomal region are many and include improved diagnostic capability, the identification of high-risk individuals, and the potential to use molecular genetic techniques to identify the TS vulnerability gene itself and to unravel the pathophysiology of these related disorders from their source. Thus far, more than 80% genome has been excluded [27] and the search continues.

Epigenetic and Environmental Factors

Studies of monozygotic twins indicate that epigenetic or environmental factors play an important role in mediating the extent to which the genetic vulnerability to TS and related disorders is expressed [24, 25]. Prenatal factors, particularly those that influence intrauterine growth, have been repeatedly implicated in TS. In discordant monozygous twin pairs, the more severely affected individual usually has a lower birth weight [25, 28]. Indeed, Hyde et al. [25] found that the greater the intrapair birth weight difference, the greater the observed difference in tic severity. Although the mechanisms underlying this association between birth weight and tic severity have not been established, differing degrees of oxygen and nutrient delivery to developing brain structures such as the basal ganglia may play an important role.

Circumstantial evidence also suggests that psychological maternal stress during pregnancy may influence the later severity of tic disorders [see ref. 29 for a review]. Specifically, ratings of maternal emotional stress during pregnancy have been found to be associated with later tic severity [30]. In addition, some of the behavioral, neuroendocrine, and neurochemical characteristics of TS patients parallel those seen in adult animals who have previously been exposed to high levels of maternal stress in utero. For example, some investigators have reported significantly higher levels of plasma adrenocorticotropin and corticosterone in the prenatally stressed animals. Similarly, we have found that a significant portion of TS patients show a heightened responsivity to stressful stimuli – with elevations in plasma adrenocorticotropin and urinary NE excretion observed in anticipation of a lumbar puncture to collect CSF [19]. In summary, maternal stress during pregnancy may be an important mediator of the TS phenotype later in life. If severe, these stresses are likely to have enduring effects on the activity of monoaminergic systems and the hypothalamic-pituitary-adrenal axis, as well as alter the development and extent of CNS lateralization.

Other possible epigenetic factors that can influence the natural history of TS include: exposure to chronic intermittent psychosocial stress in the postnatal period, chronic medical conditions such as asthma; and exposure to cocaine or other CNS stimulants (see above); and recurrent streptococcal infections.

The prospects for being able to identify specific or nonspecific risk and/or protective factors will be greatly enhanced when the TS vulnerability gene has been mapped. By selecting biologically homogeneous high-risk carriers, investigators will be able to control for a significant portion of the biological variance so that detecting effects of a risk or protective factor will be more readily accomplished. The succesful identification of risk and/or protective factors may lead directly to early interventions that will limit, if not prevent, clinically significant forms of TS and OCD.

Promising Areas for Clinical and Basic Research

TS and related disorders are likely to be associated with detectable alterations in brain function. Systematic and sustained investigation of possible pathobiological mechanisms in TS has led to promising new treatment strategies. Advances in the neurosciences are continuing at a rapid pace. Although the enormous complexity of the neurochemical, neuroendocrine, and neuropharmacological data in TS and OCD does not yield readily to simplistic explanations focused on a single neurotransmitter or neuromodulatory systems, it is likely that significant progress will be made over the next decade with regard to the biological causes and determinants of TS and related disorders and the development of safe and effective treatments.

Acknowledgments

This work was supported in part by NIH grants MH44843, MH49351, MH00508, NS16648, HD03008, RR00125, RR06022, MH30929, and the Tourette Syndrome Association. The author would also like to acknowledge the many important scientific and practical contributions made by Drs. Donald J. Cohen, David L. Pauls, Wayne K. Goodman, Mark A. Riddle, George M. Anderson, Phillip B. Chappell, Bradley S. Peterson, James Duncan, John T. Walkup, Dorothy E. Grice, Paul J. Lombroso, Yanki Yazgan, Bruce Wexler, and Robert A. King; and Ms. Sharon I. Ort, Mr. Lawrence D. Scahill, and Ms. Maureen T. McSwiggan-Hardin to this research program. Portions of this chapter are based on a chapter in a book by Leckman et al. [9] and an invited editorial by Leckman and Peterson [29].

References

1 Cohen DJ, Leckman JF: Developmental psychopathology and neurobiology of Tourette's syndrome. J Am Acad Child Adolesc Psychiatry 1994:33:2–15.
2 Leckman JF, Pauls DL, Peterson BS, Riddle MA, Anderson GM, Cohen DJ: Pathogenesis of Tourette's syndrome: Clues from the clinical phenotype and natural history; in Chase TN, Friedhoff AJ, Cohen DJ (eds): Tourette's Syndrome: Genetics, Neurobiology, and Treatment. Advances in Neurology. New York, Raven Press, 1992, vol 58, pp 15–23.
3 Braun AR, Stoetter B, Randolph C, Hsiao JK, Vladar K, Gernert J, Carson RE, Herscovitch P, Chase TN: The functional neuroanatomy of Tourette's syndrome: An FDG-PET study. I. Regional changes in cerebral glucose metabolism differentiating patients and controls. Neuropsychopharmacology 1993:9:277–291.
4 Singer HS, Reiss AL, Brown JB, Aylward EH, Shih B, Chee E, Harris EL, Reader MJ, Chase GA, Bryan RN, Denkla MB: Volumetric MRI changes in basal ganglia of children with Tourette syndrome. Neurology 1993;43:950–956.
5 Peterson BS, Riddle MA, Cohen DJ, Katz LD, Smith JC, Hardin MT, Leckman JF: Reduced basal ganglia volumes in Tourette's syndrome using 3-dimensional reconstruction techniques from magnetic resonance images. Neurology 1993;43:941–949.
6 Peterson BS, Leckman JF, Wetzles R, Duncan J, Riddle MA, Hardin MT, Cohen DJ: Corpus callosum morphology from MR images in Tourette's syndrome. Psychiatry Res, in press.
7 Alexander GE, Crutcher MD: Functional architecture of basal ganglia circuits: Neural substrates of parallel processing. Trends Neurosci 1990;13:266–276.
8 Leckman JF, Knorr AM, Rasmusson AM, Cohen DJ: Basal ganglia research and Tourette's syndrome. Trends Neurosci 1991;14:94.
9 Leckman JF, Pauls DL, Cohen DJ: Tic Disorders; in Bloom FE, Kupfer D (eds): Psychopharmacology: The Fourth Generation of Progress. New York, Raven Press, in press.
10 Devinsky O: Neuroanatomy of Gilles de la Tourette's syndrome: Possible midbrain involvement. Arch Neurol 1983;40:508–514.
11 Singer HS, Wong DF, Brown JE: Positron emission tomography evaluation of dopamine D_2 receptors in adults with Tourette syndrome; in Chase TN, Friedhoff AJ, Cohen DJ (eds): Tourette's Syndrome: Genetics, Neurobiology, and Treatment. Advances in Neurology. New York, Raven Press, 1992, vol 58, pp 233–239.
12 Seeman P, Bzowej NH, Guan HC, Bergeron C, Becker LE, Reynolds GP, Bird ED, Riederer P, Jellinger K, Watanabe S, Tourtellotte WW: Human brain dopamine receptors in children and aging adults. Synapse 1987;1:399–404.
13 Singer HS, Hahn IH, Moran TH: Abnormal dopamine uptake sites in postmortem striatum from patients with Tourette's syndrome. Ann Neurol 1991;30:558–562.
14 Cohen DJ, Young JG, Nathanson JA: Clonidine in Tourette's syndrome. Lancet 1979;ii:551.
15 Goetz CG, Tanner CM, Wilson RS: Clonidine and Gilles de la Tourette syndrome: Double-blind study using objective rating method. Ann Neurol 1987;21:307–310.
16 Leckman JF, Hardin MT, Riddle MA, Stevenson J, Ort SI, Cohen DJ: Clonidine treatment of Gilles de la Tourette's syndrome. Arch Gen Psychiatry 1991;48:324–328.
17 Leckman JF, Ort SI, Cohen DJ, Caruso KA, Anderson GM, Riddle MA: Rebound phenomena in Tourette syndrome after abrupt withdrawal with clonidine: Behavioral, cardiovascular, and neurochemical effects. Arch Gen Psychiatry 1986;43:1168–1176.
18 Leckman JF, Cohen DJ, Gertner JM, Ort S, Harcherik DF: Growth hormone response to clonidine in children ages 4–17: Tourette's syndrome vs. children with short stature. J Am Acad Child Adolesc Psychiatry 1984;23:174–181.
19 Chappell PB, Riddle MA, Anderson GM: Enhanced stress responsivity to Tourette syndrome patients undergoing lumbar puncture. Biol Psychiatry, in press.
20 McDougle CJ, Goodman WK, Leckman JF, Lee NC, Heninger GR, Price LH: Haloperidol addition in fluvoxamine-refractory obsessive compulsive disorder: A double-blind, placebo controlled study in patients with and without tics. Arch Gen Psychiatry, in press.

21 Chappell PB, Leckman JF, Riddle MA: Neuroendocrine and behavioral effects of naloxone in Tourette syndrome, in Chase TN, Friedhoff AJ, Cohen DJ (eds): Tourette's Syndrome: Genetics, Neurobiology, and Treatment. Advances in Neurology. New York, Raven Press, 1992, vol 58, pp 253–262.
22 Peterson BS, Leckman JF, Scahill L, Naftolin F, Keefe D, Charest NJ, Cohen DJ: Steroid hormones and CNS sexual dimorphisms modulate symptom expression in Tourette's syndrome. Psychoneuroendocrinology 1992;6:553–563.
23 Leckman JF, Scahill L: Possible exacerbation of tics by androgenic steroids. N Engl J Med 1990; 322:1674.
24 Price AR, Kidd KK, Cohen DJ, Pauls DJ, Leckman JF: A twin study of Tourette syndrome. Arch Gen Psychiatry 1958;42:815–882.
25 Hyde TM, Aaronson BA, Randloph C, Rickler KS, Weinberger DR: Relationship of birth weight to the phenotypic expression of Gilles de la Tourette's syndrome in monozygotic twins. Neurology 1992;42:652–658.
26 Pauls DL, Leckman JF: The inheritance of Gilles de la Tourette syndrome and associated behaviors: Evidence for autosomal dominant transmission. N Engl J Med 1986;315:993–997.
27 Pakstis AJ, Heutinik P, Pauls DL: Progress in the search for genetic linkage with Tourette syndrome: An exclusion map covering more than 50% of the autosomal genome. Am J Hum Genet 1991;48:281–294.
28 Leckman JF, Price RA, Walkup JT, Ort SI, Pauls DL, Cohen DJ: Nongenetic factors in Gilles de la Tourette's syndrome. Arch Gen Psychiatry 1987;44:100.
29 Leckman JF, Peterson BS: The pathogenesis of Tourette's syndrome: Role of epigenetic factors active in early CNS development. Biol Psychiatry 1993;34:425–427.
30 Leckman JF, Dolnansky ES, Hardin MT, Clubb M, Walkup JT, Stevenson J, Pauls DL: Perinatal factors in the expression of Tourette's syndrome: An exploratory study. J Am Acad Child Adolesc Psychiatry 1990;29:220–226.

James F. Leckman, MD, Child Study Center, Yale University, 230 South Frontage Road, PO Box 333, New Haven, CT 06510-8009 (USA)

Pathophysiology of Muscle Tone and Involuntary Movement in Young-Onset Basal Ganglia Disorders

Nobuo Yanagisawa, Takao Hashimoto

Department of Medicine (Neurology), Shinshu University School of Medicine, Matsumoto, Japan

Age-related motor symptoms in basal ganglia disorders have been known in various diseases affecting the striatum. The so-called extrapyramidal symptoms, from the hypotonic-hyperkinetic syndrome (chorea, ballism) to the hypertonic-hypokinetic syndrome (rigidity, parkinsonism), are considered as expressions of the striatal dysfunctions and they may be modified by age (table 1). On the other hand, changes in the frequency of tremor by age were asserted by Marshall [1].

In this chapter, three topics on the neuropathophysiology of age-related motor disorders in the basal ganglia diseases will be discussed. First, relations between the type of involuntary movement and the underlying muscle tone in the basal ganglia disorders will be described. Secondly, the pathophysiology of idiopathic torsion dystonia and a unique type of juvenile parkinsonism will be discussed. Then, thirdly, aging effects on the tremor frequency in Parkinson's disease and juvenile parkinsonism will be described. As all findings for those topics were obtained in EMG with surface electrodes, the principle of this method will be described first.

Surface EMG for Involuntary Movement Disorders

Involuntary movements of central origin are expressions of the abnormal excitation of motoneurons innervating skeletal muscles. EMG of involved muscles furnishes information about such movements and may contribute to their evaluation on a physiological basis, both qualitatively and quantitatively. Recently, we have tried to establish a quantitative measurement of irregular involuntary movements with EMG and have obtained some promising results [2].

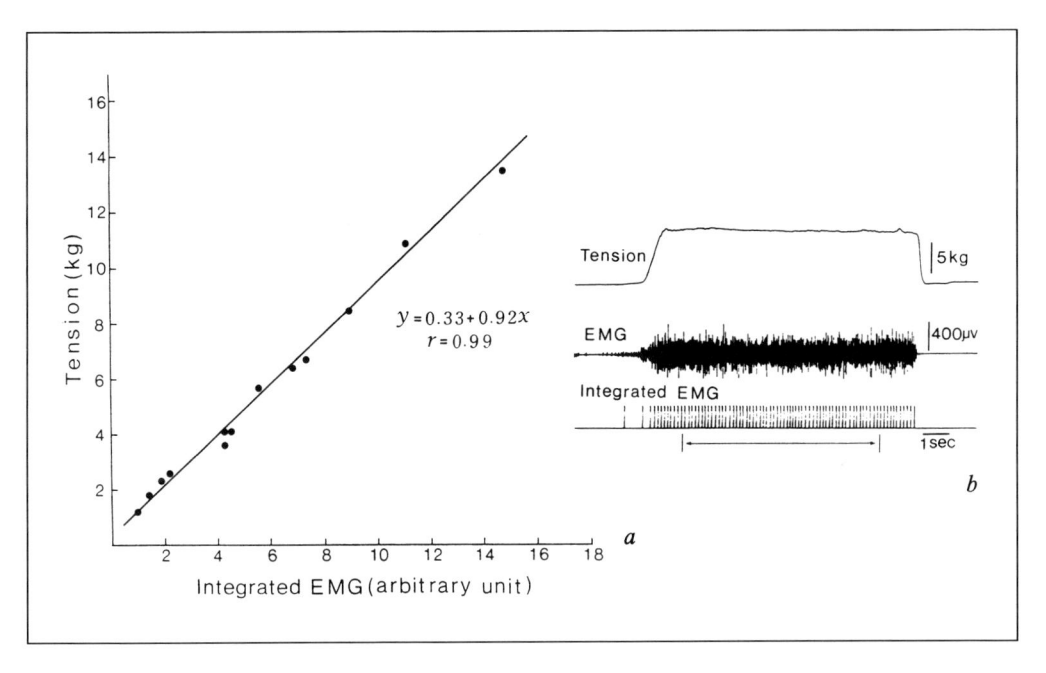

Fig. 1. Relation between integrated EMG and tension produced in isometric voluntary contraction (pretibial muscles, 30-year-old male control) [modified from Yanagisawa and Hashimoto, 2].

Table 1. Age-related motor symptoms in the basal ganglia disorders

Condition	Young	Adult
Wilson's disease	rigidity, dystonia	intention tremor
Huntington's disease	parkinsonism, dystonia	chorea
Effects of dopamine receptor blockers	dystonia	parkinsonism, chorea
Side effects of *L*-dopa on parkinsonism	chorea, ballism	chorea

Bipolar recording with surface electrodes on the skin which covers muscles reflects the activity of the muscles as a whole [2–4]. In limb muscles, the amount of EMG after rectification and integration of action potentials recorded with bipolar electrodes has quantitative correlations with the tension produced in isometric voluntary contraction and the size of movement in isotonic ballistic contraction [2] (fig. 1). Simultaneous recording of many muscles may provide infor-

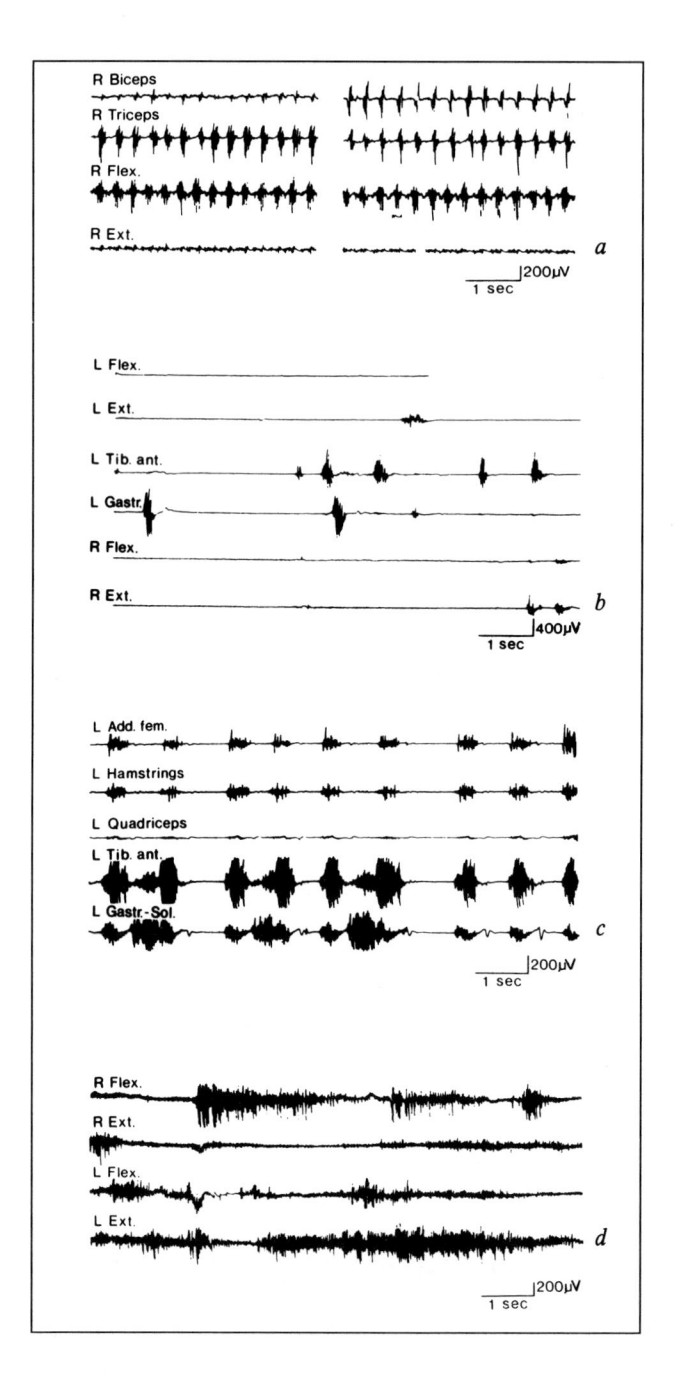

Fig. 2. EMG patterns of different types of involuntary movements [2]. *a* Tremor. *b* Chorea. *c* Ballism. *d* Athetosis-dystonia.

Table 2. Correspondence between clinical observation and EMG in involuntary movements [2]

Clinical observations	EMG
Affected parts of the body	Simultaneous recording from many muscles
Pattern of movement	Pattern of EMG
Briskness	Duration of discharges
Irregularity	Irregularity of discharges
Amount of movement	Amount of action potentials
Underlying muscle tone	Tonic discharges at rest
	Responses to passive stretch
Provoking or inhibiting factors	Recording under different physical or mental conditions

Table 3. EMG patterns of basal ganglia diseases [5]

Diseases	Involuntary discharges phasic	Involuntary discharges tonic	Stretch reflexes phasic	Stretch reflexes tonic	Disturbance of reciprocal innervation in voluntary contraction
Parkinson's disease	tremor		–	++	+ ~ ++
Huntington's disease	++	(+)	–	(+)	– ~ (+)
Hemiballism	++	–	–	–	–
Double athetosis	+	++	+	++	++
Postencephalitic dystonia	–	+++	–	+++	+++
Dystonia musculorum deformans	+	++	–	(+)	– ~ (+)

mation on the nature of movements, and correspondence between clinical observations and EMG is obtained as listed in table 2. Different patterns of EMG in various involuntary movements are shown in figure 2.

For a routine examination of central motor disorders, EMG at rest, response to the passive stretch of muscles and isometric voluntary contraction around a joint are recorded. EMG under postural stress, such as standing, or of voluntary movement used in a routine neurological examination, such as a finger-to-nose test, or alternating pronation and supination are also recorded, if necessary. A summary of EMG in basal ganglia disorders is given in table 3.

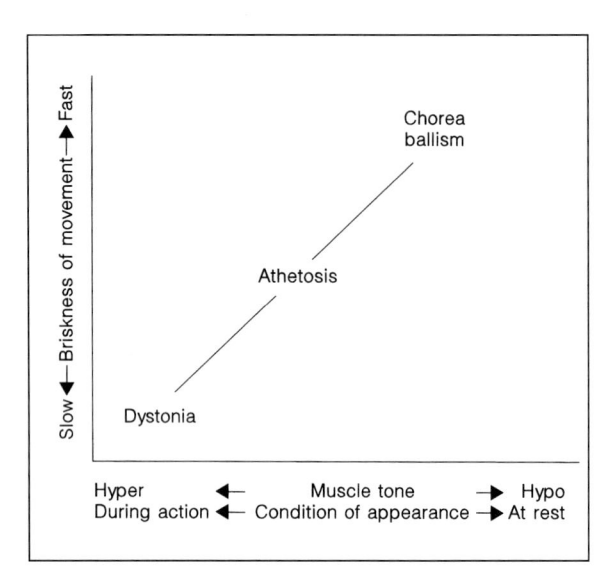

Fig. 3. Relations between briskness of involuntary movements, underlying muscle tone and conditions for appearance.

Relationships between Involuntary Movements and Underlying Muscle Tone

The briskness of involuntary movements is determined by the duration of bursts of action potentials in EMG and the underlying muscle tone [5]. Furthermore, brisk involuntary movements, such as chorea or ballism, develop while the subjects is at rest, and slow movements, such as dystonia, may develop during the execution of purposive motion or the maintenance of a posture. Such relationships between the patterns of involuntary movements and the underlying muscle tone or posture were also true for different types of movements in Huntington's disease from the common form to the rigid form or Westphal's variant [6]. In Huntington's disease, the rigid form appears only among young subjects in their teens. Historically responsible lesions for Westphal's variants invited arguments, and involvement of the large-size neurons in the striatum or affection of the neurons in the globus pallidus were discussed [7]. However, aging of the striatum may be responsible for the development of dystonia in Huntington's disease, as has been claimed in hereditary dystonia with diurnal fluctuation [8]. Relations between involuntary movements with different degrees of briskness in movement and underlying muscle tone are shown in figure 3.

Motor Disorders in Idiopathic Torsion Dystonia and Juvenile Parkinsonism

Clinical features of motor disorders in idiopathic torsion dystonia were first established by Oppenheim [9]. After confusion in the concept of dystonia since 1920s, reestablishment of the nosological concept in the 1950s [10, 11] promoted a clarification of motor disorders in idiopathic dystonia. Clinically, patients develop a peculiar posture in general (generalized dystonia) or in limited parts of the body (focal dystonia). The dystonic posture is aggravated by efforts to maintain certain postures and disappears in a recumbent position. Another motor disorder is action dystonia, in which an exaggeration of tonic muscular contraction appears in related muscles, disturbing purposive action. Tremor or irregular shaking of affected body parts is not infrequent. In the neck, trunk and extremities, the muscle tone is normal or even hypotonic while at rest, and rigidity is not encountered in most cases. EMG of idiopathic dystonia further disclosed characteristics of the motor disorder [12, 13].

The motor disorders in idiopathic dystonia disclosed by our previous study [13] may be summarized as follows:

(1) Tonic nonreciprocal, involuntary activity appeared in agonists and antagonists. This dystonia was directly stimulated by postural efforts, such as attempted sitting or standing, or by an attempted voluntary movement.

(2) Regular or irregular grouping of action potentials in EMG appeared in most cases, corresponding to tremor or myoclonus or other trembling motions.

(3) Simple voluntary contraction developed smoothly in spite of the dystonic disorder. Weakness and disorders of the reciprocal innervation were observed in some muscles.

(4) Regular grouping of action potentials at around 10 Hz appeared frequently on voluntary contraction.

(5) Passive stretch of dystonic muscles, at stages preceding absolute fixity of posture, seldom provoked tonic contraction reflexly. Paradoxical contraction of Westphal or delayed enhancement of dystonia may be provoked by muscle stretch. These characteristics of motor disorders in idiopathic dystonia were not modified by age in our experiences and they usually gave sufficient clues for differentiation from other hypertonic motor disorders.

Juvenile parkinsonism sometimes develops trunkal or foot dystonia [14]. Although juvenile parkinsonism is a heterogeneous condition characterized by akinesia, rigidity and dystonia, with a relatively young onset, this syndrome can be differentiated from idiopathic torsion dystonia by akinesia and rigidity, both of which are seldom observed in torsion dystonia except at the most advanced stage of illness.

Hereditary progressive dystonia with marked diurnal fluctuation (HPD) [15] and dopa-responsive dystonia (DRD) [16] are dystonias, mainly affecting the lower extremities. The dystonia in these conditions was a dystonic posture, and action dystonia was not characteristic. The marked diurnal fluctuation and the efficacy of *L*-dopa, both of which are characteristic for HPD and DRD, are not observed in torsion dystonia.

Aging Effects on Tremor Frequency in Parkinson's Disease

Tremor is a most predominant initial symptom and a major symptom throughout the course in Parkinson's disease. Tremor at rest is unique for Parkinson's disease. Its frequency ranges from 3 to 7, mostly 4–6 Hz [17]. Our colleague Nezu [18] studied age effects on the frequency of parkinsonian resting tremor. Patients were divided into two groups by age of onset. A group of early-onset parkinsonism was composed of 14 patients whose illness started before 40 years of age. Ages at study ranged from 14 to 48 years with a mean of 30.1. The course of illness was 1–10 years. The group of ordinary Parkinson's disease was composed of 23 patients. Age at study ranged from 44 to 72 years with a mean of 59.6 years. The duration of illness was 1–11 years. The frequency of tremor calculated in EMG in relation to the age is shown in figure 4.

There was a tendency for the frequency of tremor to decrease with advancing age. It is noteworthy that 3 patients in the early-onset group showed tremor with a high frequency of over 7 Hz, which was not observed in cases with ordinary Parkinson's disease, beginning after middle age.

Change in the frequency of tremor by age was described by Marshall [1] as physiological tremor. Marshall observed that the frequency of postural physiological tremor was around 9–11 Hz between the ages of 20 and 40 and there was no decline in the frequency during this period. Then the frequency of the tremor declined with age, steadily reaching 6.5 Hz at the ages of 70–74 and remained at about the same level between 70 and 90 years.

The neural circuitry involved in tremor production includes the rhythm generator in the brain, descending pathways and the reflex circuitry involving spinal motoneurones, peripheral nerves and muscle [19]. Rhythm formation by certain neural circuitries in the brain may be influenced by nerve conduction velocities, which change with age [20]. The reflex circuitry involving the spinal motoneurones, peripheral nerves and muscle is known as influencing the rhythm of centrally originating tremor and stabilizing the rhythm [21]. The slowing of nerve conduction velocities or of synaptic transmission with age may be factors to slow down the rhythm with age in parkinsonian tremor and physiological tremor.

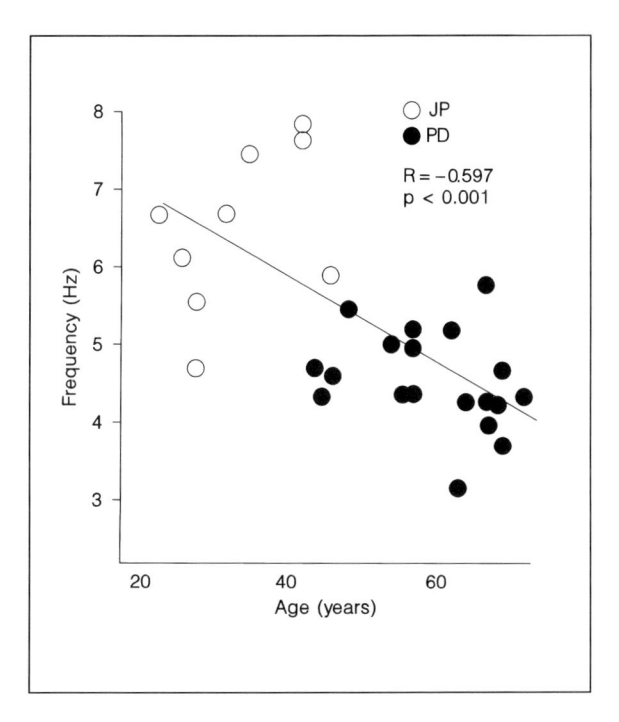

Fig. 4. Frequency of resting tremor by age in Parkinson's disease [18]. JP = Juvenile parkinsonism; PD = Parkinson's disease.

Liability to Develop Postural or Action Tremor among Young Patients with Basal Ganglia Disorders

The high-frequency tremors ranging from 8 to 12 Hz were frequently observed among young subjects with basal ganglia disorders, including hereditary progressive dystonia [8], early onset Parkinson's disease [18], torsion dystonia [13], Hallervorden-Spatz disease [22] and juvenile parkinsonism with pallidal posture [23]. This high-frequency tremor seems to be a characteristic feature among young patients with basal ganglia disorders. However, it is not pathognomonic for young patients, but adults with Parkinson's disease or torsion dystonia may develop high-frequency postural or action tremor as well.

References

1. Marshall J: The effect of ageing upon physiological tremor. J Neurosurg Psychiatry 1961;24:14–17.
2. Yanagisawa N, Hashimoto T: Quantitation of involuntary movements with electromyography; in Clifford-Rose F (ed): Parkinson's Disease and the Problems of Clinical Trials. London, Smith-Gordon, 1992, pp 131–142.
3. Lippold OGJ: The relation between integrated action potentials in a human muscle and its isometric tension. J Physiol (Lond) 1952;117:492–499.
4. Inman VT, Ralston HJ, Saunders JBGM, et al: Relation of human electromyogram to muscular tension. Electroencephalogr Clin Neurophysiol 1952;4:187–194.
5. Yanagisawa N: EMG characteristics of involuntary movements; in Bruyn GW, Roos RAC, Buruma OJS (eds): Dyskinesias, Uden, Sandoz, 1984, pp 142–159.
6. Yanagisawa N: The spectrum of motor disorders in Huntington's disease. Clin Neurol Neurosurg 1992;94(suppl):182–184.
7. Yanagisawa N: Extrapyramidal disease. Clinical and pathological aspects. Jap J Neuropharmacol 1981;3:765–783.
8. Segawa M, Nomura Y, Kase M: Diurnally fluctuating hereditary progressive dystonia; in Vinken PJ, Bruyn GW, Klawans HL (eds): Handbook of Clinical Neurology, Amsterdam, Elsevier, 1986, vol 5, pp 529–539.
9. Oppenheim H: Über eine eigenartige Krampfkrankheit des kindlichen und jugendlichen Alters (Dysbasia lordotica progressiva, Dystonia musculorum deformans). Neurol Cbl 1911;25:378–392.
10. Zeman W, Kaelbling R, Pasamanick B: Idiopathic dystonia musculorum deformans. 1. The hereditary pattern. Am J Genet 1959;11:188–202.
11. Ribera AB, Cooper IS: The natural history of dystonia musculorum deformans. A clinical study. Arch Pediatr 1960;77:55–71.
12. Herz E: Dystonia. 1. Historical review; analysis of dystonic symptoms and physiologic mechanisms involved. Arch Neurol Psychiatry 1994;51:305–318.
13. Yanagisawa N, Goto A: Dystonia musculorum deformans. Analysis with electromyography. 1971;13:39–65.
14. Yokochi M: Juvenile parkinsonism. I. Clinical features. Adv Neurol Sci (Tokyo) 1979;23:1048–1059.
15. Segawa M, Hosaka A, Miyagawa F, Nomura Y, Imai H: Hereditary progressive dystonia with marked diurnal fluctuation. Adv Neurol 1976;14:215–233.
16. Nygaard TG, Marsden CD, Duvoisin RC: Dopa-responsive dystonia. Adv Neurol 1988;50:377–384.
17. Goto A: Parkinsonism. Clinical problems. Adv Neurol Sci (Tokyo) 1968;12:957–962.
18. Nezu A: Study on tremor in parkinsonism – comparison of juvenile parkinsonism and Parkinson's disease with EMG. Shinshu Med J 1986;34:625–639.
19. Stein RB, Oguztöreli MN: Reflex involvement in the generation and control of tremor and clonus; in Desmedt (ed): Physiological Tremor, Pathological Tremors and Clonus. Basel, Karger, 1975, pp 28–50.
20. Mayer RF: Nerve conduction studies in man. Neurology 1963;13:1021–1030.
21. Pollock LJ, Davis L: Muscle tone in parkinsonian state. Arch Neurol Psychiatry 1930;23:303–331.
22. Yanagisawa N, Shiraki H, Minakawa M, Narabayashi H: Clinico-pathological and histochemical studies of Hallervorden-Spatz disease with torsion dystonia with special reference to diagnostic criteria of the disease from the clinico-pathological view point; in Tokizane T, Schadé JP (eds): Correlative Neurosciences. B. Clincal Studies. Amsterdam, Elsevier, 1969, Prog Brain Res vol 21B, pp 373–425.
23. Yanagisawa N: Juvenile parkinsonism with pallidal posture and spastic paraplegia; in Segawa M (ed): Hereditary Progressive Dystonia with Marked Diurnal Fluctuation. New York, Parthenon Publishing, 1993, pp 205–214.

Nobuo Yanagisawa, MD, Department of Medicine (Neurology),
Shinshu University School of Medicine, Asahi 3-1-1, Matsumoto 390 (Japan)

Voluntary Saccades in Normal and in Dopamine-Deficient Subjects

Okihide Hikosaka[a,b], Hideki Fukuda[a,c], Masaya Segawa[a], Yoshiko Nomura[a]

[a] Segawa Neurological Clinic for Children, Tokyo,
[b] Department of Physiology, Juntendo University, School of Medicine, Tokyo, and
[c] National Institute of Industrial Health, Kawasaki, Japan

The need for devising a method to evaluate the active brain functions becomes urgent when we deal with neurological disorders affecting so-called higher brain functions. The basal ganglia system, for example, is no longer viewed as a simple motor structure; we have begun to think that the basal ganglia are also crucial for procedural learning, problem-solving, and emotional behaviors [1].

We propose in this article that voluntary eye movement is a ideal paradigm to evaluate motor and cognitive functions, and their disorders. We measure voluntary saccades while a subject catches up a jumping light spot and detects its dimming. The task is noninvasive and even creates enthusiasm among young children.

The task was originally designed for monkeys. One of the authors (O.H.) and his colleagues have studied normal monkeys, muscimol-injected monkeys [2], and dopamine-deficient monkeys [3]. Comparison of the monkey data with the human data will be very useful to nail down the nature of behavioral changes in different diseases, as we will indicate.

Context-Dependent Oculomotor Control by the Basal Ganglia

We make a saccadic eye movement because output neurons in the superior colliculus emit a burst of spikes [4], and the afferents from the substantia nigra terminate exactly on such output neurons, among other inputs from various corti-

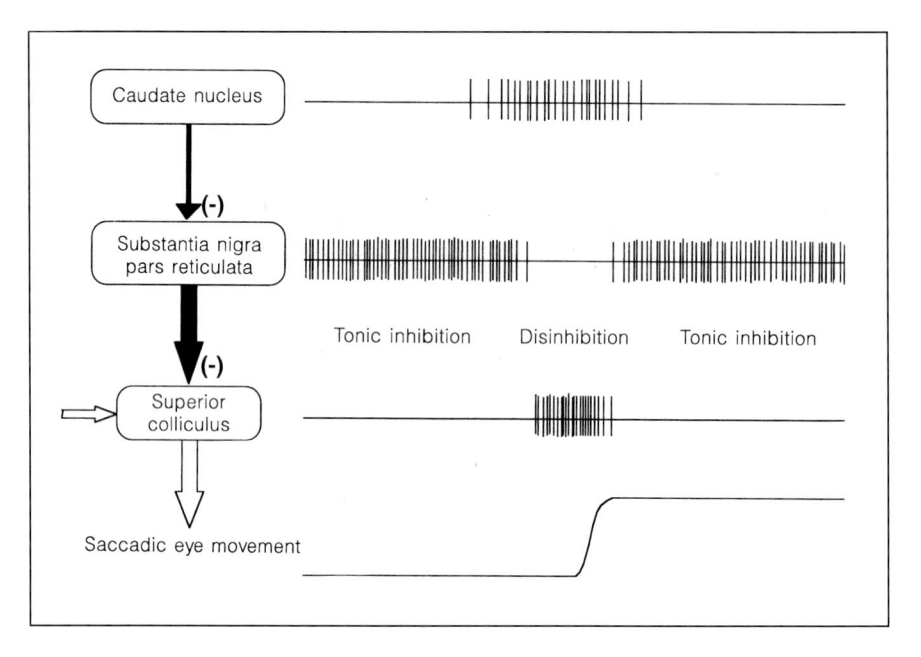

Fig. 1. The basal ganglia contribute to the initiation of saccades by disinhibition. A group of caudate neurons (top) shows spike activity before saccades to task-specific targets. The transient signal would inhibit the tonic spike activity of substantia nigra pars reticulata neurons (center), thereby removing the inhibition on saccade burst neurons in the superior colliculus (bottom).

cal regions [5]. These facts provided a perfect background for us to study the basal ganglia using eye movement paradigms.

Based on our results and other related findings, we have shown that disinhibition is a key feature of basal ganglia function [2] (fig. 1). The basal ganglia normally exert tonic inhibitory influences over their target structures, not only the superior colliculus but also other brainstem areas and part of the thalamus. The inhibition is maintained by the extremely high spontaneous activity of the output neurons in the substantia nigra pars reticulata and the internal segment of the globus pallidus. The signals from the striatum – caudate and putamen – truncate the inhibition, thus yielding a powerful facilitatory effect [6].

An important feature here is that this mechanism of disinhibition works in a context-dependent manner [6]. Such context dependency was revealed by training monkeys to make saccades in different behavioral situations [7]. The monkey first fixates a central spot of light and, if it steps away, refixates it by making a saccade. We call this type of saccade visually guided saccade. One useful variation

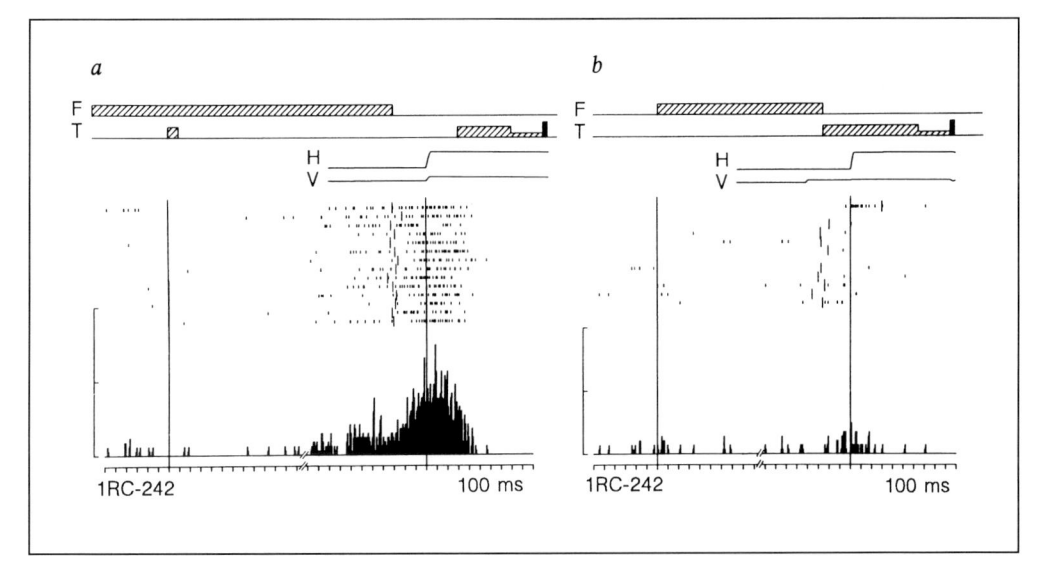

Fig. 2. Memory-specific saccadic activity in the monkey caudate. A group of caudate neurons show spike activity before a saccade to a selected area in the contralateral hemifield, when it is guided by memory *(a)*, but not when guided by vision *(b)*. Spike activity of a single caudate neuron are shown by dot raster displays. Each raster indicates the spike activity for a single task trial which was aligned on the onset of the saccade, each dot indicating a single action potential. F = Fixation point; T = target point; H and V = horizontal and vertical eye positions.

of this task is to ask the monkey to remember the location of a light flash and a few seconds later make a saccade to the remembered location. We call this memory-guided saccade.

A group of caudate neurons, which were otherwise silent, showed spike activity before a saccade to a selected area in the contralateral hemifield. Interestingly, the saccadic activity switched on and off, depending on how the saccade was initiated. Typically, as shown in figure 2, a burst of spike activity occurred before a memory-guided saccade, but the same neuron remained silent before a visually guided saccade even though these saccades were similar in amplitude and direction. Nearly one third of saccadic neurons in the caudate and the substantia nigra were selective for memory-guided saccades; another one third were selective for visually guided saccades [6, 7].

We expected therefore that basal ganglia dysfunction would lead to deficits in initiation of saccades, especially those guided by memory. This is in fact what we observed in dopamine-deficient monkeys and parkinsonian patients.

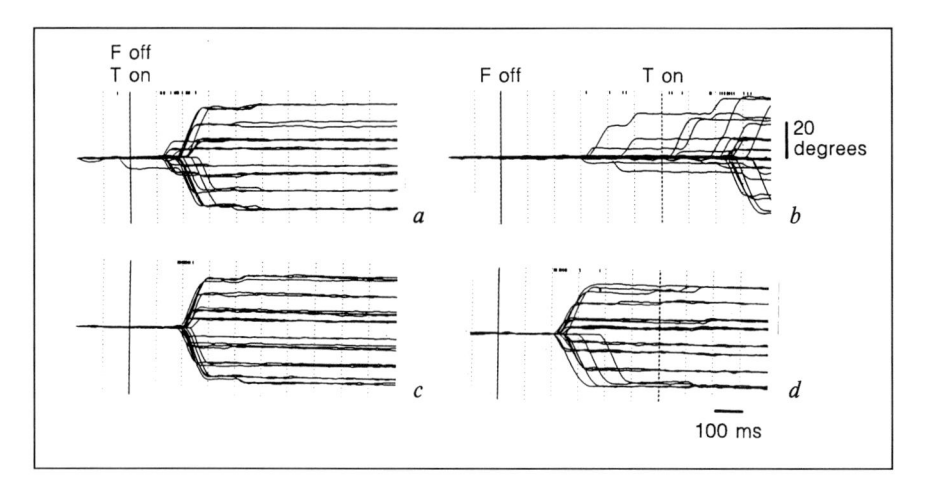

Fig. 3. Visually guided saccades *(a, c)* and memory-guided saccades *(b, d)* in a patient with Parkinson's disease *(a, b)* and a normal control subject *(c, d)*. Horizontal eye positions, aligned on the offset of the fixation point (F), are superimposed. The target (T) was chosen randomly from 8 positions – 5, 10, 20 and 30 degrees horizontally from the central fixation point.

Saccadic Eye Movements in Dopamine-Deficient Monkeys

We used MPTP to produce permanent dopamine deficiency. However, MPTP is so powerful that, if it is given systemically, the monkey will be immobile with severe parkinsonism. So we infused a small amount of MPTP locally into the monkey caudate nucleus on one side [8]. Such monkeys remained active without obvious parkinsonism, but their eye movements became deficient.

We compared visually guided saccades and memory-guided saccades before and after unilateral caudate MPTP infusion. Compared with the pre-MPTP data, visually guided saccades were relatively unimpaired. In contrast, memory-guided saccades to the contralesional side became hypometric, and slow with long latencies, their directions being sometimes wrong.

Memory-Guided Saccades Are Deficient in Parkinsonian Patients

The monkey's saccade tasks are perfectly applicable to humans; you can enjoy it as a computer game. Electrooculography (EOG) was used for eye movement recording. The subject pressed a button to initiate a trial and released it by detecting dimming of a light target. For the light targets, LEDs were embedded in a dome-shaped black screen, the light from which was seen through one of pinholes which were otherwise unseen.

Figure 3 shows examples of visually guided saccades and memory-guided saccades taken from a normal adult and a patient with Parkinson's disease. The target was chosen randomly from 8 positions – 5, 10, 20 and 30 degrees horizontally from the central fixation point. The horizontal component of EOG record was aligned on the offset of the fixation point.

In case of the visually guided saccades the target came on simultaneously. In case of memory-guided saccades the target did not appear until 600 ms later. The saccades which occurred during this time gap were thus guided by memory, not by currently available visual information. The memory-guided saccades predict the appearance of the target. When a saccade did not occur during the 600-ms gap period, we determined that the memory-guided saccade was absent in this trial, which was often the case in parkinsonian patients.

Using the saccade tasks, it became clear that parkinsonian patients do have deficits in initiation of saccades. And the deficits are often selective, apparent only in the task that requires memory-guided saccades. Latencies of visually guided saccades can even be shorter than in normal controls.

Age-Dependent Nature of Saccadic Eye Movement

A big problem here was that we did not have enough data of normal subjects. This would have been serious because we were going to deal with both young children and aged people. We therefore decided to collect data from normal subjects at various ages, which might give an insight into development and aging. So far, we have collected data from 80 normal subjects, and have found interesting tendencies.

The data were quantified for different parameters, some of which are shown in figures 4–7. Let us first examine the normal control subjects (indicated by open circles) in relation to their ages, and then compare them with dopamine-deficient patients (other symbols).

Figure 4 shows the mean latency or reaction time of visually guided saccades and memory-guided saccades against the age of the subject. In the normal control subjects, the latency of visually guided saccades was fairly constant across age, whereas the latency of memory-guided saccades was longer in young children and people over 50.

In figure 5, the saccade amplitude was normalized as the ratio to the eccentricity of the target which was either 20 or 30 degrees. There is the tendency that even normal subjects undershot the visual target up to about 10%. In con-

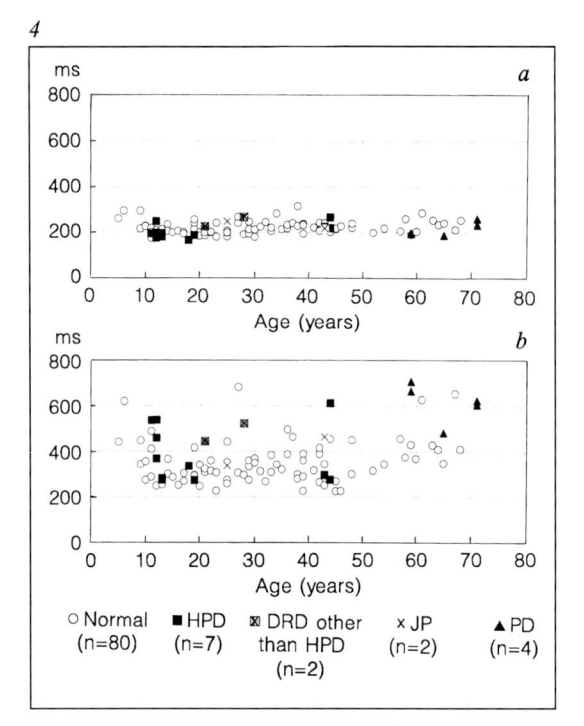

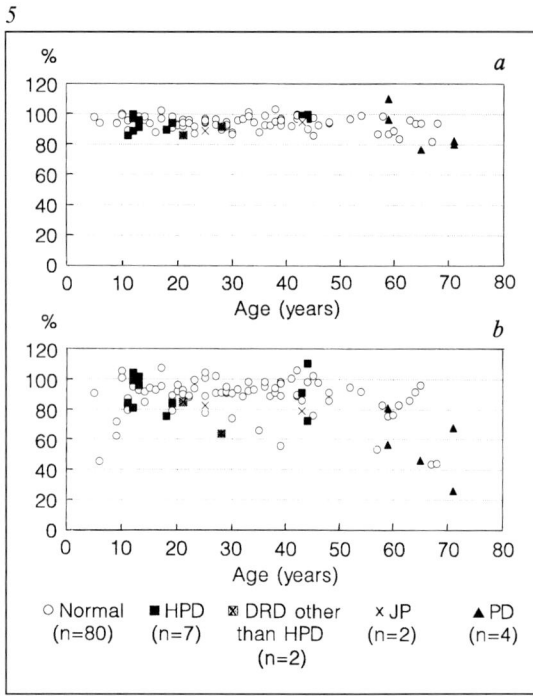

Fig. 4. Mean latency of visually guided saccades *(a)* and memory-guided saccades *(b)* in normal control subjects and patients with dopamine deficiency (HPD, DRD, JP, PD), plotted against the age of the subject. Each data point indicates the average of all successful trials (n = 50) with the target at 5, 10, 20, or 30 degrees on either side. The same convention applies to figures 5–7.

Fig. 5. Mean amplitude of visually guided saccades *(a)* and memory-guided saccades *(b)* in normal control subjects and patients with dopamine deficiency, plotted against the age of the subject. The amplitude was normalized to the eccentricity of the target as a percent value. The data include only the trials with the target at 20 and 30 degrees (n = 25 for each data point), because the signal-to-noise ratio may, in some subjects, be insufficient for smaller saccades.

trast, memory-guided saccades could be grossly hypometric; in some of the young children and older people, the amplitudes could be only half of the required values.

These data suggest that memory-guided saccade may be a good indicator of insufficient brain functions due to either immaturity or aging.

Note that, in the delayed-saccade task, the subject is required to suppress a saccade and then initiate a saccade. When a cue stimulus is presented, the subject must maintain fixation while remembering the location of the cue, and when the

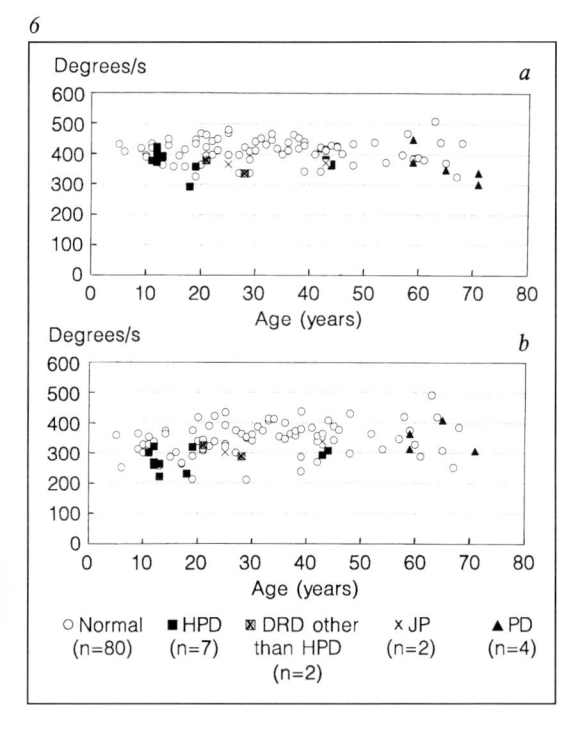

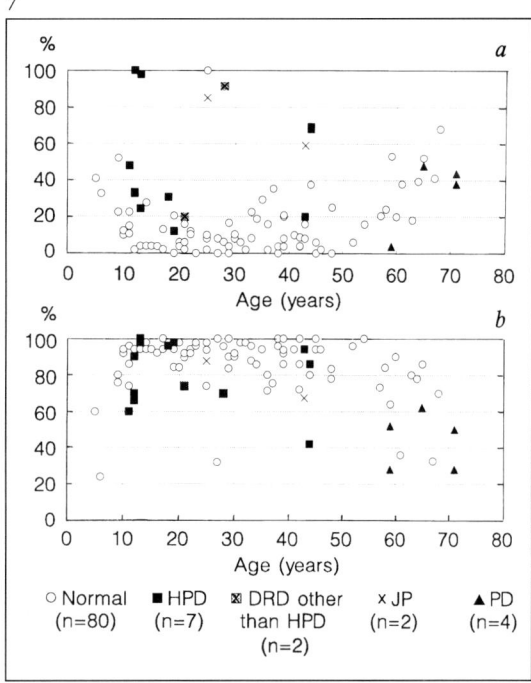

Fig. 6. Mean peak velocity of visually guided saccades *(a)* and memory-guided saccades *(b)* in normal control subjects and patients with dopamine deficiency, plotted against the age of the subject. The data include only the trials in which the amplitude of the saccade was between 17.5 and 22.5 degrees (note: the peak velocity is generally dependent on the amplitude). The number of trials for each data point thus varies among subjects (mean: 6.8).

Fig. 7. Frequency of saccades to the target cue *(a)* and the frequency of memory-guided saccades *(b)* in normal control subjects and patients with dopamine deficiency, plotted against the age of the subject. The saccade to the target cue was judged to be present when a saccade occurred toward the cue stimulus within 1 s after the presentation of the cue. The memory-guided saccade was judged to be present when a saccade started toward the remembered target within 600 ms after the offset of the fixation point (i.e., before the onset of the target). Each data point was based on 50 trials.

fixation point goes out he has to make a saccade promptly to the remembered location.

As shown in figure 7, these requirements were rather difficult for young children and older people. They tended to break fixation toward the cue stimulus but could not make a saccade to the remembered location when the saccade was required.

Saccadic Deficits in Dopamine-Deficient Diseases

We now compare these normal data with the performance of dopamine-deficient patients. In figures 4–6, on top of the normal data are superimposed the data from dopamine-deficient patients: hereditary progressive dystonia with marked diurnal fluctuation (HPD), dopa-responsive dystonia (DRD), juvenile parkinsonism (JP), and Parkinson's disease (PD).

Interestingly, these dopamine-deficient patients were quick to respond to visual targets (fig. 4). HPD and PD groups may even be quicker than normal subjects. However, the situation was reversed in memory-guided saccades. It took more time for the patients with HPD, DRD, and PD to initiate memory-guided saccades.

A similar difference was seen in saccade amplitude (fig. 5). In the dopamine-deficient patients, the amplitude of visually guided saccades was nearly normal, but memory-guided saccades tended to be hypometric. The decrease in amplitude was unclear in the HPD patients, however, which might give the impression that the saccades in the HPD patients are normal. But this is not the case.

The saccades of the HPD patients were slower in memory guided saccades, and perhaps in visually guided saccades as well (fig. 6). Such slowness might also be present in patients with DRD or JP, but not PD.

The saccadic deficits seen in the dopamine-deficient patients were reminiscent of our experimental results obtained after an injection of muscimol, a GABA agonist, int the monkey superior colliculus [9]. The saccades to the side contralateral to the muscimol injection became delayed, hypometric, and slower. The effects were stronger in memory-guided saccades; the monkey eventually became unable to make a saccade to the contralateral direction. The injected muscimol was thought to mimic overaction of the nigrocollicular GABAergic inhibition.

It might be suggested therefore that the saccadic deficits observed in the dopamine-deficient patients are due to the hyperactive nigrocollicular inhibition. Interestingly, however, the three saccade parameters – latency, amplitude and velocity – were, to some extent, differently affected in these patients.

As indicated previously, the delayed-saccade task required both suppression of saccades to a flashed cue and prompt initiation of saccades to a remembered target, which is difficult even for normal subjects, especially when they are young or old.

The two tasks were even more difficult for the dopamine-deficient patients (fig. 7). The suppression of unnecessary saccades was especially difficult for younger patients, whereas the initiation of memory-guided saccades was especially difficult for older patients.

These two kinds of deficits were even more prominent in young dystonic patients who had lesions in the striatum and older patients with multisystem atro-

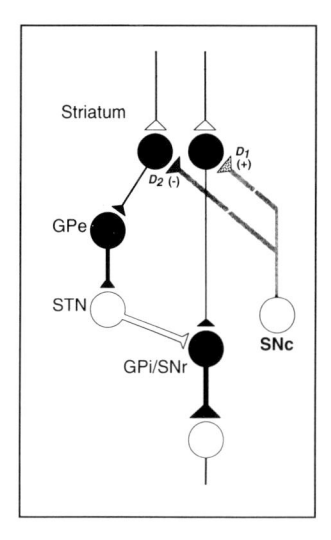

Fig. 8. Dual mode of basal ganglia action and dopaminergic modulation. Excitatory and inhibitory neurons are shown by open and filled symbols, respectively. GPe = Globus pallidus, external segment; STN = subthalamic nucleus. GPi = Globus pallidus, internal segment; SNr = substantia nigra pars reticulata. Neurons in the substantia nigra pars compacta (SNc) are thought to modulate the information processing in the striatum via D_1 and D_2 receptors.

phy (not shown). The performance of the dopamine-deficient patients was comparatively better. Most of them had been treated with dopaminergic drugs; it will be important to examine how the drug treatment changes their performance.

Such difficulty of maintaining eye fixation is similar to the observations made after injection of muscimol into the monkey substantia nigra pars reticulata [10]. The monkey became unable to maintain fixation on the central spot of light, and made saccades repeatedly to the side contralateral to the injection. The injected muscimol was thought to block the nigrocollicular inhibition, because muscimol would bind to GABA receptors on substantia nigra cells, hyperpolarize them, and stop their firing.

The suggestion drawn from this comparison might appear paradoxical: these patients have difficulty in initiating saccades because the nigrocollicular inhibition is overactive; in addition, they have difficulty in suppressing unnecessary saccades because the nigrocollicular inhibition is underactive or blocked.

We think that this apparent paradox reflects the dual mode of basal ganglia function (fig. 8).

Dual Mode of Basal Ganglia Action and Dopaminergic Innervation

One of the mechanisms would facilitate the target structure by disinhibition through the direct connection from the striatum to the output areas of the basal ganglia (SNr or GPe). Another mechanism would enhance the inhibition on the target, for the following reason. The striatal outputs are mediated by the external segment of the globus pallidus (GPe), which is inhibitory, and the subthalamic nucleus (STN), which is excitatory. Through this indirect pathway, the inputs to the striatum would lead to enhancement of the basal ganglia inhibitory outputs, because this pathway contains two inhibitions, as opposed to one inhibition.

Using trained monkeys, we found that there are clusters of visuooculomotor neurons both in the external pallidum [11] and the subthalamic nucleus [12] whose activities are suitable for suppression of saccades.

Our hypothesis is as follows. While the subject is fixating, the indirect mechanism may be more active so that saccades to a cue stimulus are effectively suppressed, but once a trigger signal occurs (in this case the offset of the fixation point), the direct mechanism will take over to disinhibit the saccade mechanism in the superior colliculus. The dopaminergic mechanism may be important in switching between these two modes of action.

References

1 Hikosaka O: Basal ganglia – possible role in motor coordination and learning. Curr Opin Neurobiol 1991;1:638–643.
2 Hikosaka O, Wurtz RH: The basal ganglia; in Wurtz RH, Goldberg ME (eds): The Neurobiology of Saccadic Eye Movements. Amsterdam, Elsevier, 1989, vol 3, pp 257–281.
3 Miyashita N, Matsumura M, Usui S, et al: Deficits in task-related eye movements induced by unilateral infusion of MPTP in the monkey caudate nucleus. Soc Neurosci Abstr 1990;16:235.
4 Sparks DL: Translation of sensory signals into commands for control of saccadic eye movements: Role of primate superior colliculus. Physiol Rev 1986;66:118–171.
5 Karabelas AB, Moschovakis AK: Nigral inhibitory termination on efferent neurons of the superior colliculus: an intracellular horseradish peroxidase study in the cat. J Comp Neurol 1985;239:309–329.
6 Hikosaka O, Sakamoto M, Usui S: Functional properties of monkey caudate neurons. I. Activities related to saccadic eye movements. J Neurophysiol 1989;61:780–798.
7 Hikosaka O, Wurtz RH: Visual and oculomotor functions of monkey substantia nigra pars reticulata. I. Relation of visual and auditory responses to saccades. J Neurophysiol 1983;49:1230–1253.
8 Imai H, Nakamura T, Endo K, Narabayashi H: Hemiparkinsonism in monkeys after unilateral caudate nucleus infusion of 1-methyl-4-phenyl-1,2,3,6-tetrahydropyridine (MPTP): Behavior and histology. Brain Res 1988;474:327–332.
9 Hikosaka O, Wurtz RH: Modification of saccadic eye movements by GABA-related substances. I. Effect of muscimol and bicuculline in the monkey superior colliculus. J Neurophysiol 1985;53:266–291.
10 Hikosaka O, Wurtz RH: Modification of saccadic eye movements by GABA-related substances. II. Effects of muscimol in the monkey substantia nigra pars reticulata. J Neurophysiol 1985;53:292–308.
11 Kato M, Hikosaka O: Function of the indirect pathway in the basal ganglion oculomotor system. Visuo-oculomotor activities of external pallidum neurons. Monogr Neural Sci. Basel, Karger, 1995, vol 14, pp 178–187.
12 Matsumura M, Kojima J, Gardiner TW, Hikosaka O: Visual and oculomotor functions of monkey subthalamic nucleus. J Neurophysiol 1992;67:1615–1632.

Dr. O. Hikosaka, Department of Physiology, Juntendo University School of Medicine,
2-1-1 Hongo, Bunkyo-ku, Tokyo 113 (Japan)

Phasic Activity during REM Sleep in Movement Disorders

Hideki Fukuda[a,b], *Masaya Segawa*[a], *Yoshiko Nomura*[a], *Kyoko Nishihara*[c], *Yukio Ono*[c,d]

[a] Segawa Neurological Clinic for Children, Tokyo,
[b] Department of Industrial Physiology, National Institute of Industrial Health, Ministry of Labor, Kawasaki,
[c] Department of Psychophysiology, Psychiatric Research Institute of Tokyo, and
[d] Division of Total Health Evaluation, Tsukuba Medical Center, Tsukuba, Japan

To study the pathophysiology of movement disorders, it is important and useful to analyze phasic activities during sleep such as twitch movements (TMs), gross movements and rapid eye movements (REMs) [1, 2]. Since movements and postures during sleep are not under the voluntary modulation of the cortex, these phasic components reflect activities of the subcortical neurons which are directly modulated by amine neurons and the basal ganglia [1, 3].

In previous studies, we observed that phasic activities during sleep showed abnormalities of the basal ganglia, which are specific for each disorder [1]. TMs, particularly those in the REM stage, are modulated by the nigrostriatal·dopamine neuron (NS·DA neuron), and the rate of occurrence in this stage reflects activities of the NS·DA neuron [1, 3]. In the present study, we evaluated the TMs of the mentalis muscle during sleep in various kinds of diseases of the basal ganglia modulated by the NS·DA neuron, including hereditary progressive dystonia with marked diurnal fluctuation (HPD) and chronic tic syndrome (CTS), and suggest a hypofunction of the NS·DA neuron in CTS.

Methods

Subjects in this study were as follows: (a) 11 normal children as controls (8 males and 3 females) with ages ranging from 6 to 13 years; (b) 7 cases with disorders of the basal ganglia, aged 5–17 years, who had symptomatic torsion dystonia, 3 unilateral and 4 bilateral lesions

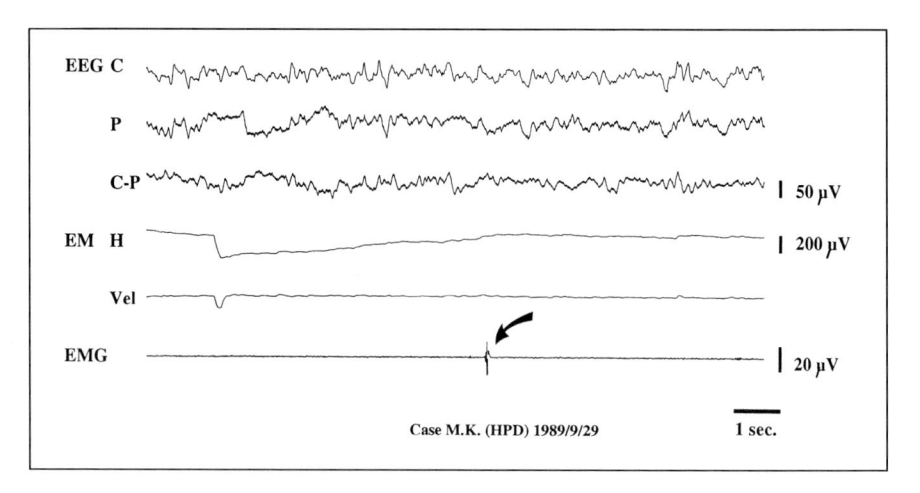

Fig. 1. TMs observed in the mentalis muscle during the REM stage. The arrow indicates the TMs counted in this study observed in the mentalis muscle. C = Central area; P = parietal area; EM H = horizontal eye movement; Vel = velocity of eye movement; EMG = electromyogram of the mentalis muscle.

in the striatum with or without lesions in other nuclei of the basal ganglia, detected by computerized tomography or magnetic resonance imaging; (c) 9 females with HPD, aged 5–14 years; and (d) 10 cases with CTS, chronic vocal or motor tics and Gilles de la Tourette syndrome (GTS) (9 males and 1 female), aged 6–14 years.

Polysomnography for normal children was performed on 3 consecutive nights, while for clinical cases with diseases of the basal ganglia, HPD or CTS, it was done for 1 night before administration of *L*-dopa. The polysomnographic parameters recorded of normal children were electroencephalogram (EEG) for the bilateral frontal, central, occipital and temporal areas, electrooculogram (EOG) for horizontal and vertical eye movements, electromyogram (EMG) for the mentalis muscle, electrocardiogram (ECG) and respiration; for the clinical cases, they were EEG for the unilateral central and parietal areas, EOG for horizontal eye movement, EMG from the mentalis muscle and 12 muscles, 6 of each side, from the trunk and extremities; sternocleidomastoid, proximal and distal muscles of both upper and lower extremities and abdominal rectus, ECG and respiration.

Sleep stages were scored in 20-second epochs according to the standard criteria [4] and TMs of the mentalis muscle during sleep were counted throughout the sleep. TMs, as defined previously [1, 3], are a short EMG activity localized to one muscle and lasting less than 0.5 s, but in this study they were defined as EMG activity of an amplitude of more than 20 µV (fig. 1). TMs occurring in each stage were counted per minute. Regarding statistical analysis, a one-way analysis of variance and a t test were used.

As demonstrated in figure 2, the average numbers of TMs per minute per sleep stage observed in normal children in the first night record showed a significant difference among sleep stages. The values were highest in REM sleep, next in sleep stage 1 and low in other NREM stages with the lowest in sleep stage 4. This pattern is identical to that observed in

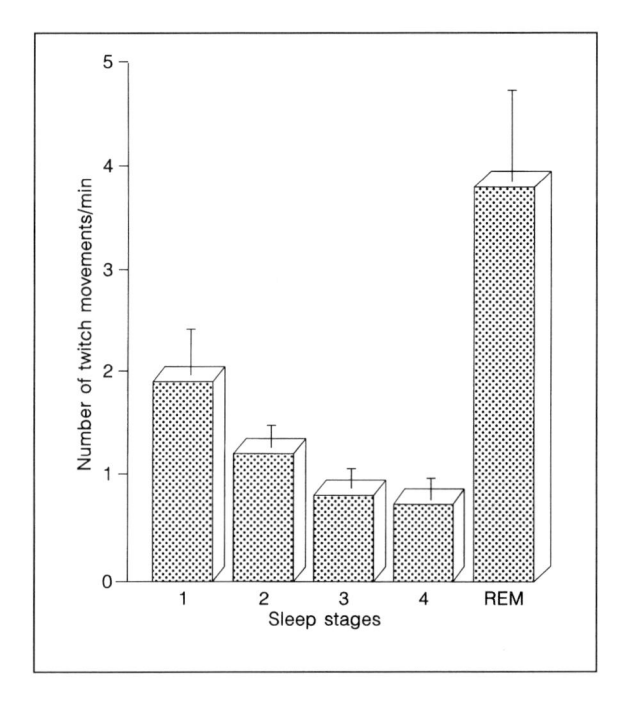

Fig. 2. Number of TMs per minute (mean ± SEM) in each sleep stage for normal children. S1–4: sleep stage 1–4. REM = REM stage.

previous studies [3] and revealed the stability of this paramter without the first-night effects. Thus, we compared the number of TMs per minute during REM sleep between normal and diseased children and among the three disorders to evaluate the mode of involvement of the NS·DA neuron or the basal ganglia. Furthermore, the age variation of this index was evaluated by examining the values of normal children and patients of various ages. The nocturnal variation of TMs was examined by comparing the number of TMs during REM sleep in the first two sleep cycles with those in the last two sleep cycles.

Results

Figure 3 shows a significant reduction in the mean number of TMs per minute in all of the affected cases compared with normal children. Furthermore, when lesions of the basal ganglia were divided into unilateral and bilateral lesions, the decrement was more remarkable in the latter. In 3 cases with unilateral lesions, the mean number of TMs per minute was 1.2, 1.5 and 0.4, respectively, while in 4 cases with bilateral lesions, it was 0.0, 0.2, 0.1 and 0.3, respectively.

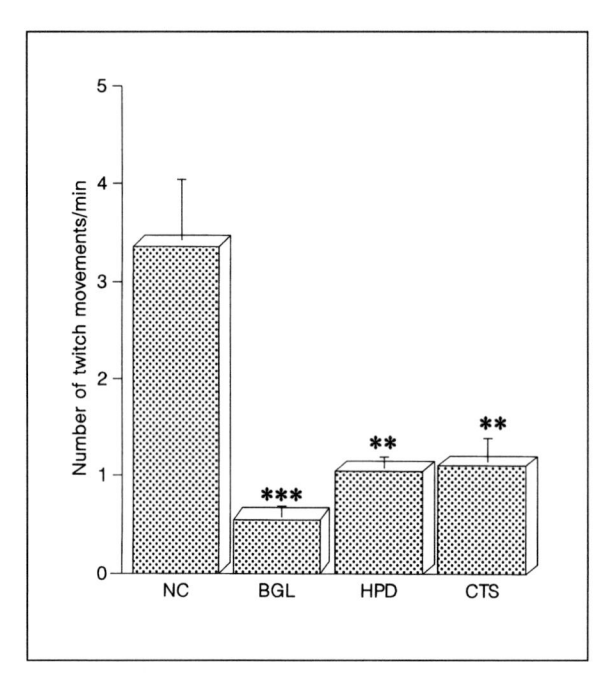

Fig. 3. Number of TMs per minute (mean ± SEM) in REM sleep compared among normal children, cases with lesions in the basal ganglia (BGL), cases with HPD and CTS. ** $p<0.01$, *** $p<0.001$: statistical analysis of the values of diseased patients against those of normal children.

As shown in figure 4, the number of TMs in normal children gradually decreased with age. Such age-dependent decrements of TMs were also observed in cases with HPD and with CTS, except for 1 case with CTS aged 14 years. This age variation was not observed in cases with lesions of the basal ganglia.

The nocturnal variation of the number of TMs is shown in figure 5. All of the normal children consistently showed a stable nocturnal variation, that is, the number of TMs was larger in the last cycle of sleep than in the first. This difference was statistically highly significant ($p < 0.001$). Regarding patients with lesions of the basal ganglia, no significant nocturnal variation was observed, though 3 cases with unilateral lesions demonstrated a slight, nonsignificant nocturnal variation. In both HPD and CTS, TMs significantly increased in the last cycle ($p<0.02$ for both diseases), though this variation was unremarkable in older cases with these disorders.

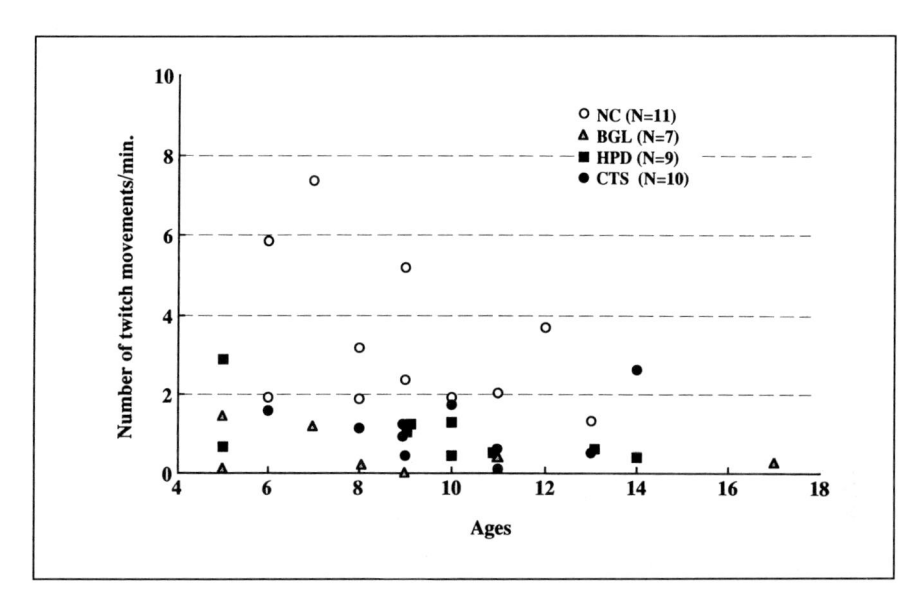

Fig. 4. Variation of the number of TMs during REM sleep in the four groups according to age. Abbreviations are the same as for figure 3.

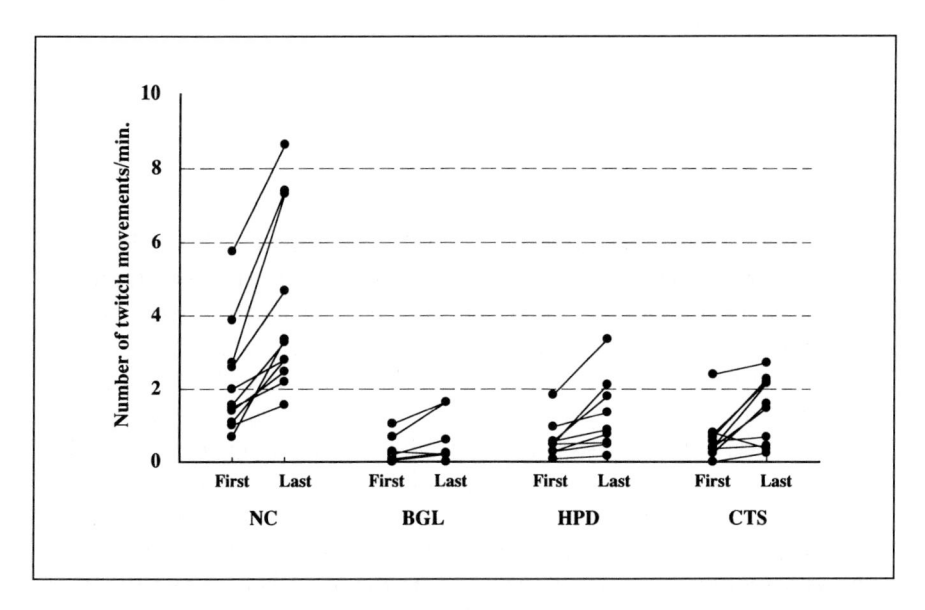

Fig. 5. Difference in the number of TMs per minute during REM sleep between the first two sleep (First) and last two sleep cycles (Last) are compared among normal children and the three movement disorders. Abbreviations are the same as for figure 3.

Discussion

Polysomnographic studies on three kinds movement disorders: symptomatic torsion dystonia with lesion of the basal ganglia, HPD and CTS, revealed a significant decrease in the rate of occurrence of TMs in the REM stage in all of these disorders.

These results verified our previous results on HPD and symptomatic disorders of the basal ganglia [1, 5], and also revealed a significant decrease of TMs in REM sleep in CTS. Furthermore, the present study showed particular age and nocturnal variations of the phasic components in normal children. As the number of TMs in the REM stage reflects the activities of the NS·DA neuron [1-3], the results of the present study revealed a decrease in DA activities in CTS, as well as in HPD.

Studies on autopsied cases showed a marked decremental age variation of the levels of tyrosine hydroxylase in the striatum, in the first two decades of life [6] and an animal experiment revealed the circadian oscillation of DA content in the striatum [7]. So the age and nocturnal variations of TMs in normal children are considered as the reflection of these variations of activity in the NS·DA neuron. Preservation of age and nocturnal variations in HPD and CST implied that in these disorders the NS·DA neuron, though reduced in its activities, preserved these neuronal functions, without any morphological abnormalities, either in the neuron or in the basal ganglia. The lack of nocturnal variation in older cases of both HPD and CTS may be due to the marked decrement of DA activities with age or the advancement of the disease. Absence of age and nocturnal variations in cases with symptomatic torsion dystonia with bilateral lesions of the basal ganglia may be due to masking of the functional variation of the NS·DA neuron by striatal lesions.

Our previous polysomnographic studies on GTS revealed a significant increase in TMs of the axial and limb muscles in all sleep stages, particularly in the REM stage [8], while there was no difference in the results in HPD between previous [1, 5] and present studies.

In our previous studies, we counted as TMs all muscle activities that were localized to one muscle and lasted less than 0.5 s, irrespective of the amplitude.

As observed in figure 6, there are numerous TMs with amplitudes of less than 20 μV. In HPD, these small TMs are rarely detected (fig. 1). Thus, the discrepancy found between previous and the present studies as concerns tic disorders might be due to the fact that in our previous study all TMs were counted, including those with low amplitude. Previous studies also showed particular abnormality in the sleep-stage-dependent modulation of GMs [8], suggesting the development of receptor supersensitivity.

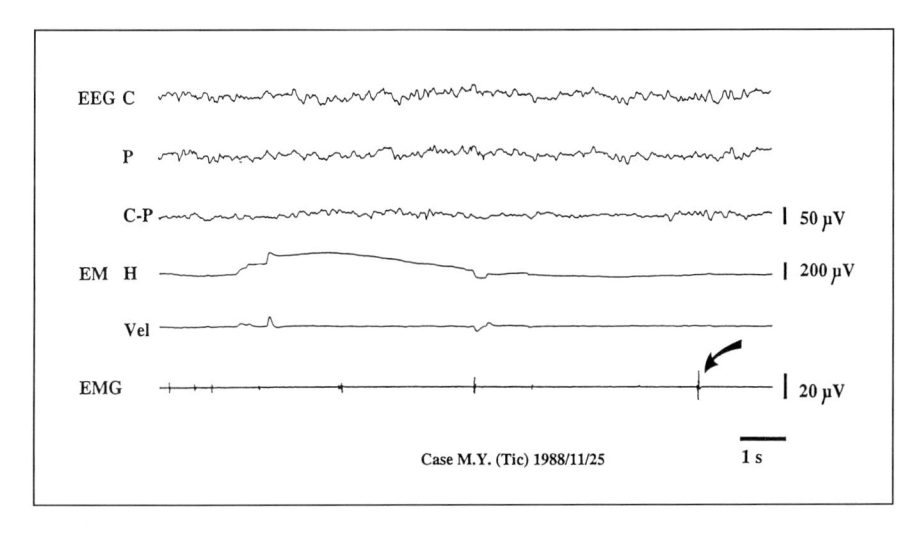

Fig. 6. TMs during REM sleep in a case with CTS. The arrow indicates TMs of an amplitude higher than 20 μV. Note the presence of a number of TMs of lower amplitudes. Abbreviations are the same as for figure 1.

The existence of a moderate number of TMs with a lower amplitude in GTS or CTS might be due to receptor supersensitivity. Thus past and present polysomnographic studies have revealed the pathophysiologies of CTS as a hypofunctioning of the NS·DA neuron with receptor supersensitivity. In contrast, there is no evidence suggesting receptor supersensitivity in HPD [9, this volume].

The direct projection of the striatum has shown important roles in dystonia [10]. In PET studies, D_2 receptors are shown to be spared or not involved in dopa-responsive dystonia or HPD [11, this volume] and GTS [12, this volume]. Thus the pathophysiologies underlying the reduction in the number of TMs in the disorders considered in this study are interpreted as follows:

In HPD, the decrease in the activity of the DA neuron leads to disfaciliation of the striatal neurons, which disinhibits the output pathways of the basal ganglia from the medial segment of the globus pallidus and the pars reticulata of the SN, and may finally suppress TMs. The striatal lesion in symptomatic torsion dystonia may cause dysfunction of the GABAergic neurons projecting to the internal segment of the globus pallidus and to the pars reticulata of the substantia nigra and disinhibit the output pathways from these nuclei, consequently causing dystonia and a decrease in TMs during the REM stage. These processes explain the modulation of horizontal REMs in these symptomatic dystonias well [2]. In CTS, a pathophysiology similar to HPD might exist. However, with receptor supersensitivity, incomplete facilitation of striatal direct projection may cause irregular and incomplete inhibition of the output pathways of the basal ganglia to appear as increased TMs with low amplitude. As REMs are modulated differently from TMs by the NS·DA neuron [1, 2], correlative studies of TMs with REMs in the REM stage are necessary for further investigation of the pathophysiology of the NS·DA neuron and the basal ganglia in CTS and GTS.

References

1 Segawa M, Nomura Y, Hikosaka O, Soda M, Usui S, Kase M: Roles of the basal ganglia and related structures in symptoms of dystonia; in Carpenter MB, Jayaraman A (eds): Basal Ganglia II: Structure and Function. New York, Plenum Press, 1987, pp 489–504.
2 Segawa M, Nomura Y: Rapid eye movements during REM stage are modulated by nigrostriatal dopamine neuron; in Bernardi G, Carpenter MB, Di Chiara G, Morelli M, Stanzione P (eds): Basal Ganglia III. New York, Plenum Press, 1991, pp 663–671.
3 Segawa M: Body movement during sleep: Its significance in neurology. Shinkei Naika (Tokyo) 1985;22:317–325.
4 Rechtschaffen A, Kales H: A Manual of Standardized Terminology, Techniques and Scoring System for Sleep Stages of Human Subjects. Washington, US Government Printing Office, 1968.
5 Segawa M, Nomura Y, Tanaka S, Hakamada S, Nagata E, Soda M, Kase M: Hereditary progressive dystonia with marked diurnal fluctuation: Consideration on its pathophysiology based on the characteristics of clinical and polysomnographical findings; in Fahn S, Marsden CD, Calne DB (eds): Advances in Neurology. New York, Raven Press, 1988, vol 50, Dystonia 2, pp 367–376.
6 McGeer EG, McGeer PL: Some characteristics of brain tyrosine hydroxylase; in Mandel AJ (ed): New Concept in Neurotransmitter Regulation. New York, Plenum Press, 1973, pp 53–68.
7 Phillips AG: Presented at the 3rd International Basal Ganglia Society Meeting. Cagliari, 1989.
8 Nomura Y: Tics (Including Gilles de la Tourette syndrome). Adv Neurol Sci (Tokyo) 1985;29: 265–275.
9 Hikosaka O, Fukuda H, Segawa M, Nomura Y: Voluntary saccades in normal and in dopamine-deficient subjects. Monogr Neural Sci. Basel, Karger, 1995, vol 14, pp 59–68.
10 Mitchell IJ, Luquin R, Boyce S, Clarke CE, Robertson RG, Sambrook MA, Crossman AR: Neural mechanisms of dystonia: Evidence from a 2-deoxyglucose uptake study in a primate model of dopamine agonist-induced dystonia. Mov Disord 1990;5:49–54.
11 Leenders KL, Antonini A, Meinck HM, Weindl A: Striatal dopamine D_2 receptors in dopa-responsive dystonia and Parkinson's disease. Monogr Neural Sci. Basel, Karger, 1995, vol 14, pp 95–100.
12 Turjanski N, Weeks R, Sawle GV, Brooks DJ: PET studies of the dopaminergic and opioid function in dopa-responsive dystonic syndromes and Tourette's syndrome. Monogr Neural Sci. Basel, Karger, 1995, vol 14, pp 77–86.

Hideki Fukuda, PhD, Segawa Neurological Clinic for Children, 2–8 Surugadai Kanda, Chiyoda-ku, Tokyo 101 (Japan)

Segawa M, Nomura Y (eds): Age-Related Dopamine-Dependent Disorders.
Monogr Neural Sci. Basel, Karger, 1995, vol 14, pp 77–86

PET Studies of the Dopaminergic and Opioid Function in Dopa-Responsive Dystonic Syndromes and Tourette's Syndrome

N. Turjanski, R. Weeks, G.V. Sawle, D.J. Brooks

MRC Cyclotron Unit, Hammersmith Hospital, London, UK

Positron emission tomography (PET) has been widely used to explore in vivo disturbances in cerebral function associated with extrapyramidal disorders. In particular, the dopaminergic system has been intensively studied, because of the availability of highly specific ligands which can reveal pre- and postsynaptic function. Most PET work has concentrated on determining the nature of the dopaminergic deficit in Parkinson's disease (PD).

Endogenous *L*-dopa is formed from tyrosine by the enzyme tyrosine hydroxylase, then converted to dopamine by the action of the aromatic amino acid decarboxylase which is subsequently stored in the presynaptic vesicles until stimulation of release occur. The activity of dopamine is terminated by re-uptake into the synaptic terminal or degradation by monoamine oxidase B and catechol-O-methyltransferase. Dopamine receptors fall into two main classes: D_1 and D_5 which are coupled to adenyl cyclase, and D_2, D_3 and D_4 receptors which either have no effect on, or inhibit this enzyme. D_3 and D_4 receptor subtypes are only weakly expressed in the basal ganglia [1]. All these receptors subtypes can exist in high- and low-agonist affinity conformation.

Presynaptic Dopaminergic PET Tracers. The functional integrity of dopaminergic nigrostriatal synapses has been assessed with PET using a variety of radiotracers. [^{11}C]nomifensine and [^{11}C]WIN 35,428 are antagonists of striatal dopamine re-uptake [2–4]. [^{11}C]tetrabenazine inhibits monoamine vesicular transport [5], while [11C]deprenyl labels monoamine oxidase B [6]. [18F]6-fluorodopa ([18F]dopa) is the tracer most frequently used to study the functional integrity of nigrostriatal projections as it is metabolised in a similar way to endogenous L-dopa [7]. Specific striatal [18F]dopa uptake can be described by an influx constant (K_i) that represents the product of the volume of distribution of the free tracer in the striatum and its decarboxylation rate constant [8]. K_i therefore reflects a combination of the density of functioning terminals and the concentration of aromatic amino acid decarboxylase [7]. The validity of this approach has been recently confirmed by Snow et al. [9] who reported a correlation between [18F]dopa influx constants and subsequent post-mortem nigral cell counts.

Postsynaptic Dopaminergic PET Tracers. The striatal density of postsynaptic dopaminergic receptors has been measured with PET using a variety of dopamine antagonists [7, 10]. Until recently, only D_2 tracers have been available for human studies. These are antagonists which bind reversibly or irreversibly to the receptors. Those binding irreversibly, such as [11C]methylspiperone and [18F]fluoroethylspiperone, are characterized by stronger affinity. The advantage of using more weakly binding reversible antagonists, such as [76Br]bromospiperone and [11C]raclopride, is that when tissue levels are in equilibrium with plasma, the striatal:cerebellar uptake ratio directly reflects the striatal receptor binding potential as the cerebellum contains few dopamine receptors [10]. Alternatively, formal kinetic modelling can be used to compute total tracer volumes of distribution [7, 11]. D_1 receptor binding can now be assessed with a benzazepine [11C]SCH 23390. This is a reversible antagonist of D_1 receptors and gives equilibrium striatum:cerebellum ratios of around 2 [12].

Opioid Receptor PET Tracers. The basal ganglia contain high concentrations of μ, κ and δ opioid receptor subtypes [13]. All three receptor subtypes are found on post-synaptic striatal neurons located on both intrinsic striatal neurons and striatal efferent neurons [14]. Opioid receptors are also found on striatal presynaptic neurons but there is uncertainty over which receptor subtype is expressed [15, 16]. There are a number of high-affinity positron-emitting ligands which can be used to demonstrate striatal opioid receptor density. [11C]carfentanil is a μ-selective opioid agonist [17], [18F]cyclofoxy is μ- and κ-selective [18] and [11C]-etorphine is a non-selective ligand with potent agonist properties [19]. [11C]diprenorphine is a synthetic opioid ligand which has equal affinity to μ-, κ- and δ-receptor subtypes [20]. Using the occipital area as a non-specific reference tissue, the striatal to occipital ratio reflects the specific striatal receptor binding potential.

Clinical Studies

Dopa-Responsive Dystonic Syndromes

PD, dopa-responsive dystonia (DRD) and adult-onset dystonia parkinsonism (DYS-P) are all clinical entities that manifest dystonia and parkinsonism and may present diagnostic confusion.

Idiopathic PD is characterized by an akinetic-rigid syndrome and gait disturbances. Some patients with otherwise classical disease present with dystonic posturing of the hand or foot even before introduction of dopaminergic therapy [21, 22]. The loss of dopamine terminal function in idiopathic PD has been extensively studied with [^{18}F]dopa PET [10]. The first reports of the use of this tracer in early PD described reduced putamen uptake that was greater contralateral to the more affected limbs [23]. [^{18}F]dopa K_i values have subsequently been shown to correlate with bradykinesia, gait alterations and overall disease severity [24]. Patients with PD have mean [^{18}F]dopa K_i values reduced to 50% of normal levels in the putamen, and about 80% of normal levels in the caudate [25, 26]. Preclinical detection of PD is consequently possible [27], and the progression of PD can be objectively monitored with [^{18}F]dopa PET [28, 29].

Presynaptic tracers such as [^{11}C]nomifensine and [^{11}C]WIN 35, 428 show large reductions in dopamine transporter binding in the striatum of patients with PD [2–4], while [^{11}C]deprenyl shows normal striatal uptake in PD [6].

PET studies on striatal D_2 receptor status in PD using the above-mentioned tracers have produced variable results, but overall, they suggest that D_2 receptor densities are normal or up-regulated in untreated PD, while they appear to be normal or down-regulated in treated PD [10]. [^{11}C]SCH 23390 studies suggest that striatal D_1 receptor binding potential remains normal in both treated and untreated PD [30, 31].

Familial DRD characteristically presents in childhood with a diurnally fluctuating dystonic gait; parkinsonism tends to appear later in the course of the disease. Adult-onset parkinsonism may also be a phenotypic expression of this disorder. Usually, DRD shows a striking and sustained response to treatment with low doses of *L*-dopa [32–38]. Sawle et al. [39] performed [^{18}F]dopa PET in 6 classical cases of DRD with childhood onset. They reported a mild but significant uniform reduction of mean putamen and caudate tracer uptake compared with normal subjects, a different pattern from that seen in sporadic PD where putamen is targeted. Other studies with [^{18}F]dopa PET have found striatal tracer uptake to be normal in both classical DRD cases [40] and their relatives presenting with adult-onset parkinsonism [41].

Some adult patients with *DYS-P* develop florid dystonic symptoms preceding or appearing concomitantly with parkinsonism, especially when the disease presents before the age of 40 [42, 43]. The nosological status of DYS-P is currently uncertain. Few autopsies have been performed in this condition, though nigral Lewy bodies were found in 2 cases [44]. This suggests that DYS-P may occur as a phenotypic variant of PD. Two previous [^{18}F]dopa PET studies on patients with DYS-P [40, 45] reported reduced striatal tracer uptake, but the low resolution of the cameras used prevented separation of caudate and putamen signals.

In this review we compare striatal [^{18}F]dopa uptake in adult-onset DYS-P, typical PD cases without dystonia, and patients with classical childhood-onset DRD. These findings have been previously reported [46].

Patients. We studied 6 adult-onset DYS-P patients with a mean age of 51 (range 27–63) years. Mean age at onset of the disease was 46 (23–58) years and mean Hoehn and Yahr score at the time of PET was 2.5 (1–4). All our patients fulfilled the following criteria: (1) florid dystonic symptoms, witnessed by us, preceding, or occurring at the onset of parkinsonism, before starting *L*-dopa treatment; (2) parkinsonism, indistinguishable from idiopathic PD, which was at least moderately *L*-dopa responsive; (3) intact basal ganglia on cerebral computed tomography or magnetic resonance imaging, and (4) no clinical evidence of pyramidal tract dysfunction.

In 3 of the 6 DYS-P cases, *L*-dopa exacerbated their dystonia, but in 2 of these 3 cases reduction of the dose reversed this. At the time of PET, 4 cases had a fluctuating therapeutic response of their parkinsonism to *L*-dopa and 2 had associated dyskinesias. One patient developed myoclonic jerks after therapy.

The PET findings of these patients were compared with 6 DRD cases previously described by Sawle et al. [39]. The DRD patients' mean age at the time of PET was 39 (18–66) years and at disease onset was 8 (3–12) years. Five had a positive family history of *L*-dopa-responsive dystonia, and had developed parkinsonism. All had a marked response to *L*-dopa medication without long-term treatment complications.

Results were compared with 12 age-matched patients with sporadic PD, mean age 52 (27–75) years, and mean Hoehn and Yahr stage 2.1 (1–3). Mean age of onset of PD was 47 (25–73) and none had pretreatment dystonia. Their clinical details have been previously reported [25]. Thirty-four normal subjects, mean age 58 (20–77), were scanned as controls. The methodology of [^{18}F]dopa PET analysis has been described elsewhere [2, 7].

PET Findings. Compared to normals, the DYS-P patients had significantly reduced mean putamen and caudate [^{18}F]dopa influx constants (table 1). Mean putamen tracer uptake was reduced to approximately 50% of normal in both DYS-P and PD. Putamen [F]dopa uptake was roughly half that of caudate in both conditions. Mean caudate uptake was 24% lower in DYS-P than in the PD group

Table 1. Summary of caudate and putamen [^{18}F]dopa PET uptake in normals, DRD, PD, DYS-P and TS groups; mean striatal [^{11}C]raclopride BP and [^{11}C]diprenorphine caudate and putamen:occipital ratio in normals and TS groups

[^{18}F]dopa [modified from ref. 33]	Age mean	Caudate K_i, min^{-1}	Putamen K_i, min^{-1}
Normal (34)	58	0.0105 ± 0.0018	0.0097 ± 0.00108
DRD (6)	38	0.0098 ± 0.0005	0.0080 ± 0.0007[a]
PD (12)	51	0.0095 ± 0.0016	0.0047 ± 0.0010[a]
DYS-P (6)	51	0.0072 ± 0.0018[a]	0.0044 ± 0.0014[a,b]
TS (10)	30	0.0097 ± 0.0010	0.0092 ± 0.0013

[^{11}C]raclopride	Age mean	Caudate BP	Putamen BP
Normal (9)	50	2.29 ± 0.16	2.35 ± 0.14
TS (5)	33	2.29 ± 0.34	2.31 ± 0.22

[^{11}C]diprenorphine	Age mean	Caudate:occipital	Putamen:occipital
Normal (6)	36	2.60 ± 0.25	2.48 ± 0.37
TS (6)	34	2.71 ± 0.23	2.68 ± 0.36

Results are means ± SD. BP = Binding potential (B_{max}/K_d).
Comparisons were performed using Student's paired two-tailed t test.
[a] p < 0.01 compared to normals; [b] p < 0.01 compared to DRD.

but this difference was not statistically significant. The DYS-P group showed a significantly greater reduction of putamen function than the DRD group.

In conclusion, our findings were compatible with adult-onset *L*-dopa-responsive DYS-P being a phenotypic variant of PD and strongly suggest that DYS-P has a different biochemical defect from childhood-onset DRD.

Gilles de la Tourette's Syndrome

Gilles de la Tourette's syndrome (TS) is a familial disorder characterized by simple and complex motor and vocal tics and obsessive-compulsive behavioral manifestations [47]. Isolated obsessive-compulsive disorders and multiple tic disease may well be an expression of the clinical spectrum of this disorder [48, 49].

The genetic and biochemical defects underlying TS have not been characterized, but the following observations suggested an abnormality of the dopaminergic pathway may be relevant: (1) TS generally responds to dopamine receptor antagonists (neuroleptics) and dopamine-depleting agents (reserpine, tetrabenazine); (2) dopamine re-uptake inhibitors, such as cocaine [50, 51] and amphetamines exacerbate tics [52]; (3) neuroleptic withdrawal can lead to tics [53]; (4) baseline levels of dopamine metabolites have been reported to be reduced in the cerebrospinal fluid of TS patients [54, 55]; (5) a post-mortem study using [^{3}H]mazindol has reported increased dopamine re-uptake site density in TS striatum though levels of striatal dopamine and its metabolites were normal [5, 6]. Striatal D_1 and D_2 receptor density were also normal [56]. A preliminary [^{11}C]methylspiperone PET study in 4 TS patients suggested increased D_2 receptor density [57], although more recently a group of 19 TS patients studied by the same authors showed no overall mean differences compared to controls [58].

Patients. Ten patients were selected for [^{18}F]dopa PET, 9 satisfied DSM-III criteria for TS [47], while the 10th had multiple-tic disease. Their mean age at the time of PET scanning was 30 (range 18–48) years, and at disease onset was 6 (2–13) years. Seven TS patients were taking neuroleptic drugs at the time of PET while 3 were drug naive; 2 had TS and 1 had multiple-tic disease. All patients had simple motor tics, all except the patient with multiple tics had simple vocal tics and obsessive-compulsive symptoms. Complex motor tics were present in 8 and 7 had complex vocalizations. These findings have been previously reported [59]. The patients' data were compared with those previously shown for 34 normal controls (table 1).

Five further TS patients with a mean age of 33 (18–46) years at the time of PET, and a mean onset of disease at 8 (5–12) years were studied with [^{11}C]raclopride. Three were drug naive and 2 had received no neuroleptic treatment for at least 3 months before PET. All 5 patients had simple motor and vocal tics. One had complex motor tics and 2 had complex vocal tics, obsessions and compulsions.

The patients' PET results were compared with those of 9 control subjects with a mean age of 50 (24–74) years.

PET Findings. The mean caudate and putamen K_i values of the TS and normal groups were similar, all patients' values lying within 2 standard deviations of the normal mean. [^{18}F]dopa uptake was not significantly different between treated and untreated TS groups. Table 1 summarizes the mean caudate and putamen K_i values for TS patients and controls.

The mean caudate and putamen [^{11}C]raclopride binding potentials in TS were similar to normals. However, 1 of the TS patients' values lay 2 standard deviations below the normal mean. Results are summarized in table 1.

Opioid Results in TS: [¹¹C]diprenorphine PET

As the opioid system may regulate dopamine release [60], an alternative and interesting hypothesis for the pathogenesis of TS is a dysfunction of the opioid system. This proposal arose following a post-mortem study of TS cases that showed a reduction of dynorphin-like immunoreactivity throughout the brain, especially in the external segment of the globus pallidus [61]. Against a dynorphin deficit underlying TS, however, is the finding of elevated levels of dynorphin in the cerebrospinal fluid of TS patients [62].

Patients. We performed [¹¹C]diprenorphine PET studies in 6 patients with TS with a mean age of 34 (range 18–60). All patients exhibited simple motor and vocal tics, 4 patients had onset of disease in childhood, whilst 2 patients had disease onset in their mid twenties. Five patients were drug free for over a year before scanning and 1 patient was off neuroleptic medication for 1 month before scanning. None of the patients were in pain during the scans or taking opiate-like medication.

The patients' results were compared with those of 6 normal subjects with a mean age of 36 years.

PET Findings. The striatal activity of [¹¹C]diprenorphine in the patients with TS was similar to that found in the normal subjects.

In conclusion, our [¹⁸F]dopa PET results do not suggest a reduction in number, or a dysfunction in dopa metabolism of the pre-synaptic dopaminergic terminals in TS. However, we cannot exclude an alteration in the production of endogenous *L*-dopa as [¹⁸F]-dopa utapke reflects AADC rather than TH activity. Our results are also unable to support the presence of an abnormality in the density of striatal D_2 receptor binding in TS, but cannot exclude an altered proportion of striatal D_2 receptors in high and low agonist affinity configuration, as [¹¹C]raclopride binds equally to these receptor conformations [52]. Preliminary results with [¹¹C]diprenorphine do not support the hypothesis that the striatal opioid system is involved in the pathophysiology of TS.

References

1   Kandel ER: Disorders of thought: Schizophrenia; in Kandel ER, et al (eds): Principles of Neural Science. New York, Elsevier, 1991, pp 853–868.
2   Brooks DJ, Salmon EP, Mathias CJ, et al: The relationship between locomotor disability, autonomic dysfunction, and the integrity of the striatal dopaminergic system, in patients with multiple system atrophy, pure autonomic failure, and Parkinson's disease, studied with PET. Brain 1990; 113:1539–1552.
3   Leenders KL, Salmon EP, Tyrell P, et al: The nigrostriatal dopaminergic system assessed in vivo by positron emission tomography in healthy volunteer subjects and patients with Parkinson's disease. Arch Neurol 1990;47:1290–1298.

4   Frost JJ, Rosier AJ, Reich SG, et al: Positron emission tomographic imaging of the dopamine transporter with ^{11}C-WIN 35,428 reveals marked declines in mild Parkinson's disease. Ann Neurol 1993;34:423–431.
5   DaSilva JN, Kilbourn MR, Domino EF: In vivo imaging of monoaminergic nerve terminals in normal and MTPT-lesioned primate brain using positron emission tomography (PET) and [^{11}C]tetrabenazine. Synapse 1993;14:128–131.
6   Bench C, Lammertsma AA, Dolan RJ, Brooks DJ, Frackowiak RSJ: Cerebral monoamine oxidase B (MAO-B) activity in normal subjects, Alzheimer's disease and Parkinson's disease (abstract). J Cereb Blood Flow Metab 1993;13:S246.
7   Sawle GV, Brooks DJ: Positron emission tomography studies of neurotransmitter systems. J Neurol 1990;237:451–456.
8   Brooks DJ: Studies in cerebral pharmacology with PET; in Rose F (ed): Advances in Neuropharmacology. Smith-Gordon, 1993, pp 1–12.
9   Snow BJ, Tooyama I, McGeer EG, et al: Human positron emission tomographic [^{18}F]Fluorodopa studies correlate with dopamine cell counts and levels. Ann Neurol 1993;34:324–330.
10 Brooks DJ: Functional imaging in relation to parkinsonian syndromes. J Neurol Sci 1993;115:1–17.
11 Swart JA, Korf J: In vivo dopamine receptor assessment for clinical studies using positron emission tomography. Biochem Pharmacol 1987;36:2241–2250.
12 Halldin C, Stone-elander S, Farde L, Ehrin E, Langstrom B, Sedvall G: Preparation of ^{11}C-labelled SCH 23390 for the in vivo study of dopamine D_1 receptors using positron emission tomography. Int J Radiat Appl Instrum [A] 1986;37:1039–1043.
13 Goodman RR, Adler BA, Pasternak GW: Regional distribution of opioid receptors; in Pasternak GW (ed): The Opiate Receptors. Clifton, Humana Press, 1988, pp 197–228.
14 Cross AJ, Hille C, Slater P: Subtraction autoradiography of opiate receptor subtypes in human brain. Brain Res 1987;418:343–348.
15 Trovero F, Herve D, Desban M, Glowinski J, Tassin JP: Striatal opiate µ-receptors are not located on dopaminergic nerve endings in the rat. Neuroscience 1990;39:313–320.
16 Murrin LC, Coyle JT, Kuhar MJ: Striatal opiate receptors: Pre- and postsynaptic localisation. Life Sci 1980;27:1175–1183.
17 Frost JJ, Wagner HN, Dannals RF, et al: Imaging opiate receptors in the human brain with positron tomography. J Cereb Blood Flow Metab 1985;9:231–236.
18 Kawai R, Channing MA, Rice KC, Newman AH, Blasberg RG: Opiate receptor subtype discrimination in vivo using cyclofoxy and the site specific alkylation agent beta-FNA. J Cereb Blood Flow Metab 1991;11(suppl 2):S872.
19 Maziere M, Berger G, Godot JM, Comar D. Etorphine [11]C: A new tool of the in vivo study of the brain opiate receptors. J Label Compounds Radiopharmacol 1981;18:15, 16.
20 Frost JJ, Sadzot B, Mayberg HS, et al: Estimation of receptor number and affinity for ^{11}C-diprenorphine binding to opiate receptors in man by PET. J Cereb Blood Flow Metab 1989;9(Suppl 1):S192.
21 Poewe WH, Lees AJ, Stern GM: Dystonia in Parkinson's disease. Clinical and Pharmacological features. Ann Neurol 1988;23:73–78.
22 Kidron D, Melamed E: Forms of dystonia in patients with Parkinson's disease. Neurology 1987;37:1009–1011.
23 Nahmias C, Garnett ES, Firnau G, Lang A: Striatal dopamine distribution in parkinsonian patients during life. J Neurol Sci 1985;69:223–230.
24 Eidelberg D, Moeller JR, Dhawan V, et al: The metabolic anatomy of Parkinson's disease: Complementary [^{18}F]fluorodeoxyglucose and [^{18}F]fluorodopa positron emission tomographic studies. Mov Disord 1990;5:203–213.
25 Brooks DJ, Ibanez V, Sawle GV, et al: Differing patterns of striatal ^{18}F-dopa uptake in Parkinson's disease, multiple system atrophy, and progressive supranuclear palsy. Ann Neurol 1990;28:547–555.
26 Martin WRW, Stoessl AJ, Adam MJ, et al: Positron emission tomography in Parkinson's disease: Glucose and Dopa metabolism. Adv Neurol 1986;45:95–98.

27 Sawle GV: The detection of preclinical Parkinson's disease: What is the role of positron emission tomography? Mov Disord 1993;8:271–277.
28 Bhatt MH, Snow BJ, Wayne Martin WR, Pate BD, Ruth TJ, Calne DB: Positron emission tomography suggest that the rate of progression of idiopathic parkinsonism is slow. Ann Neurol 1991;29: 673–677.
29 Sawle GV, Turjanski N, Brooks DJ, Frackowiak RSJ: The rate of disease progression in Parkinson's disease: PET findings in patients receiving medical treatment or following foetal mesencephalic transplantation. Neurology 1992;42(suppl 3):295–290.
30 Rinne JO, Laihinen A, Nagren K, et al: PET demonstrates different behaviour of striatal dopamine D_1 and D_2 receptors in early Parkinson's disease. J Neurosci Res 1990;27:494–499.
31 Shinotoh H, Hirayama K, Tateno Y: Dopamine D_1 and D_2 receptors in Parkinson's disease and striatonigral degeneration determined by PET. Adv Neurol 1993;60:488–493.
32 Nygaard TG, Marsden CD, Fahn S: Dopa-responsive dystonia: Long-term treatment response and prognosis. Neurology 1991;41:174–181.
33 Nygaard T, Marsden CD, Duvoisin RC: Dopa-responsive dystonia. Adv Neurol 1988;50:377–384.
34 Yamamura Y, Sobue I, Ando K, Iida M, Yanagi T, Kono C: Paralysis agitans of early onset with marked diurnal fluctuation of symptoms. Neurology 1973;23:239–244.
35 Nygaard TG: Dopa-responsive dystonia: Delineation of the clinical syndrome (abstract). Proc 10th Int Symp on Parkinson's Disease, 1991, pp 44–40.
36 Nygaard TG, Trugman JM, De Yebenes JG, Fahn S: Dopa-responsive dystonia. The spectrum of clinical manifestations in a large North American Family. Neurology 1990;40:66–69.
37 Segawa M, Nomura Y, Yamashita S, et al: Long-term effects of L-dopa on hereditary progressive dystonia with marked diurnal fluctuation; in Berardelli A, et al (eds): Motor Disturbances. II. New York, Academic Press, 1990, pp 305–318.
38 Segawa M, Nomura Y, Tanaka S, et al: Hereditary progressive dystonia with marked diurnal fluctuation. Consideration of its pathophysiology based on the characteristics of clinical and polysomnographical findings. Adv Neurol 1988;50:367–376.
39 Sawle GV, Leenders KL, Brooks DJ, et al: Dopa-responsive dystonia: [^{18}F]dopa positron emission tomography. Ann Neurol 1991;30:24–30.
40 Snow BJ, Okada A, Martin W, Duvoisin RC, Calne DB: PET scanning in dopa-responsive dystonia, parkinsonism-dystonia, and young-onset parkinsonism; in Segawa M (ed): Proc of the Hereditary Progressive Dystonia Symp, Tokyo 1990. London, Parthenon Press, 1992, pp 181–186.
41 Takahashi H, Snow B, Nygaard TG, Calne DB: Clinical heterogeneity of Dopa-responsive dystonia: PET observations. Proc 10th Int Symp Parkinson's Disease, 1991, p 155.
42 Rivest J, Quinn N, Marsden CD: Dystonia in Parkinson's disease, multiple system atrophy, and progressive supranuclear palsy. Neurology 1990;40:1571–1578.
43 Golbe LI: Young-onset Parkinson's disease: A clinical review. Neurology 1991;41:168–173.
44 LeWitt PA, Hiner BC, David WS, The Parkinson's Study Group (DATATOP). Parkinsonism with dystonia: Clinical and neurochemical investigations (abstract). Proc 10th Int Symp on Parkinson's Disease, 1991, p 45.
45 Leenders KL, Quinn N, Frackowiak RS, Marsden CD: Brain dopaminergic system studied in patients with dystonia using positron emission tomography. Adv Neurol 1988;50:243–247.
46 Turjanski N, Bhatia K, Burn DJ, Sawle GV, Marsden CD, Brooks DJ: Comparison of striatal ^{18}F-dopa uptake in adult-onset dystonia-parkinsonism, Parkinson's disease, and dopa-responsive dystonia. Neurology 1993;43:1563–1568.
47 Diagnostic and Statistical Manual of Mental Disorders, ed 3, rev. Washington, American Psychiatric Association, 1987.
48 Kurlan R: Tourette's syndrome: Current concepts. Neurology 1989;39:1625–1630.
49 Singer HS, Walkup JT: Tourette's syndrome and other tic disorders. Diagnosis, pathophysiology, and treatment. Medicine 1991;70:15–31.
50 Factor SA, Sanchez-Ramos JR, Weiner W: Cocaine and Tourette's syndrome. Ann Neurol 1988;4: 423–424.
51 Mesulam MM: Cocaine and Tourette's syndrome. N Engl J Med 1986;315:398.

52 Pollack MA, Cohen NL, Friedhoff AJ: Gilles de la Tourette's syndrome. Familial occurrence and precipitation by methylphenidate therapy. Arch Neurol 1977;34:630–632.
53 Klawans HL, Falk DK, Nausieda PA, Weiner WJ: Gilles de la Tourette's syndrome after long-term chlorpromazine therapy. Neurology 1978;28:1064–1068.
54 Butler IJ, Koslow SH, Seifert WE, Caprioli RM, Singer HS: Biogenic amine metabolism in Tourette's syndrome. Ann Neurol 1979;6:37–39.
55 Goetz CG, Tanner CM: Gilles de la Tourette's syndrome in twins: Clinical and neurochemical data. Mov Disord 1990;5:173–175.
56 Singer HS, Hahn IH, Moran TH: Abnormal dopamine uptake sites in postmortem striatum from patients with Tourette's syndrome. Ann Neurol 1991;30:558–562.
57 Wong DF, Pearlson GD, Young LT, et al: D_2 dopamine receptors are elevated in neuropsychiatric disorders other than schizophrenia. J Cereb Blood Flow Metab 1989;9(Suppl 1):S593.
58 Singer HS, Wong DF, Brown JE, et al: Positron emission tomography evaluation of dopamine D_2 receptors in adults with Tourette's syndrome. Adv Neurol 1992;58:233–239.
59 Turjanski N, Sawle GV, Playford ED, et al: PET studies of the pre- and post-synaptic dopaminergic system in Tourette's syndrome. J Neurol Neurosurg Psychiatry, in press.
60 Manzanares J, Lookingland KJ, Morre KE: Kappa opioid receptor-mediated regulation of dopaminergic neurons in the rat brain. J Pharmacol Exp Ther 1991;256:500–505.
61 Haber SN, Wolfer D: Basal ganglia peptidergic staining in Tourette's syndrome. A follow-up study. Adv Neurol 1992;58:145–150.
62 Leckman JF, Riddle MA, Berrettini WH, et al: Elevated CSF dynorphin A [1–8] in Tourette's syndrome. Life Sci 1988;43:2015–2023.

David J. Brooks, MD, MRC Cyclotron Unit, Hammersmith Hospital, Du Cane Road,
GB–London W12 OHS (UK)

Fluorodopa PET Scans of Juvenile Parkinsonism with Prominent Dystonia in Relation to Dopa-Responsive Dystonia[1]

Hirohide Takahashi[a,d], Barry Snow[a], Torbjoern Nygaard[c], Masayuki Yokochi[b], Donald Calne[a]

[a] Neurodegenerative Disorder Centre, University of British Columbia, Vancouver, Canada;
[b] Tokyo Metropolitan Institute for Neuroscience, Tokyo, Japan;
[c] Columbia University, New York, N.Y., USA, and
[d] Showa University Fujigaoka Hospital, Yokohama, Japan

Although the nosological concept of early-onset idiopathic parkinsonism (EOIP; idiopathic parkinsonism starting before the age of 40 years) has not been fully established, cumulative evidence has shown that it differs clinically in many aspects from late-adulthood-onset idiopathic parkinsonism (PD; Parkinson's disease) [1]. Among EOIP cases, it is noted that the frequency of dystonia and familial incidence increase as the age of onset becomes younger [2]. Although the incidence of EOIP starting in childhood or adolescence is very low, they may have prominent dystonia accompanying or preceding parkinsonism (JPD; juvenile parkinsonism with prominent dystonia). Diurnal fluctuation of symptoms has also been apparent in some cases [3].

Patients with JPD may therefore be confused with another distinct disorder of childhood-onset dystonia: dopa-responsive dystonia (DRD) or hereditary progressive dystonia (HPD). DRD/HPD is typically characterized by hereditary, female dominant, diurnally fluctuating, childhood-onset dystonia which shows dramatic and sustained response to small doses of *L*-dopa [3–5]. A tendency for parkinsonism to predominate over the dystonic elements, increasing *L*-dopa dose requirements, fluctuation in response, and drug-induced dyskinesias, frequent in JPD or

---

[1] This work was supported by the Dystonia Medical Research Foundation and the Medical Research Council of Canada.

Table 1. Juvenile parkinsonism with prominent dystonia

Case	Age years	Sex	Age of onset years	Family history	L-dopa mg/day	Initial signs	Diurnal fluctuation	Dys-kinesia	Wearing-off
1	27	F	8	yes	800 (DCI) (100 × 8)	foot dystonia, hand dystonia, action hand tremor	yes	yes	yes
2	22	F	16	yes	400 (DCI) (50 × 8)	hand tremor, then dystonic hand and voice, gait difficulties	no	yes	yes
3	45	M	6	yes	300 (DCI) (100 × 3)	foot dystonia, then generalized dystonia	yes	no	no

EOIP, usually allow the clinical distinction of these disorders [1, 4, 5]. However, it may often be difficult to distinguish these disorders by clinical observations alone, especially in their early course. In fact, a few cases of JPD had been misclassified as DRD/HPD, and the appropriate diagnosis only became obvious after several years of follow-up [4, 6]. Pathologically, EOIP comprises a group of heterogeneous conditions. Some cases have a pathology typical for PD: degenerative nigrostriatal neuronal loss with Lewy bodies, while few were without Lewy bodies or had unusual pathology [2]. DRD has no evidence of nigrostriatal neuronal loss, but shares a profound striatal dopaminergic deficit [7] with degenerative PD.

Positron emission tomography (PET) using the tracer 6-L-[^{18}F]fluorodopa ([^{18}F]dopa) allowed us to investigate the function of the nigrostriatal dopaminergic pathway in vivo [8]. Following administration, the tracer [^{18}F]dopa crosses the blood-brain barrier, is decarboxylated to fluorodopamine by L-aromatic acid decarboxylase, and remains in the nerve terminals for the 2 h of the scan. PET can image this accumulation and thereby display the anatomical distribution of the functionally intact nerve endings [9]. In addition, tracer uptake data, gathered during the scan, can be used to derive an [^{18}F]dopa uptake rate constant, an index of the dopaminergic activity of nigrostriatal nerve terminals [8]. This [^{18}F]dopa uptake rate constant correlates well with numbers of surviving dopaminergic cells in the substantia nigra, demonstrated in monkeys [10] and humans [11]. [^{18}F]dopa PET is sufficiently sensitive to detect an asymptomatic dopaminergic deficit [12–14]. In DRD/HPD, we found normal striatal [^{18}F]dopa uptake [15, 16]. This contrasts with reduced uptake found in patients with PD, consistent with nigrostriatal neuronal loss [17].

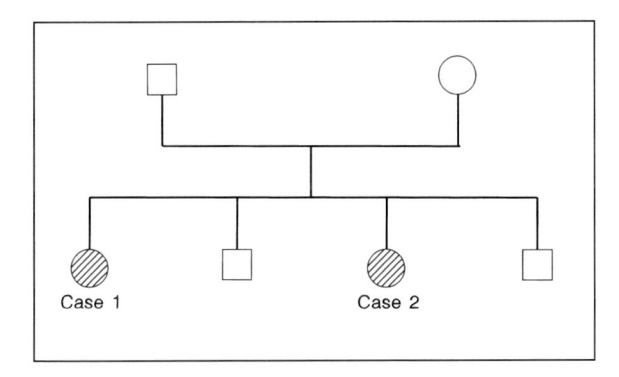

Fig. 1. Pedigree of the family. Among 4 siblings, both females were affected.

We used FD PET to study the integrity of the nigrostriatal system in 3 cases of hereditary JPD and compared the results with those of other cases of EOIP and DRD/HPD.

Subjects and Method

Subjects from the United States, Canada and Japan were invited to the University of British Columbia, Vancouver, for the PET scan.

Juvenile Parkinsonism with Prominent Dystonia (table 1)

Case 1: 27-Year-Old Female. She was well until the age of 8 years when foot dystonia on the right became manifest and she started to stumble. Bilateral hand dystonia and action tremor presented at the same time. Six months later, her left leg became affected with dystonia as well. The symptoms were worse in the evening and premenstrually. At age 10 years, she was started on trihexyphenidyl hydrochloride the effect of which was immediate. The dose was built up to 40 mg and she became 'almost normal'. At age 24 years, she was started on *L*-dopa (300 mg with a decarboxylase inhibitor; DCI) and responded dramatically. However, the effect started to wear off 6 months later. Currently, she requires 800–900 mg of *L*-dopa (with DCI) and she has to take the medication almost every 3 h. On this regimen, she notices occasional 'wearing-off' and 'muscle twitching' at night [previously described in references 4 and 16, in part].

Case 2: 22-Year-Old Female, Younger Sister of Case 1 (fig. 1). She was well until the age of 16 years when she noticed a very fine tremor in both hands. Few years later, her voice became dystonic and she started experiencing gait difficulties (stiff leg on one side) and muscle cramps of the left arm on writing. *L*-dopa improved her symptoms; however, within several years, she noted wearing-off after 3–4 h [previously presented in reference 16, in part].

Case 3: 45-Year-Old Male. Birth and early motor development was normal. Around the age of 6 years, he developed foot dystonia which caused him to stumble. Dystonia was

Table 2. Early-onset idiopathic parkinsonism

Case	Age years	Sex	Age of onset, years	Family history	L-dopa mg/day	Wearing-off
1	27	F	12	no	400 (DCI)	yes
2	53	M	20	no	500 (DCI)	yes
3	45	M	27	no	300 (DCI)	yes
4	35	M	29	no	600 (DCI)	yes
5	36	M	30	no	0	
6	46	F	30	no	600 (DCI)	yes
7	38	M	31	no	0	
8	33	M	32	no	0	
9	34	M	33	no	600 (DCI)	yes
10	36	F	34	no	250 (DCI)	yes
11	45	M	34	no	2,500 (DCI)	yes
12	41	M	35	no	500 (DCI)	no
13	43	M	35	no	1,400 (DCI)	yes
14	43	M	36	no	400 (DCI)	no
15	39	M	37	no	450 (DCI)	yes
16	47	M	39	no	1,500 (DCI)	yes

Table 3. Dopa-responsive dystonia/hereditary progressive dystonia

Case	Age years	Sex	Age of onset years	Family history	L-dopa mg/day	Dystonia
1	42	M	5	yes	150 (DCI)	arm (L → R)
2	34	F	5	yes	300	leg (L → R), then generalized
3	42	F	5	no	500 (DCI)	foot (L), then foot & hand (R)
4	17	M	6	yes	100 (DCI)	foot (R & L), then legs
5	37	F	5	yes	100 (DCI)	foot (R & L), arm (L → R)
6	41	F	7	yes	100 (DCI)	foot & hand (R), then generalized
7	47	F	9	no	600	foot (R → L), then arm (L)
8	15	F	10	yes	300 (DCI)	foot (L → R)
9	63	F	11	yes	100 (DCI)	foot (R), then hand (L)
10	63	F	11	yes	100 (DCI)	foot (R → L)

worse in the evening or when he was tired. Later on, the dystonia spread to his arm, trunk and neck. He also developed postural tremor of the hands. The dystonia progressed until his late teens, but remained stationary thereafter. At the age of 25 years, he was started on L-dopa, which markedly improved both his dystonia and his parkinsonism. He is currently on 300 mg of L-dopa with DCI, and wearing-off reactions or dyskinesias are not clinically evident. His elder sister was similarly affected, with milder dystonia. However, she developed wearing-off reactions and severe dyskinesia soon after the initiation of L-dopa therapy [both siblings were previously described in reference 18].

Early-Onset Idiopathic Parkinsonism (table 2)
Sixteen patients with EOIP (13 males, 3 females, mean age of onset; 30.9 years) were scanned [previously presented in reference 16]. None of them were familial cases.

Dopa-Responsive Dystonia/Hereditary Progressive Dystonia (table 3)
Ten patients who met the published clinical criteria [2] for DRD (2 males, 8 females, mean age of onset; 7.4 years) were scanned [previously presented in reference 16].

Normals
Thirty three neurologically normal volunteers with a similar age distribution (mean 31.7, range 18–57 years) were also selected as control subjects. None were taking medications at the time of scanning [previously presented in reference 16].

[^{18}F]Dopa PET
The UBC/TRIUMF PETT VI system was operated in the high-resolution mode. All subjects received 100 mg of carbidopa 1 h before scanning. [^{18}F]dopa (2.0–3.5 mCi) was administered intravenously at the start of scanning. A graphic method was used to calculate the steady-state [^{18}F]dopa uptake rate constant for the whole striatum [8]. This method incorporates both the measured blood radioactivity and the corrected striatal radioactivity.

Results and Discussion

Figure 2 shows the striatal [^{18}F]dopa uptake rate constants for normals, patients with DRD/HPD, JPD and other EOIP. Patients with JPD and EOIP had significantly reduced uptake rate constants. This contrasted with normal [^{18}F]dopa uptake rate constants found in DRD/HPD patients.

Interpretation of the [^{18}F]dopa PET data needs a comment. Although it does not completely trace the endogenous dopamine, injected [^{18}F]dopa is metabolized in a way similar to therapeutic L-dopa; it is taken up across the blood-brain barrier into the dopaminergic nerve terminals, then decarboxylated to fluorodopamine, and stored, probably in vesicles. Therefore [^{18}F]dopa PET is dependent on the structural as well as biochemical integrity of the nigrostriatal dopaminergic neurons and their striatal terminals. This is in agreement with our recent findings that the [^{18}F]dopa uptake rate constant correlates well with the surviving nigral dopaminergic neurons in both MPTP-intoxicated monkeys [10] and humans [11].

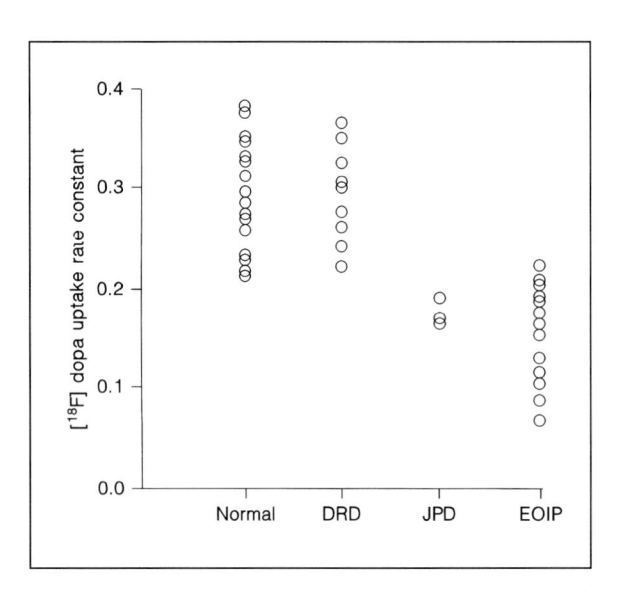

Fig. 2. Scatter plots showing striatal [^{18}F]dopa uptake rate constants (ml striatum^{-1} min^{-1}) for each patient group.

In PD, the [^{18}F]dopa uptake rate constant is reduced, reflecting degenerative nigrostriatal neuronal loss. The reduced [^{18}F]dopa uptake may also explain the development of 'wearing-off', as it may reflect loss of nigrostriatal neurons and a defect in dopamine storage. In contrast, the [^{18}F]dopa uptake rate constant is normal in DRD/HPD, suggesting intact dopa uptake, decarboxylation and storage mechanisms. This explains well why DRD/HPD patients continue to respond to small doses of L-dopa without developing motor fluctuations, as intact vesicular storage mechanism permit gradual release and steady levels of dopamine in the synaptic cleft [16]. Our PET result is also in accord with neuropathological findings in the brain of a woman with DRD. In her brain, striatal dopamine levels were severely reduced with a normal number of nigral neurons; these neurons, however, were hypopigmented. There was no evidence of a degenerative process. Tyrosine hydroxylase (TH) protein levels in the substantia nigra were normal while TH protein levels and TH activity were reduced in the striatum [17]. As noted above, [^{18}F]dopa PET measures neither the activity of TH nor the intrinsic striatal pool of dopamine. The endogenous dopamine pathway differs from that of [^{18}F]dopa in that it begins with the metabolism of phenylalanine to tyrosine, then to dopamine by TH. Therefore, a defect at, or before, the hydroxylation of tyrosine would result in a reduced pool of striatal dopamine but may not affect [^{18}F]dopa PET.

Most of our EOIP patients had reduced [^{18}F]dopa uptake rate constants similar to those found in PD. This had been expected because most of them had clinical motor fluctuations (table 2), therefore, it was most likely that they had lost substantial numbers of intact dopaminergic nerve terminals.

Like these EOIP patients, our JPD patients had reduced [^{18}F]dopa uptake rate constants in spite of their prominent dystonic features. This suggests that JPD and EOIP patients share a common pathophysiology, namely, reduced numbers of nigrostriatal neurons.

Clinically, our JPD patients could initially have been misclassified as DRD/HPD, because they had prominent dystonia which responded well to *L*-dopa. However, subsequent development of wearing-off and dyskinesias, seen in case 1 and 2, and in case 3's affected sister, suggested that they had a disease separate from DRD/HPD. Our results provided further evidence that our JPD patients have a pathophysiology distinct from that of DRD/HPD, in spite of their similar dystonic features. The question as to why very young subjects with EOIP manifest prominent dystonia and less parkinsonism in their early course remains unanswered. Both EOIP and DRD/HPD may display substantial clinical heterogeneity depending on the age of onset, as older relatives of DRD patients may present with pure parkinsonism and normal [^{18}F]dopa PET [17].

We conclude that [^{18}F]dopa PET is capable of distinguishing JPD from DRD/HPD.

References

1 Yokochi M: Nosological concept of juvenile parkinsonism with reference to the dopa-responsive syndrome. Adv Neurol 1993;60:548–552.
2 Yokochi M: Clinicopathological identification of juvenile parkinsonism in reference to dopa-responsive disorders; in Segawa M (ed): Hereditary Progressive Dystonia. London, Parthenon Publishing, 1992, pp 37–48.
3 Nygaard TG, Snow BJ, Fahn S, Calne DB: Dopa-responsive dystonia: Clinical characteristics and definition; in Segawa M (ed): Hereditary Progressive Dystonia. London, Parthenon Publishing, 1992, pp 21–35.
4 Nygaard TG, Marsden CD, Fahn S: Dopa-responsive dystonia: Long-term treatment response and prognosis. Neurology 1991;41:174–181.
5 Segawa M, Nomura Y: Hereditary progressive dystonia with marked diurnal fluctuation; in Segawa M (ed): Hereditary Progressive Dystonia. London, Parthenon Publishing, 1992, pp 3–17.
6 Nomura Y, Segawa M: Intrafamilial and interfamilial variations of symptoms of Japanese hereditary progressive dystonia with marked diurnal fluctuation; in Segawa M (ed): Hereditary Progressive Dystonia. London, Parthenon Publishing, 1992, pp 73–96.
7 Rajput AH, Gibb WRG, Zhong XH, Shannak KS, Kish S, Chang LG, Hornykiewicz O: Dopa-responsive dystonia: Pathological and biochemical observations in a case. Ann Neurol 1994;35:396–402.
8 Martin WRW, Palmer MR, Patlak CS, Calne DB: Nigrostriatal function in man studied with positron emission tomography. Ann Neurol 1989;26:535–542.
9 Garnett ES, Firnau G, Nahmias C: Dopamine visualized in the basal ganglia of living man. Nature 1983;305:137–138.

10 Pate BD, Kawamata T, Yamada T, McGeer EG, Hewitt KA, Snow BJ, Ruth TJ, Calne DB: Correlation of striatal fluorodopa uptake in MPTP monkey with dopaminergic indices. Ann Neurol 1993; 34:331–338.
11 Snow BJ, Tooyama I, McGeer EG, Yamada T, Calne DB, Takahashi H, Kimura H: Human positron emission tomographic [^{18}F]fluorodopa studies correlate with dopamine cell counts and levels. Ann Neurol 1993;34:324–330.
12 Calne DB, Langston JW, Martin WRW, Stossl AJ, Ruth TJ, Adam MJ, Pate BD, Schulzer M: Positron emission tomogrpahy after MPTP: Observations relating to the cause of Parkinson's disease. Nature 1985;317:246–248.
13 Snow BJ, Peppard RF, Guttman M, Okada J, Martin WRW, Steel J, Eisen A, Carr G, Schoenberg B, Calne DB: Positron emission tomographic scanning demonstrates a presynaptic dopaminergic lesion in Lytico-Bodig (the ALS-PD complex of Guam). Arch Neurol 1990;47:870–874.
14 Guttman M, Yong VW, Kim SU, Calne DB, Martin WRW, Adam MJ, Ruth TJ: Asymptomatic striatal dopamine depletion: PET scans in unilateral MPTP monkeys. Synapse 1988;2:469–473.
15 Okada A, Nakamura K, Snow BJ, Bhatt MH, Nomoto M, Osame M, Calne DB: PET scan study on the dopaminergic system in a Japanese patient with hereditary progressive dystonia (Segawa's disease). Adv Neurol 1993;60:591–594.
16 Snow BJ, Nygaard TG, Takahashi H, Calne DB: PET studies of dopa-responsive dystonia and early-onset idiopathic parkinsonism. Ann Neurol 1993;34:733–738.
17 Takahashi H, Snow BJ, Nygaard TG, Calne DB: Clinical heterogeneity of dopa-responsive dystonia: PET observations. Adv Neurol 1993;60:586–590.
18 Yanagisawa N: Juvenile parkinsonism with pallidal posture and spastic paraplegia; in Segawa M (ed): Hereditary Progressive Dystonia. London, Parthenon Publishing, 1992, pp 205–214.

Hirohide Takahashi, MD, Division of Neurology, Department of Medicine,
Showa University Fujigaoka Hospital, 1–30, Fujigoaka, Midori-ku, Yokohama 227 (Japan)

Striatal Dopamine D_2 Receptors in Dopa-Responsive Dystonia and Parkinson's Disease

K.L. Leenders[a], A. Antonini[a], H.-M. Meinck[b], A. Weindl[c]

[a] PET Department, Paul Scherrer Institute, Villigen, Switzerland;
[b] Department of Neurology, University of Heidelberg, and
[c] Department of Neurology, Technical University of Munich, FRG

The clinical aspects of dopa-responsive dystonia (DRD) syndromes are extensively described elsewhere in this volume. Patients suffering from DRD are known to respond dramatically and over many years to *L*-dopa therapy. The clinically effective dose of this drug is often low and the long-term therapeutic complications seen in Parkinson's disease (PD) are not encountered in DRD. The question therefore is in which aspects these two dopamine deficiency conditions, DRD and PD, differ at the level of the nigrostriatal dopaminergic neurotransmitter system. Since most patients with DRD are young, hardly any postmortem data are available.

The influence of normal aging or disease on the nigrostriatal system in man can be directly addressed in vivo by positron emission tomography (PET) using radiolabeled tracers. The presynaptic striatal capacity to convert dopa to dopamine and to contain dopamine can be measured using the ^{18}F-labeled analogue of dopa, ([^{18}F]fluorodopa). Several publications demonstrate that DRD patients show normal or only slightly decreased striatal [^{18}F]fluorodopa uptake, whereas patients with dystonia-parkinsonism and early-onset parkinsonism have significant decreases similar to those seen in PD [1–3]. Also patients with benign forms of parkinsonism in a family with DRD patients showed normal striatal [^{18}F]fluorodopa uptake in contrast to patients with PD at whatever age it occurs [4]. The postsynaptic part of the striatal dopaminergic neurotransmitter system in DRD has not been investigated before. We have used [^{11}C]raclopride as a PET tracer to investigate cerebral D_2 dopamine receptors in 7 patients with DRD. The aim was to determine whether the increased binding of [^{11}C]raclopride seen in PD patients [5] is absent in DRD. The results will have important implications for interpreting the pathophysiological differences within the dopamine deficiency syndromes.

Methods and Subjects

Methods

For tracer synthesis, scanning procedure and data analysis, the reader is referred to previous reports [7, 8]. [^{11}C]raclopride is a radioligand whose uptake by the brain can be measured by PET. It is a neuroleptic of the benzamide class and binds, as an antagonist, selectively to dopamine D_2 receptors [6]. As index for specific tracer uptake, the following ratio was calculated for each subject:

$$\frac{\text{target ROI activity} - \text{cerebellum activity}}{\text{cerebellum activity}}$$

Values of the left and right hemispheres of each subject were pooled.

Subjects

The values of the healthy volunteers (n = 32) used in this chapter are described in a previous publication [8].

The DRD (n = 7) and PD (n = 33) patients were compared with the appropriate age group of healthy volunteers. The PD patients were in an early phase of their disease and recently diagnosed. They were investigated wit PET before *L*-dopa therapy was started. They all responded to apomorphine and to subsequent *L*-dopa treatment. The results of the PD patients are partially taken from Leenders et al. [5] and will be described in more detail in a forthcoming publication.

The age of the 7 DRD patients at the time of PET investigation was between 18 and 30 years (mean 25 ± 4; median 27). The disease duration was between 14 and 26 years. All patients were on *L*-dopa therapy (dose mainly around 300 mg daily). Of the 7 patients, 3 were from one family (cousins) and 2 other patients were brothers.

Results

Mean putamen uptake index for the DRD patients was 3.16 ± (SD) 0.68. The control values of the age-matched group (n = 16) were 2.97 ± 0.54. For the caudate nucleus the values were 2.79 ± 0.60 and 2.60 ± 0.60, respectively. The differences were statistically not significant. In controls putamen uptake index was on average 7–16% higher compared to caudate nucleus. In the DRD patients this value was 14%.

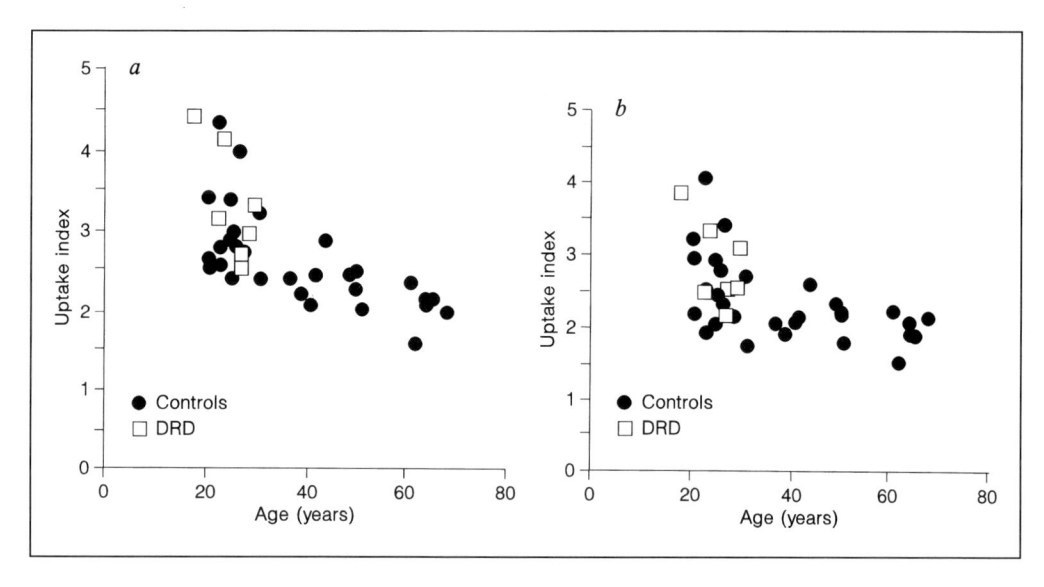

Fig. 1. Scatter diagram of the index for specific [^{11}C]raclopride uptake in caudate nucleus (*a*) and putamen (*b*) versus age in 32 healthy controls and 7 DRD patients.

In figure 1a, b, the index of specific uptake in putamen and caudate nucleus of controls and DRD patients are plotted as a function of age. The total decline from 37 to 68 years in controls was 19% in the putamen and 12% in the caudate nucleus. This corresponds to a decrease of uptake of 0.6% per year in the putamen and 0.4% per year in the caudate nucleus in the age range between 37 and 68 years. See also Antonini et al. [8]. No side-to-side difference of mean caudate nucleus and putamen uptake was found in the healthy controls or the DRD patients.

These findings are in contrast with those of the PD patients where an increase [^{11}C]raclopride uptake is seen. Mean of right and left putamen index in PD was significantly higher than in controls ($p < 0.0001$) (fig. 2). Uptake values were also increased in the mean of right and left caudate nucleus ($p < 0.01$). The ratio putamen over caudate was 31% (SD ± 16) higher in the patients compared to controls ($p < 0.0001$). Right putamen index correlated with left body side mean clinical score ($p < 0.01$; Spearman rank). Similarly, left putamen index correlated with right body side mean score ($p < 0.02$; Spearman rank). The best correlation was seen with akinesia ($p < 0.009$ Spearman rank). Percentage side-to-side differences of tracer uptake in putamen were highly significantly correlated with percentage clinical side-to-side differences in putamen ($p < 0.0001$). For the caudate nucleus this correlation was only marginally significant ($p < 0.03$).

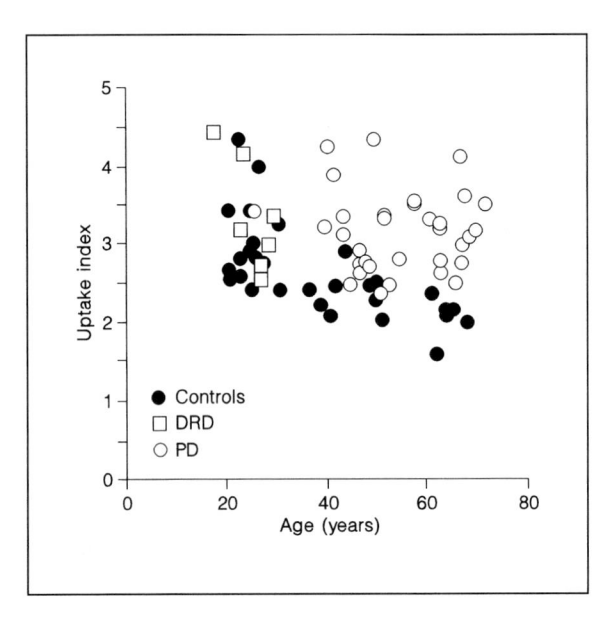

Fig. 2. The same scatter diagram as in figure 1b, but now also showing the values of the 33 PD patients.

Discussion

In the striatum, dopamine D_2 receptors are for the vast majority located on striatal projection neurons (to pallidum or substantia nigra) or on interneurons. In vivo investigation of binding by radioligands (as in our case using the dopamine D_2 antagonist [^{11}C]raclopride) allows the study of diseases in which these neurons are either directly affected (like in Huntington's disease or multi-system atrophy), or where these neurons, whilst themselves not pathologically affected, may respond with cell membrane receptor density or affinity changes due to specific presynaptic neurotransmitter input.

In *DRD* there is no difference in striatal dopamine D_2 receptor density compared to age-matched controls. This is in line with the normal density of presynaptic nerve terminals, as indicated by PET [^{18}F]fluorodopa studies [1–3]. [^{18}F]fluorodopa uptake does not reflect endogenous dopamine concentration directly, but rather dopa-decarboxylase and dopamine retention capacity [9]. No PET studies using dopamine D_2 receptor tracers on a larger group of DRD patients have been published to date.

In *healthy controls* there is a significant decrease with age of specific [^{11}C]raclopride binding in human striatum. This means that comparison of patients with a control group requires appropriate age matching. The reduction of striatal dopamine D_2 receptor binding sites with age may be associated either with a loss of intrinsic striatal (projection) neurons or, alternatively, with a decrease of receptor expression in the cell membranes. The latter could possibly be due to impairment of cellular regulatory systems of the striatal projection neurones [for a review, see ref. 8].

In PD, the postsynaptic dopaminergic receptor response contrasts with that in normal aging [6, 8] and DRD. In PD, significant increases of dopamine D_2 receptor binding are found, particularly in the putamen (figure 2). These increases correspond with laterality and severity of clinical signs. In PD, increased D_2 receptor density is most likely a balanced response to decreased dopaminergic nigrostriatal input.

Why do DRD and PD, both endogenous dopamine deficiency syndromes, show such different patterns of neurotransmitter system behavior? We suggest that abnormal reduction (in addition to that caused by normal aging) in the number of functional dopaminergic synaptic density, i.e. loss of cellular elements (nerve terminals) in striatum due to loss of nigro-striatal dopaminergic neurons, results in pathological post-synaptic compensation (increased number of receptors) if the projection and interneurons are not affected by disease. This seems to be the case in PD. Here, the endogenous dopamine content is only a few percent of normal, whereas the dopaminergic nigrostriatal neurons are on average only reduced between 40 and 50% of normal. Upregulation of dopamine D_2 receptors appears to correlate quantitatively more to the decreased number of striatal dopaminergic nerve terminals than to the extremely low content of endogenous dopamine. The upregulation of dopamine D_2 receptor density in PD is often called receptor 'supersensitivity', which is an unfortunate misnomer: the receptors are not more sensitive, but more in number. In most cases of multisystem atrophy, dopamine D2 receptor density is reduced, often severely, because of outright loss of projection and possibly interneurons.

In DRD, no or only very little loss of nerve terminals is apparent, although the endogenous striatal dopamine content is very low as judged from the few postmortem data available (see elsewhere in this volume). This explains the normal or near-normal striatal [^{18}F]fluorodopa uptake and normal postsynaptic dopamine D_2 receptor density. It also explains why L-dopa therapy is so successful usually at low doses: the nerve terminal pool is large, probably hardly reduced from normal, and easily absorbs L-dopa, thus restoring the intrinsic striatal dopaminergic drive. In PD, this phenomenon is also present, but less so, since here the nerve terminals are reduced (30–50% of normal), which probably explains why in PD larger L-dopa doses are required and why they usually need to be given at shorter intervals than in DRD. Apart from the pharmacokinetic influence of the condition and number of dopaminergic nerve terminals on L-dopa efficiency, naturally the therapeutic effect also depends on whether the postsynaptic system is intact. In PD and DRD, this is clearly the case, whereas in multisystem atrophy the dopamine D_2, and probably also other receptor systems are markedly reduced. In the latter case, *L*-dopa therapy is usually of not much benefit or not effective at all.

Finally, the fact that striatal endogenous dopamine content reduction by itself does not result in increased [^{11}C]raclopride binding, as shown by our studies of DRD patients, implies that the increases in tracer binding found in PD indeed reflect increased receptor density and do not express larger availability of tracer binding sites due to a lack in endogenous dopamine.

References

1 Sawle GV, Leenders KL, Brooks DJ, Harwood G, Lees AJ, Frackowiak RSJ, Marsden CD: Dopa-responsive dystonia: [^{18}F]dopa positron emission tomography. Ann Neurol 1991;30:24–30.
2 Snow BJ, Nygaard TG, Takahashi H, Calne DB: Positron emission tomographic studies of dopa-responsive dystonia and early-onset idiopathic parkinsonism. Ann Neurol 1993;34:733–738.
3 Turjanski N, Bhatia K, Burn DJ, Sawle GV, Marsden CD, Brooks DJ: Comparison of striatal ^{18}F-dopa uptake in adult-onset dystonia-parkinsonism, Parkinson's disease, and dopa-responsive dystonia. Neurology 1993;43:1563–1568.
4 Nygaard TG, Takahashi H, Heiman GA, Snow BJ, Fahn S, Calne DB: Long-term treatment response and fluorodopa positron emission tomographic scanning of parkinsonism in a family with dopa-responsive dystonia. Ann Neurol 1992;32:603–608.
5 Leenders KL, Antonini A: Pathophysiology of Parkinson's disease and positron emission tomography (PET); in Bignami A (ed): New Issues in Neurosciences. Basic and Clinical Approaches. Basal Ganglia and Movement Disorders. Stuttgart, Thieme, 1991, vol 3/2, pp 143–148.
6 Farde L, Ehrin E, Eriksson L, Greitz T, Hall H, Hedström CG, Litton JE, Sedvall G: Substituted benzamides as ligands for visualization of dopamine receptor binding in the human brain by positron emission tomography. Proc Natl Acad Sci USA 1985;82:3863–3867.
7 Leenders KL, Antonini A, Thomann R, Locher J Th, Maitre L, Gerebtzoff A, Beer H-F, Ametamey S, Weinreich R, Gut A, Gnirss F, Ofner S, Schilling W, Waldmeier PC: Savoxepine: Striatal dopamine-D_2 receptor occupancy in human volunteers measured using positron emission tomography (PET). Eur J Clin Pharmacol 1993;44:135–140.
8 Antonini A, Leenders KL, Reist H, Thomann R, Beer HF, Locher JH: Effect of age on D_2 dopamine receptors in normal human brain measured by positron emission tomography and ^{11}C-raclopride. Arch Neurol 1993;50:474–480.
9 Leenders KL, Salmon EP, Turton D, Tyrrell P, Perani D, Brooks DJ, Sagar H, Jones T, Marsden CD, Frackowiak RSJ: The nigrostriatal dopaminergic system assessed in vivo by positron emission tomography in healthy volunteer subjects and patients with Parkinson's disease. Arch Neurol 1990; 47:1290–1298.

Prof. Dr. K.L. Leenders, PET Department, Paul Scherrer Institute,
CH–5232 Villigen (Switzerland)

Striatal Dopamine in Dopa-Responsive Dystonia: Comparison with Idiopathic Parkinson's Disease and Other Dopamine-Dependent Disorders

Oleh Hornykiewicz

Institute of Biochemical Pharmacology, University of Vienna, Austria

Parkinson's Disease versus Other Conditions with Parkinsonian Symptomatology

Neurochemically, idiopathic Parkinson's disease (iPD) is a striatal dopamine (DA) deficiency syndrome [1]. In addition to iPD, reduced striatal DA concentrations have been found in many other degenerative brain disorders accompanied by parkinsonian symptomatology (table 1) [2]. Despite this similarity, iPD is set apart from the other parkinsonian conditions by three major neurochemical distinguishing marks: (a) The degree of DA deficiency in the striatum of patients with iPD is always severe, with individual patients' values being clearly below the lowest control values in every instance [3]; (b) in every patient with iPD, the loss of DA in the putamen is regularly more severe than in the caudate nucleus [4–6]; (c) in iPD there exists a rostrocaudal pattern of striatal DA loss that is characteristic for each of the two striatal subdivisions [6, 7]. This pattern of DA loss is particularly pronounced in the caudate nucleus where the rostral portion is substantially more severely affected (>90% DA loss) than in the caudal subdivision (<70% DA loss); in the putamen, the rostrocaudal DA pattern goes in the opposite direction [6] (see fig. 1).

Judging by the above neurochemical criteria, it is obvious that degenerative brain disorders with parkinsonian symptomatology other than iPD differ sub-

Table 1. Dopamine loss in caudate nucleus versus putamen in some neurodegenerative brain disorders: comparison with iPD

Brain disorder	n	DA, % of control CN	DA, % of control PUT	Ratio CN/PUT
Idiopathic Parkinson's disease	8	18	2.2	8.18
Jacob-Creutzfeldt disease	1	35	48	0.73
Progressive supranuclear palsy	5	20	27	0.74
Pick's disease	4	6.4	8.9	0.72
Rett syndrome	1	6.0	12	0.50
Cortico-basal ganglionic degeneration	1	4.2	3.7	1.41
Postencephalitic parkinsonism	6	1.5	0.6	2.50
Hallervorden-Spatz disease	1	1.4	0.9	1.56
Neuronal intranuclear inclusion body disorder	1	0.6	0.4	1.50
Striatonigral degeneration	1	0.4	0.4	(1.0)

For references, see Hornykiewicz [2]. CN = Caudate, PUT = putamen.

Table 2. Clinical history of the patient with dopa-responsive dystonia [9]

19-year-old female (English-Irish ancestry) with inconspicuous birth record and no family history of a similar disorder

- Normal development until age 5, when she started walking on tiptoes

- At age 7: intermittent involuntary flexion of right arm wrist and a tendency to fold the right arm across the chest while walking

- When examined at age 8: right foot tended to turn in and plantarflex (equinovarus), making her walk on the right tiptoe and occasionally with the right arm folded across the chest; dystonic hypertonicity at the right ankle and the right wrist; forearm pronation/supination, piano playing finger movements and foot tapping slower on the right than the left side; no other neurological abnormalities

- Drug treatment: On L-dopa (since age 8), 250 mg 3 × daily: complete improvement with posture, gait and tone normal; no effects other than occasional abdominal cramps; patient finished high school and held 3 different jobs

- Cause of death: car accident.

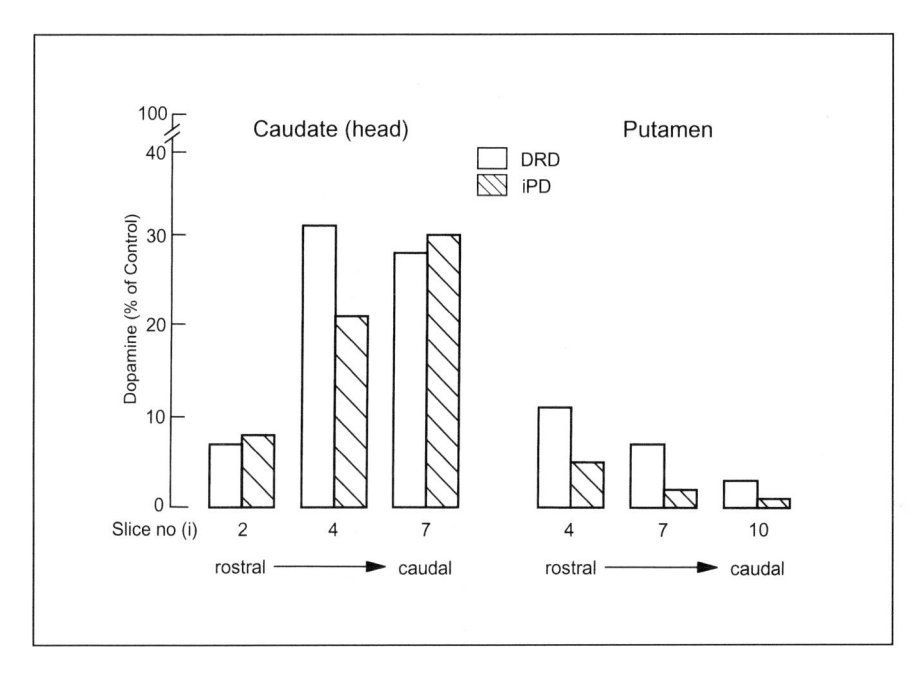

Fig. 1. Rostrocaudal gradients of striatal DA loss in a case with DRD compared with iPD. The slice numbers refer to the intermediate subdivisions of the corresponding slices shown in the insert of figure 1 in Kish et al. [6]. Actual data can be found in Kish et al. [6] and Rajput et al. [9].

stantially from iPD proper. This is true in respect to both the interregional caudate versus putamen DA difference (table 1) and the subregional rostrocaudal DA patterns [2, 3]. These differences justify the conclusion that iPD is indeed a separate disease entity with, most likely, a specific etiology.

Dopa-Responsive Dystonia

In contrast to the non-iPD neurodegenerative conditions with parkinsonian symptomatology, analysis of basal ganglion DA in a case with typical DOPA-responsive dystonia (DRD; Segawa dystonia [8]) disclosed intriguing analogies to iPD. A detailed pathological and biochemical study of this case was published elsewhere [9].

This 19-year-old female patient (table 2), who died in a car accident, had dystonia since age 5. She showed a complete response to a low dose of *L*-dopa

Table 3. Striatal DA in a case with DRD – comparison with iPD

	CN	PUT	References
DRD (n = 1)			
DA, µg/g	1.09	0.66	9
DA, % of control	18	8	
iPD (n = 8)			
DA, $\bar{X} \pm$ SEM, µg/g	0.62±0.11	0.12±0.03	3, 6
DA, range, µg/g	0.25–1.07	0.03–0.26	
DA, % of control	20	2.6	
Controls (n = 10)			
DA, $\bar{X} \pm$ SEM, µg/g	3.07±0.28	4.67±0.34	3, 6
DA, range, µg/g	2.01–4.46	3.49–6.79	

CN = Caudate; PUT = putamen.

(250 mg without a decarboxylase inhibitor, thrice daily); the response had been stable for 11 years until her sudden death. Histopathological examination revealed no abnormalities in the striatal nuclei, a normal number and pattern of substantia nigra compacta (DA) neurones which, however, contained distinctly less melanin than age-matched controls.

Despite the normal cellularity of the compact zone of the substantia nigra, neurochemical analysis revealed subnormal DA levels in the substantia nigra, caudate nucleus and putamen. The magnitude of the striatal DA loss was not as severe as in iPD (in the preparkinsonian range), but clearly below the lower limit of the control range (table 3).

Surprisingly, the interregional caudate/putamen pattern and the subregional rostrocaudal patterns of DA loss in this DRD case were very similar to the corresponding DA patterns in iPD. Thus, as in iPD, DA loss in the putamen was distinctly greater (8% DA remaining) than in the caudate nucleus (18% DA remaining) (table 3). Also in analogy to iPD, the rostral portion of the caudate nucleus suffered a distinctly greater DA loss than the caudal part of the nucleus; this gradient was opposite to the DA gradient in the putamen, where it was the caudal subdivision that was found to contain less DA than the rostral subdivision (fig. 1).

In addition to striatal DA reduction, both the tyrosine hydroxylase activity and the enzyme protein concentration were subnormal in our case with DRD, whereas the level of striatal DA transporter (DA uptake sites), measured by the specific [^{3}H]GBR 12935 binding, was at the lower limit of the normal range [9].

Pathophysiological Considerations

At present, the cause of the iPD-like DA deficiency in our DRD case remains unexplained. It is possible that a defect in producing adequate amounts of tetrahydrobiopterin, the cofactor of tyrosine hydroxylase, was responsible for the reduced striatal DA levels [10, 11], although this possibility would hardly explain the very specific, iPD-like, pattern of the striatal DA loss found in this DRD case. Another possibility is an impairment, in the DRD striatum, of the normal function of a neurotrophic factor with a subregionally uneven striatal distribution pattern. It is known that in the brain both establishment, during the embryonic development, and postnatal preservation of an intact terminal field of innervation (arborization) is crucially dependent on trophic factor activity [12]. Thus, an inborn, or developmental, impairment of such a trophic factor in the DRD brain might prevent, possibly in a subregionally uneven manner, the establishment of a normal complement of striatal DA terminals, without resulting in death of the nigral perikarya. In this respect, the striking similarity of the patterns of striatal DA loss in the DRD case and in iPD suggests the idea that similar, possibly age-related, striatal neurotrophic factor impairments might, in part, play a role in the etiopathology of iPD.

The above data raise another important question: Why did the striatal DA reduction in the DRD case produce symptoms of dystonia rather than (mild) parkinsonism? One possibility would be overactivity of cholinergic striatal mechanisms as a consequence of the marked DA loss; it is well known that removal of the inhibitory dopaminergic influence in the striatum results in overactivity of the cholinergic neurones. In this respect, young individuals seem to be especially sensitive as evidenced by the observation that in children DA-receptor-blocking neuroleptics frequently produce dystonic reactions (rather than, as in adults, parkinsonism) [13]; these dystonic reactions easily resolve upon administration of anticholinergic drugs. It is, therefore, possible that the actual loss of striatal DA (that in older individuals would produce clinical signs of [mild] parkinsonism), in young individuals produces signs of dystonia. This might explain why, besides L-dopa, also anticholinergic medication has a favorable effect in DRD patients. (The fact that the therapeutic response to anticholinergics is not as complete as with L-dopa is not surprising, considering that, in contrast to L-dopa, anticholinergics do not influence the primary neurochemical defect.)

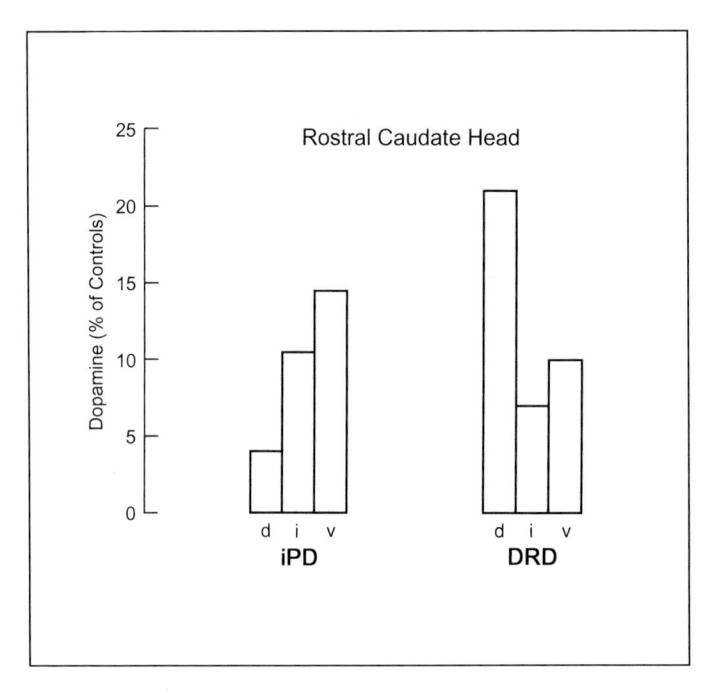

Fig. 2. Dorsoventral DA gradient in the rostral part of the caudate head (slice No. 2 in Kish et al. [6] and Rajput et al. [9]) in iPD: comparison with the (opposite) DA gradient in a case with DRD. d = Dorsal; i = intermediate; v = ventral subdivision.

An alternate possibility for the association between the observed striatal DA reduction and dystonia is offered by one potentially crucial neurochemical difference between the analyzed DRD case and iPD. In iPD, the subregional striatal DA patterns also include a dorsoventral gradient of DA loss, with the dorsal subdivisions of the caudate and putamen losing more DA than the ventral subdivisions [6]. In striking contrast to this gradient, in the analyzed DRD case, the ventral subdivision of the most rostral caudate head suffered a greater DA loss than its dorsal counterpart (fig. 2), containing only about one-half (0.52 µg/g) the DA measured in the dorsal portion of the rostral caudate (1.08 µg/g) [9]. (In the caudal part of the caudate head and in the putamen, the dorsoventral DA gradient was similar as in iPD.) The neurophysiological role of the ventral vs. dorsal subdivision of the rostral caudate head for the functioning of the basal ganglia is not known. However, in studies on the compartmental substructure of the human striatum it has been noticed that within the rostral caudate in particular, the matrix compartment was more homogeneous in its dorsal/lateral portions than in the medial/ventral portions of the nucleus, where the (acetylcholinesterase-poor) striosomes/patches were more numerous [14]. Viewed against this compartmental substructure of the rostral caudate, the DA loss in iPD caudate head would appear to be more pronounced in the matrix, and in DRD in the striosome/patch compartment. The role of the various striatal compartments for the functioning of the striatum is still hypothetical [15, 16], but it may be significant that recently Burke and Baimbridge [17] found a preferential neurone loss in the striosomal compartment in the neonatal hypoxia-ischemia model of injury to the rat striatum. This developmental brain injury is considered to represent an experimental model of one important cause of dystonia in childhood. The intriguing hypothesis suggests itself that preferential damage to the striosomal DA of the rostral caudate nucleus may be specifically related to the dystonic symptomatology of DRD, thus both clinically and neurochemically distinguishing this striatal DA deficiency condition from iPD.

References

1 Hornykiewicz O: Parkinson's disease: From brain homogenate to treatment. Fed Proc 1973;32: 183–190.
2 Hornykiewicz O: Parkinson's disease and the adaptive capacity of the nigrostriatal dopamine system: Possible neurochemical mechanisms. Adv Neurol 1993;60:140–147.
3 Hornykiewicz O, Kish SJ, Rajput AH: Neurochemical aspects of Parkinson's disease and the dementing brain disorders: Relation to brain aging; in Nagatsu T, Fisher A, Yoshida M (eds): Basic, Clinical, and Therapeutic Aspects of Alzheimer's and Parkinson's Diseases. New York, Plenum, 1990, vol 1, pp 445–452.
4 Ehringer H, Hornykiewicz O: Verteilung von Noradrenalin und Dopamin (3-Hydroxytyramin) im Gehirn des Menschen und ihr Verhalten bei Erkrankungen des extrapyramidalen Systems. Klin Wochenschr 1960;38:1236–1239.
5 Bernheimer H, Birkmayer W, Hornykiewicz O, Jellinger K, Seitelberger F: Brain dopamine and the syndromes of Parkinson and Huntington. J Neurol Sci 1973;20:415–455.
6 Kish SJ, Shannak K, Hornykiewicz O: Uneven pattern of dopamine loss in the striatum of patients with idiopathic Parkinson's disease. N Engl J Med 1988;318:876–880.
7 Nyberg P, Nordberg A, Wester P, Winblad B: Dopaminergic deficiency is more pronounced in putamen than in nucleus caudatus in Parkinson's disease. Neurochem Pathol 1983;1:193–202.
8 Segawa M, Nomura Y, Kase M: Diurnally fluctuating hereditary progressive dystonia; in Vinken PJ, Bruyn GW, Klawans HL (eds): Handbook of Clinical Neurology: Extrapyramidal Disorders. Amsterdam, Elsevier, 1986, vol 5, pp 529–547.
9 Rajput AH, Gibb WRG, Zhong XH, Shannak KS, Kish S, Chang LG, Hornykiewicz O: Dopa-responsive dystonia: Pathological and biochemical observations in a case. Ann Neurol, 1994 in press.
10 Lewitt PA, Miller LP, Levine RA, Lovenberg W, Newman RP, Papavasiliou A, Rayes A, Eldridge R, Burns RS: Tetrahydrobiopterin in dystonia: Identification of abnormal metabolism and therapeutic trials. Neurology 1986;36:760–764.
11 Furukawa Y, Nishi K, Kondo T, Mizuno Y, Narabayashi H: CSF biopterin levels and clinical features of patients with juvenile parkinsonism. Adv Neurol 1993;60:562–567.
12 Cowan WM, Fawcett JW, O'Leary DDM, Stanfield BB: Regressive events in neurogenesis. Science 1984;225:1258–1265.
13 Swett C: Drug-induced dystonia. Am J Psychiat 1975;132:532–534.

14 Graybiel AM, Ragsdale CW Jr: Histochemically distinct compartments in the striatum of human, monkey, and cat demonstrated by acetylthiocholinesterase staining. Proc Natl Acad Sci 1978;75: 5723–5726.
15 Graybiel AM: Neurotransmitters and neuromodulators in the basal ganglia. Trends Neurosci 1990; 13:244–254.
16 Gerfen CR: The neostriatal mosaic: Multiple levels of compartmental organization in the basal ganglia. Annu Rev Neurosci 1992;15:285–320.
17 Burke RE, Baimbridge KG: Relative loss of the striatal striosome compartment, defined by calbindin-D_{28k} immunostaining, following developmental hypoxic-ischemic injury. Neuroscience 1993; 56:305–315.

Prof. Dr. O. Hornykiewicz, Institute of Biochemical Pharmacology, University of Vienna, Borschkegasse 8a, A–1090 Vienna (Austria)

Dopa-Responsive Dystonia: Clinical, Pathological, and Genetic Distinction from Juvenile Parkinsonism[1]

Torbjoern G. Nygaard

Department of Neurology, Columbia-Presbyterian Medical Center, New York, N.Y., USA

Three years ago, a meeting at this site served as a forum to synthesize our understanding of the relationships and distinctions between what may be viewed as two major disease categories giving rise to dystonia-parkinsonism in childhood and early adolescence. These categories are dopa-responsive dystonia, within which we include hereditary progressive dystonia with marked diurnal fluctuation (HPD) and juvenile parkinsonism. This chapter will summarize our current knowledge about the pathophysiologies of these disorders. We will explore the clinical rationale for considering HPD as part of a single disorder, dopa-responsive dystonia (DRD). And finally, there are new molecular genetic data which might allow us to take another step forward in determining the relationship of HPD to DRD.

Historical Background

Beck [1] made the first apparent clinical description of DRD in 1947 with her report of a young girl and her paternal uncle affected with dystonia. Corner [2] subsequently reported the diurnal nature of symptoms and a dramatic response to trihexyphenidyl in the girl. Proof of the dopa-responsiveness of symptoms in this family came in 1976 following their treatment with *L*-dopa.

---

[1] This work was supported in part by the Dystonia Medical Research Foundation, the Parkinson's Disease Foundation, and NIH Grant NS32035.

The introduction of *L*-dopa for the treatment of Parkinson's disease in 1967 [3] spurred therapeutic trials in many neurologic conditions. Chase [4] probably provided the first report of a patient with a dramatic *L*-dopa response in a small trial of *L*-dopa use in dystonia, but the first attention to the special clinical characteristics of *L*-dopa-responsive cases came from two subsequent reports. In 1971, Castaigne et al. [5] reported two brothers with a 'progressive extrapyramidal disorder', and Segawa et al. [6] described two cousins with 'hereditary basal ganglia disease with marked diurnal fluctuation', who experienced a remarkable response to *L*-dopa therapy.

Two factors undoubtedly blunted the impact of these observations on Western neurologists. First, Cooper [7] cautioned that *L*-dopa rendered patients less responsive to thalamotomy; a major treatment modality for dystonia at the time. The second reason was alluded to by Eldridge et al. [8], who found that only about 5% of patients with dystonia reported their greatest therapeutic benefit from *L*-dopa. This low likelihood of finding an *L*-dopa-responsive case may have quickly dimmed enthusiasm for this treatment in dystonia; while the frequency estimates for DRD, or HPD, are similar at 1 in 2,000,000 in both Japan and England [9], the relative contributions that these diagnoses make to total number of children with idiopathic dystonia differ. Current estimates are that 5–10% of children with idiopathic dystonia in America or Europe may have DRD while 50% of children with idiopathic dystonia in Japan may have DRD, or HPD.

Juvenile parkinsonism (JP, here including only childhood- and adolescent-onset parkinsonism) has occupied the attention of neurologists for a longer period of time. Consideration of JP as a specific entity has been a topic of much debate. Analyses as long ago as those by Willage [10], in 1911, and more recently by Quinn et al. [11], have concluded that age 20 (or possibly 18) should be the lower age limit accepted for the onset of Parkinson's disease (PD). Parkinsonism beginning before this age represents clinical entities other than PD. Review of currently available neuropathological data supports this notion as, perhaps, only 1 case with onset this young has been reported with 'typical Lewy body pathology' [12], while the rest comprise many other pathologies [13]. Appreciation of the parkinsonian features in DRD has led to increased awareness of the potential clinical overlap between DRD and JP [14–16].

Clinical Definitions

Dopa-Responsive Dystonia

To date, with the inclusion of the cases presented by Markova and Ivanova-Smolenskaya [17, this volume], over 200 cases of DRD have appeared in the literature. The typical clinical features in these cases include: onset with dystonia in childhood, usually affecting gait; the concurrent or later development of signs of parkinsonism in most affected individuals; and a dramatic therapeutic response to *L*-dopa [18]. Signs and symptoms in DRD often worsen later in the day ('diurnal fluctuation') or increase following exertion.

The mean age at onset is about 6–7 years, and the largest number of cases present at about these ages. Following onset of dystonia in one or both legs or a gait disorder, there is increasing disability in the legs and appearance of postural instability. Only a small number of patients have onset with other symptoms (arm dystonia [19, 20], torticollis [20], retrocollis [16], 'poor coordination' [19] or slowness in dressing [21], however, all eventually manifest leg involvement. Symptomatic arm involvement occurs in most patients and a majority develop axial manifestations (increased lumbar lordosis, scoliosis, or torticollis) before treatment. Most patients will progress to generalized involvement; however, the rate of progression may vary considerably; generalization occurred within a year in several patients. A small number of patients experience only exercise-induced dystonia that resolves following 1–2 h rest (without sleep). Increased awareness of DRD has resulted in earlier treatment and shorter periods of deficit.

Juvenile Parkinsonism

JP has a somewhat more heterogeneous presentation. Among JP cases for which data are available, the presenting complaint was foot dystonia in less than one third of cases, while larger numbers had initial complaints including rigidity, bradykinesia or rest tremor [13]. The age-at-onset spectrum in JP is different, as the earliest cases may present at about age 6 years while a larger number of cases present at subsequent ages. A progression of both the dystonic and parkinsonian elements may occur, but in general parkinsonian elements will predominate. This progression may parallel that in PD, though in some cases may be quite slow.

Diurnal Fluctuation

Diurnal fluctuation, the increase in severity of symptoms through the day, has attracted special attention in DRD and JP. What constitutes 'fluctuation' varies widely in degree as some consider this to mean a patient is 'normal' in the morning before experiencing symptomatic exacerbation, while others apply this to the further exacerbation of symptoms that may have been quite severe even early in the day. Using the broader of these definitions, diurnal fluctuations occurred in about 70% of DRD cases. Ethnic background appears to influence this as fluctuations are the rule in Japanese cases (cases of HPD), while 70% of North American and European cases and 60% of Russian cases have fluctuations. Importantly, several families have affected members with and without fluctuations. In our experience, patients still have obvious signs during periods when they report being symptom free. Diurnal fluctuation of symptoms occurred in 28% of JP cases. Most patients with longstanding symptoms of DRD or JP report an attenuation in the degree of variation with disease progression. Thus, diurnal fluctuation is neither specific nor invariable to either disorder, and may not be present at all stages of disease.

Non-Classical Clinical Features in DRD and JP

Among children with DRD, a few cases had an unexplained delay in attainment of early motor milestones that preceded overt dystonic manifestations in later infancy or childhood [22, 23]. Hyperreflexia, including sustained ankle clonus, or an apparent Babinski sign (the dystonic toe termed the 'pseudo-Babinski') by Hunt [24] or the 'striatal foot' [25]) occurred in about 20% of patients with DRD and several patients with JP. This has caused cerebral palsy to be an early consideration in many patients, particularly those with DRD and late walking [23].

Oculogyria has occurred in a few cases with either JP [26–29] or DRD [17, 30–32]. This symptom is a hallmark of postencephalitic parkinsonism [33]. In the JP group this tended to occur near the onset of disease, and the pathological basis for this, in at least 1 case, was neuronal intranuclear hyaline inclusion disease (NIHID) [29]. Among cases with DRD, oculogyria occurred in only 4 untreated cases with long-standing disease.

Treatment Response

L-dopa, as is necessary in our criteria, has been effective in all cases of DRD. Many cases have had varying degrees of benefit from anticholinergic agents, carbamazepine, or bromocriptine as prior treatment [16]. *L*-dopa responsiveness has been present in patients who were symptomatic for as long as 58 years before initiating treatment [16]. Doses required vary from as little as 50 mg (with a peripheral dopa-decarboxylase inhibitor) on alternate days to 2,000 mg *L*-dopa (without inhibitor). The current longest duration of treatment is 26 years; the treatment response in this case, as with all others, remains stable and unhampered by those treatment complications that are common to PD. A few cases with longstanding focal dystonias (writer's cramp or spastic dysphonia) have not had these particular elements respond to treatment [16]. We are not aware of any instance in which one family member responded to *L*-dopa and another fully affected member did not.

Among JP cases in whom *L*-dopa therapy has been reported, there has been a good response in the vast majority. But over half of these individuals experienced dyskinesias, wearing-off, or 'on-off' within an average of 2 years after beginning treatment. The treatment duration in most of the others was less than 5 years and thus may have been too short for these response fluctuations to occur. No case of DRD has developed wearing-off, 'on-off', or other complications of therapy which are common in PD and the JP group.

Genetics

Among 112 DRD index cases (a single occurrence of disease within a family which brought the family to attention), 45 (40%) had a family history of disease [34]. A segregation analysis using first-degree relatives of 21 DRD index cases found an equal risk of disease in siblings, parents, and children of these cases, consistent with autosomal dominant inheritance with reduced penetrance [35]. Additionally, there are sex-related differences in penetrance with the penetrance in women at 45% with a much lower penetrance of 15% estimated in men. This is also reflected in the ratio of reported cases of DRD with females outnumbering males by about 3:1 [34]. In the segregation study, among first-degree relatives older than 40, there was a higher frequency of parkinsonism than would be expected as the risk of PD in the general population [34]. The *L*-dopa response in these individuals resembled that seen in DRD (low dose requirements and long-term stability of treatment effect free of complications) and suggested that parkinsonism, not due to PD, might be a later occurring phenotype in DRD.

Among index cases with JP included in a recent review [13], 29 of 68 (43%) had a family history of another member affected with a similar disorder (one monozygotic twin pair [36] was counted as a 'sporadic' case). There were 8 instances of consanguineous parental marriages, strongly suggesting an autosomal recessive mode of inheritance [11, 37–43]. Two families had male-male transmission of disease suggesting autosomal inheritance [11, 38]. One multiply affected family has either an autosomal or x-linked dominant inheritance [43]. In the remaining 18 families, the affected family members were siblings; this high risk to siblings suggests that autosomal recessive inheritance may play a large role in JP. The age of onset in most of the other affected members was also in childhood or adolescence. In 6 families there were other affected members with onset between 21 and 40, one case with a brother affected at 24 and a parent at 73, 1 case with a sibling affected at 56, and another with both grandfathers (brothers) affected with dystonia or 'shaking palsy', beginning at unspecified ages. Although the total numbers of men and women affected with JP were similar, among sporadic cases men predominated (25:14).

Laboratory Studies

CSF Monoamine Metabolites
In DRD, cerebrospinal fluid levels of homovanillic acid (HVA) have been reduced in most patients studied; diurnal determinations have revealed afternoon increases [45, 46] and decreases [19, 45] in HVA. HVA levels have also been low in the majority of patients studied with JP [47–49]. The serotonin metabolite, 5-HIAA, has been variable in DRD with normal, elevated, and reduced values reported.

Biopterin Pathway Metabolites
Cerebrospinal fluid biopterin (a tyrosine hydroxylase cofactor) has been markedly reduced in all DRD patients studied [16, 45, 50, 51]. Neopterin, a biopterin metabolite, has also been reduced in DRD [45, 50–54]. JP patients also have a reduction in CSF biopterin which may overlap the levels seen in DRD [53].

Fluorodopa Positron Emission Tomography
The consensus is emerging that PET scanning with fluorodopa is normal in DRD in contrast to JP where reductions in fluorodopa uptake and retention parallel those seen in PD. At least 19 DRD patients have had fluorodopa PET and a smaller number with JP have been scanned [55–58].

Pathology
Details of the first pathological and neurochemical study of a case of DRD appear in the chapter by Hornykiewicz [59, this volume] and in a report by Rajput et al. [60]. The findings in this case suggest a defect affecting the development of striatal dopaminergic terminals. The site and nature of this defect are similar to those which have long been suspected as being the basis for HPD. A wide variety of different pathologies have been reported in JP [13]. All of these share, at their core, cell loss in the nigrostriatal pathway.

Molecular Genetics

We have utilized three DRD North American families of English or Welsh descent for linkage analyses. The clinical details of 2 of these families were presented previously [20, 57, 61]; the third had 5 affected members and, like the other 2, also had members affected with a syndrome of pure parkinsonism in late adulthood. After evaluating over 250 genetic markers, and excluding over half the genome, we detected evidence for linkage of DRD to markers on chromosome 14 [62]. Subsequent analyses restricted DRD to the region bounded by the anonymous loci D14S269 and D14S63, a genetic interval of about 15 cM. Multipoint lod scores (log of the odds of linkage) in this interval exceed 6. This region of chromosome 14 is relatively devoid of identified genes and possesses no obvious candidate gene. Chromosomal studies have not revealed any evidence of obvious deletions or rearrangements in DRD [unpubl. data].

Only limited data on linkage in JP have appeared to date; the candidate gene tyrosine hydroxylase was excluded in a study of autosomal recessive JP families [63].

Discussion

Work in the past 3 years has brought the distinction between DRD and JP into sharper focus. Among the clinical determinants that may distinguish DRD and JP [64], those that allow the highest degree of confidence in diagnosis remain the absence of a resting tremor at onset in DRD, and the presence of complications of therapy (wearing-off, on-off, and prominent dyskinesias) in JP. The differences in frequencies of either gender affected, apparent pattern of inheritance of disease (in the absence of a firm diagnosis in another affected family member), and the presence of diurnal fluctuation cannot be used in an individual case in the assignment of diagnosis. While CSF studies have revealed a significant overlap in the biochemical profiles of DRD and JP, the ability of fluorodopa PET to distinguish these diagnoses has become clearer with the application of this modality to the study of a larger number of patients. Fluorodopa PET data also provide support for the consideration of pure parkinsonism, pathogenetically distinct from PD, as a late-adult phenotype of DRD in some individuals. The new pathological and neurochemical data from a case with DRD add support to the long-held suspicion that DRD is due to a developmental event and is distinct from the myriad of pathological processes that can cause JP. The exact developmental stage at which this occurs remains unknown. In PD, it is estimated that there must be an 80% loss of striatal dopamine content before the appearance of signs or symptoms [65]. CSF analyses in several unaffected members in DRD families found reductions in monoamine metabolites and biopterin similar to those in typically affected cases [66, 67]. The low symptomatic penetrance of DRD suggests that most gene carriers, although they may have an equivalent biochemical defect, must have a functional dopaminergic state above this level. However, the appearace of a parkinsonian phenotype in several obligate gene carriers in late adult years seems to indicate that age-related factors lead to a gradual decline in dopaminergic function [68] that may unmask this deficit.

The demonstration of linkage of DRD to markers on chromosome 14q [62] offers us a new tool to resolve several of the clinical issues concerning DRD, HPD and JP. Within all 3 of the families studied in our linkage work, there were members with and without diurnal fluctuations. This suggests that the factor responsible for fluctuations (such as a modulator of the diurnal activity of tyrosine hydroxylase) is a genetic factor that segregates independently from the gene causing DRD. The explanation for the uniformity of fluctuations in HPD may be that the alleles for this second factor in the Japanese population are restricted to those resulting in fluctuations while in European and North American populations a greater diversity in alleles may include those resulting in no fluctuation; the Russian population may have even a greater representation of these 'non-fluctuating' alleles.

Among the families studied in our linkage work, the family members with the parkinsonian phenotype could be inferred to have inherited the disease-bearing chromosome in 9 of the 10 instances; the one discordant member had limited response to moderate doses of *L*-dopa and may have had PD. This observation bolsters support for considering adult-onset parkinsonism as a DRD phenotype.

The clinical, pathophysiological, and pathological data that separate DRD and JP are convincing; the application of genetic tools will now allow us to address the final separation of DRD from JP. In our closing remarks from the 1990 symposium at this site, we concluded that clinical data do not allow us to separate DRD and HPD; we suggested that the final resolution of this issue may need to await the identification of the (DRD) disease gene [34]. The way is now paved to approach studying whether DRD and HPD are genetically related.

References

1 Beck D: Dystonia musculorum deformans with another case in the same family. Proc R Soc Med 1947;40:551.
2 Corner BD: Dystonia musculorum deformans in siblings: Treated with Artane (trihexypenidyl). Proc R Soc Med 1952;45:451–452.
3 Cotzias GC, Van Woert MH, Schiffer LM: Aromatic amino acids and modification of parkinsonism. N Engl J Med 1967;276:374–379.
4 Chase TN: Biochemical and pharmacologic studies of dystonia. Neurology 1970;20(suppl 2):122–130.
5 Castaigne P, Rondot P, Ribadeau Dumas JL, Said J: Affection extrapyramidale évoluant chez deux jeunes frères: Effets remarquables du traitement par la *L*-dopa. Rev Neurol 1971;124:162–166.
6 Segawa M, Ohmi K, Itoh S, Aoyama M, Hayakawa H: Childhood basal ganglia disease with remarkable response to *L*-dopa, 'hereditary progressive basal ganglia disease with marked diurnal variation' (in Japanese). Shinryo 1971;24:667–672.
7 Cooper IS: Levodopa-induced dystonia. Lancet 1972;ii:1317–1318.
8 Eldridge R, Kanter W, Koerber T: Levodopa in dystonia. Lancet 1973;ii:1027–1028.
9 Nygaard TG: Dopa-responsive dystonia: Delineation of the clinical syndrome and clues to pathogenesis. Adv Neurol 1993;60:577–585.
10 Willage H: Über paralysis agitans in jugendlichem Alter. Z Ges Neurol Psychiat 1911;4:520–587.
11 Quinn N, Critchley P, Marsden CD: Young onset Parkinson's disease. Mov Disord 1987;2:73–91.
12 Olsson Je, Brunk U, Lindvall B, Eeg-Olofsson O: Dopa-responsive dystonia with depigmentation of the substantia nigra and formation of Lewy bodies. J Neurol Sci 1992;112:90–95.

13 Nygaard TG: Idiopathic dystonia-parkinsonism; in Stern MB, Koller WC (eds): Parkinsonia syndromes. New York, Dekker, 1993, pp 451–466.
14 Narabayashi H, Yokochi M, Iizuka R, Nagatsu T: Juvenile parkinsonism; in Vinken PJ, Bruyn GW, Klawans HL (eds): Handbook of Clinical Neurology. Amsterdam, Elsevier Science, 1986, pp 153–165.
15 Segawa M, Nomura Y, Kase M: Hereditary progressive dystonia with marked diurnal fluctuation: clinicopathophysiological identification in reference to juvenile Parkinson's disease. Adv Neurol 1987;45:227–234.
16 Nygaard TG, Marsden CD, Fahn S: Dopa-responsive dystonia: Long-term treatment response and prognosis. Neurology 1991;41:174–181.
17 Markova E, Ivanova-Smolenskaya I: Phenotypic polymorphism of dopa-responsive dystonia in Russia. Basel, Karger, 1995. Monogr Neural Sci, vol 14, pp 36–40.
18 Nygaard TG, Marsden CD, Duvoisin RC: Dopa-responsive dystonia. Adv Neurol 1988;50:377–384.
19 Iwami O, Kawamura J, Hashimoto S, Suenaga T, Nakamura M: Hereditary progressive dystonia with marked diurnal fluctuation: A report of two siblings, one of them showing age-dependent changes of symptoms (in Japanese). Rinsho Shinkeigaku 1990;30:961–965.
20 Nygaard TG, Trugman JM, de Yebenes JG, Fahn S: Dopa-responsive dystonia. The spectrum of clinical manifestation in a large North American family. Neurology 1990;40:66–69.
21 Segawa M, Hosaka A, Miyagawa F, Nomura Y, Imai H: Hereditary progressive dystonia with marked diurnal fluctuation. Adv Neurol 1976;14:215–233.
22 Allen N, Knopp W: Hereditary parkinsonism-dystonia with sustained control by L-dopa and anticholinergic medication. Adv Neurol 1976;14:201–213.
23 Nygaard TG, Waran SP, Levine RA, Naini AB, Chutorian AM: Dopa-responsive dystonia simulating cerebral palsy. Pediatr Neurol 1994;11:236–240.
24 Hunt JR: Progressive atrophy of the globus pallidus. Brain 1917;40:58–148.
25 Duvoisin RC, Yahr MD, Lieberman J, Antunes J, Rhee S: The striatal foot. Trans Am Neurol Assoc 1972;97:267.
26 Angelini L, Nardocci N, Broggi G, Moretti G, Mainini P: A perplexing case of juvenile extrapyramidal disease. Ital J Neurol Sci 1981;2:135–138.
27 Kilroy AW, Paulsen WA, Fenichel GM: Juvenile parkinsonism treated with levodopa. Arch Neurol 1972;27:350.
28 Furumoto H, Kitano K, Wang DS, Matsumoto S, Moroo I: Oculogyric crisis as an initial symptom of juvenile parkinsonism-like disease (in Japanese). Rinsho Shinkeigaku 1989;29:1287–1289.
29 Parker JC Jr, Dyer ML, Paulsen WA: Neuronal intranuclear hyaline inclusion disease associated with premature coronary atherosclerosis. J Clin Neuro Ophthalmol 1987;7:244–249.
30 Nygaard TG, Duvoisin RC: Hereditary dystonia-parkinsonism syndrome of juvenile onset. Neurology 1986;36:1424–1428.
31 Deonna T: DOPA-sensitive progressive dystonia of childhood with fluctuations of symptoms: Segawa's syndrome and possible variants. Results of a collaborative study of the European Federation of Child Neurology Societies (EFCNS). Neuropediatrics 1986;17:81–85.
32 Rajput AH: Levodopa in dystonia musculorum deformans. Lancet 1973;i:432.
33 Krusz JC, Koller WC, Ziegler DK: Historical review: Abnormal movements associated with epidemic encephalitis lethargica. Mov Disord 1987;2:137–141.
34 Nygaard TG, Snow BJ, Fahn S, Calne DB: Dopa-responsive dystonia: Clinical characteristics and definition; in Segawa M (ed): Hereditary Progressive Dystonia with Marked Diurnal Fluctuation. Carnforth, Parthenon Publishing, 1993, pp 21–36.
35 Nygaard TG: An analysis of North American families with dopa-responsive dystonia; in Segawa M (ed): Hereditary Progressive Dystonia with Marked Diurnal Fluctuation. Carnforth, Parthenon Publishing, 1993, pp 97–104.
36 Umehara F, Nomoto M, Usuki F, Matsumoto W, Osame M: Juvenile parkinsonism in monozygotic twins (in Japanese). Rinsho Shinkeigaku 1991;31:306–309.
37 Askenasy JJ, Mendelson L, Keren O, Braun Z: Juvenile Parkinson's disease and its response to L-dopa therapy. J Neural Transm Parkinson's Dis Dement Sect 1990;2:23–30.

38   Yokochi M, Narabayashi H, Iizuka R, Nagatsu T: Juvenile parkinsonism: Some clinical, pharmacological, and neuropathological aspects. Adv Neurol 1984;40:407–413.
39   Yamamura Y, Sobue I, Ando K, Iida M, Yanagi T: Paralysis agitans of early onset with marked diurnal fluctuation of symptoms. Neurology 1973;23:239–244.
40   Still CN, Herberg K: Long-term levodopa therapy for torsion dystonia. South Med J 1976;69:564–566.
41   Ujike H, Nakashima M, Kuroda S, Otsuki S: Two siblings of juvenile Parkinson's disease dystonic type (Yokochi type 3) and hereditary progressvie dystonia with marked diurnal fluctuation (Segawa) (in Japanese). Rinsho Shinkeigaku 1989;29:890–894.
42   Davison C: Pallido-pyramidal disease. J Neuropathol Exp Neurol 1954;13:50–59.
43   Ota Y, Miyoshi S, Ueda O, Mukai T, Maeda A: Familial paralysis agitans juvenilis. A clinical, anatomical, and genetic study. Folia Psychiatr Neurol Jpn 1958;12:112–121.
44   Dwork AJ, Balmaceda C, Fazzini EA, Maccollin M, Cote L, Fahn S: Dominantly inherited, early-onset parkinsonism: Neuropathology of a new form. Neurology 1993;43:69–74.
45   Fink JK, Barton N, Cohen W, Lovenberg W, Burns RS, Hallett M: Dystonia with marked diurnal variation associated with biopterin deficiency. Neurology 1988;38:707–711.
46   Maekawa N, Hashimoto T, Sasaki M, Oishi T, Tsuji S: A study on catecholamine metabolites in CSF in a patient with progressive dystonia with marked diurnal fluctuation (in Japanese). Rinsho Shinkeigaku 1988;28:1206–1208.
47   Clough CG, Mendoza M, Yahr MD: A case of sporadic juvenile Parkinson's disease. Arch Neurol 1981;38:730–731.
48   Naidu S, Wolfson LI, Sharpless NS: Juvenile parkinsonism: A patient with possible primary striatal dysfunction. Ann Neurol 1978;3:453–455.
49   Sachdev KK, Singh N, Krishnamoorthy MS: Juvenile parkinsonism treated with levodopa. Arch Neruol 1977;34:244–245.
50   Ishida A, Takada G, Kobayashi Y, Toyoshima I, Takai K: Effect of tetrahydrobiopterin and 5-hydroxytryptophan on hereditary progressive dystonia with marked diurnal fluctuation: A suggestion of the serotonergic system involvement. Tohoku J Exp Med 1988;154:233–239.
51   LeWitt PA, Miller LP, Levine RA, Lovenberg W, Newman RP, Papavasiliou A, Rayes A, Eldridge R, Burns RS: Tetrahydrobiopterin in dystonia: Identification of abnormal metabolism and therapeutic trials. Neurology 1986;36:760–764.
52   LeWitt PA, Newman RP, Miller LP, Lovenberg W, Eldridge R: Treatment of dystonia with tetrahydrobiopterin (letter). N Engl J Med 1983;308:157–158.
53   Fink JK, Ravin P, Argoff CE, Levine RA, Brady RO, Hallett M, Barton NW: Tetrahydrobiopterin administration in biopterin-deficient progressive dystonia with diurnal variation. Neurology 1989; 39:1393–1395.
54   Furukawa Y, Nishi K, Kondo T, Mizuno Y, Narabayashi H: CSF biopterin levels and clinical features of patients with juvenile parkinsonism. Adv Neurol 1993;60:562–567.
55   Eidelberg D, Moeller JF, Dhawan V, Sidtis JJ, Ginos JZ, Strother SC, Cedarbaum J, Greene P, Fahn S, Rottenberg DA: The metabolic anatomoy of Parkinson's disease: complementary [$_{18}$F]fluorodeoxyglucose and [$_{18}$F]fluorodopa positron emission tomographic studies. Mov Disord 1990;5:203–213.
56   Sawle GV, Leenders KL, Brooks DJ, Harwood G, Lees AJ, Frackowiak RSJ, Marsden CD: Dopa-responsive dystonia: [$_{18}$F]dopa positron emission tomography. Ann Neurol 1991;30:24–30.
57   Nygaard TG, Takahashi H, Heiman GA, Snow BJ, Fahn S, Calne DB: Long-term treatment response and fluorodopa positron emission tomographic scanning of parkinsonism in a family with dopa-responsive dystonia. Ann Neurol 1992;32:603–608.
58   Snow BJ, Nygaard TG, Takahashi H, Calne DB: Positron emission tomographic studies of dopa-responsive dystonia and early-onset idiopathic parkinsonism. Ann Neurol 1993;34:733–738.
59   Hornykiewicz O: Striatal dopamine in dopa-responsive dystonia: Comparison with idiopathic Parkinson's disease and other dopamine-dependent disorders. Basel, Karger, 1995. Monogr Neural Sci, vol 14, pp 101–108.
60   Rajput AH, Gibb WRG, Zhong XH, Shannak KS, Kish S, Chang LG, Hornykiewicz O: Dopa-responsive dystonia – pathological and biochemical observations in a case. Ann Neurol 1994;35: 396–402.

61 Kwiatkowski DJ, Nygaard TG, Schuback DE, Perman S, Trugman JM, Bressman SB, Burke RE, Brin MF, Ozelius L, Breakefield XO, Fahn S, Kramer PL: Identification of a highly polymorphic microsatellite VNTR within the argininosuccinate synthetase locus: Exclusion of the dystonia gene on 9q32–34 as the cause of dopa-responsive dystonia in a large kindred. Am J Hum Genet 1991;48: 121–128.
62 Nygaard TG, Wilhelmsen KC, Risch NJ, Brown DL, Trugman JM, Gilliam TC, Fahn S, Weeks DE: Linkage mapping of dopa-responsive dystonia (DRD) to chromosome 14q. Nat Genet 1993;5:386–391.
63 Tanaka H, Ishikawa A, Ginns EI, Miyatake T, Tsuji S: Linkage analysis of juvenile parkinsonism to tyrosine hydroxylase gene locus on chromosome-11. Neurology 1991;41:719–722.
64 Calne DB, Nygaard TG, Snow BJ: The distinction between early onset idiopathic parkinsonism (juvenile Parkinson disease) and dopa-responsive dystonia (hereditary progressive dystonia, Segawa dystonia); in Segawa M (ed): Hereditary Progressive Dystonia with Marked Diurnal Fluctuation. Carnforth, Parthenon Publishing, 1993, pp 215–218.
65 Bernheimer H, Birkmayer W, Hornykiewicz O, Jellinger K, Seitelberger F: Brain dopamine and the syndromes of Parkinson and Huntington: Clinical, morphological, and neurochemical correlations. J Neurol Sci 1973;20:415–455.
66 Fink JK, Ravin P, Argoff CE, Levine RA: Inheritance of neurochemical abnormalities in progressive dystonia with diurnal variation (abstract). Am J Hum Genet 1990;47:A155.
67 Takahashi H, Levine RA, Galloway MP, Snow BJ, Calne DB, Nygaard TG: Biochemical and fluorodopa positron emission tomographic findings in an asymptomatic carrier of the gene for dopa-responsive dystonia. Ann Neurol 1994;35:354–356.
68 McGeer PL, McGeer EG, Suzuki JS: Aging and extrapyramidal function. Arch Neurol 1977;34: 33–35.

Torbjoern G. Nygaard, MD, Department of Neurology, Columbia-Presbyterian Medical Center, 710 West 168th Street, New York, NY 10032 (USA)

The Gene for Hereditary Progressive Dystonia with Marked Diurnal Fluctuation Maps to Chromosome 14q

Kotaro Endo[a], Hajime Tanaka[a], Masaaki Saito[a], Shoji Tsuji[a], Torbjoern G. Nygaard[b], Daniel E. Weeks[c], Yoshiko Nomura[d], Masaya Segawa[d]

[a] Department of Neurology, Brain Research Institute, Niigata University, Niigata, Japan;
[b] Department of Neurology, Columbia Presbyterian Medical Center, New York, N.Y., USA;
[c] Department of Human Genetics, University of Pittsburgh, Pittsburgh, Penn., USA;
[d] Segawa Neurological Clinic for Children, Tokyo, Japan

Hereditary progressive dystonia with marked diurnal fluctuation (HPD) is a postural dystonia with onset in childhood, which responds well to *L*-dopa without any adverse effects. Since the original description of HPD by Segawa et al. [1–3], more than 70 cases have been reported from Japan and other countries [4]. The age of onset of clinical symptoms ranges from 1 to 11 years (mean 6.03 ± 2.82 years). In most cases, the initial symptoms are fatigability and gait disturbance due to leg dystonia with flexion-inversion of one foot. The symptoms are aggravated in the evening and markedly alleviated in the morning, after sleep. The symptoms progress gradually, spread to other extremities and, within 5 or 6 years, all limbs become involved. *L*-dopa produces remarkable effects in all cases without any adverse effects [5].

In 1988, Nygaard et al. [6] collected cases with dystonia showing beneficial response to *L*-dopa, labelled dopa-responsive dystonia (DRD), which include all dystonias that respond to *L*-dopa irrespective of their etiologies Nygaard et al. [6, 7] define DRD as childhood onset of dystonia or gait disorder, complete or near complete responsiveness of symptoms to low doses of levodopa, and maintenance of a smooth clinical response.

Although HPD and DRD share many of the characteristic clinical features including childhood onset dystonia and a remarkable beneficial effect of *L*-dopa, there are differences between these two conditions: diurnal fluctuation of dystonia is a cardinal feature in HPD, but it is observed in only 77% of cases of DRD, and while dopa-induced dyskinesia has rarely been observed in HPD, it is frequent in DRD [7]. With these differences in clinical observations, it seems to be too early to conclude that HPD and DRD are the same disease in terms of their etiologies.

Previous investigations have indicated that catecholamine metabolites including homovanillic acid are decreased in cerebrospinal fluid of HPD patients [8–10]. Interestingly, the age of onset and the clinical course, with minimal progression after the third decade in HPD, correlate well with the changes of tyrosine hydroxylase activities in the nerve terminals of the nigrostriatal neurons, which is high in infancy and shows a rapid decline during later childhood [11]. Based on this observation, Segawa et al. [1, 3] proposed that the pathophysiology of HPD is a functional abnormality in the nerve terminals of the nigrostriatal dopaminergic neurons in the caudate. These data raised the possibility that the tyrosine hydroxylase gene is the causative gene for HPD. The possibility, however, was excluded by linkage analysis using polymorphisms of tyrosine hydroxylase genes as well as other polymorphic markers including H-Ras1 and insulin genes [12], and we are left without any clues to the genetic locus for HPD.

In 1993, Nygaard et al. [13] mapped the genetic locus for DRD to the long arm of chromosome 14 using microsatellite polymorphisms. They have shown that the DRD gene is tightly linked to chromosome 14q with a maximum lod score of 4.67 at 8.6 cM for D14S63. With this background, it is now possible to test if HPD and DRD shares the same locus on chromosome 14q. To answer this question, we have performed linkage analysis using microsatellite loci on chromosome 14q.

Patients and Methods

HPD Families
A single family of HPD, which was described in the previous linkage analysis to tyrosine hydroxylase locus [12], was analyzed.

Genotype Analysis
Blood samples were obtained with informed consent from 18 family members including 4 affected individuals. High-molecular-weight genomic DNA was extracted from either peripheral mononuclear cells or lymphoblastoid cell lines as described [14]. Microsatellite loci selected in the present analysis are D14S79, D14S47, D14S52, D14S66, and D14S77 [15, 16]. Analysis of microsatellite polymorphisms was performed by polymerase chain reaction as described [17].

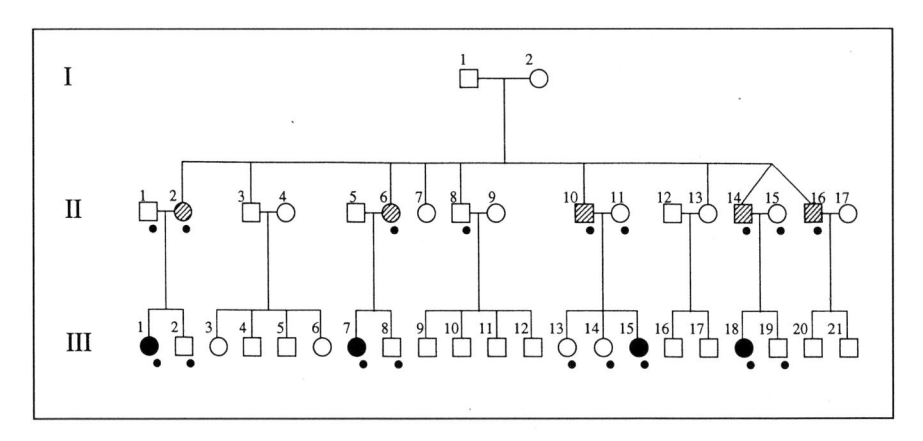

Fig. 1. Pedigree chart of an HPD family (pedigree T). Closed symbols indicate affected individuals. Although none of their parents are affected, we assumed that individuals II-2, II-6, II-10, II-14, and II-16 (shown by shaded symbols) carry the mutant gene. Small solid circles below symbols represent individuals whose genomic DNAs were analyzed.

Linkage Analysis

Linkage analysis was performed using the MLINK program in LINKAGE [18]. We assumed the same conditions of a disease allele frequency of 0.00001 and sex-dependent penetrance of 0.29 in males and 0.42 in females for calculation of pair-wise lod scores as described by Nygaard et al. [13].

Results and Discussion

The structure of the HPD pedigree analyzed in the present study is shown in figure 1. In the HPD family, 4 siblings are affected but their parents do not show any neurological signs or symptoms. As autosomal dominant inheritance with low penetrance has been postulated for HPD, we assumed that individuals II-2, II-6, II-10, II-14, and II-16 among their parents carry the mutant gene.

Pair-wise lod scores between the HPD locus and microsatellite loci at D14S79, D14S47, D14S52, D14S66, and D14S77 are given in table 1. Among the loci tested, D14S52 gave the highest maximal lod score of 2.681 at a recombination fraction of 0.0. A maximal lod score of 1.464 was also obtained with D14S79 at a recombination fraction of 0.0. Positive lod scores of 0.506–0.704 were also obtained for the other three loci, D14S47, D14S66 and D14S77. Furthermore, we did not observe any recombination events in affected individuals in the HPD family.

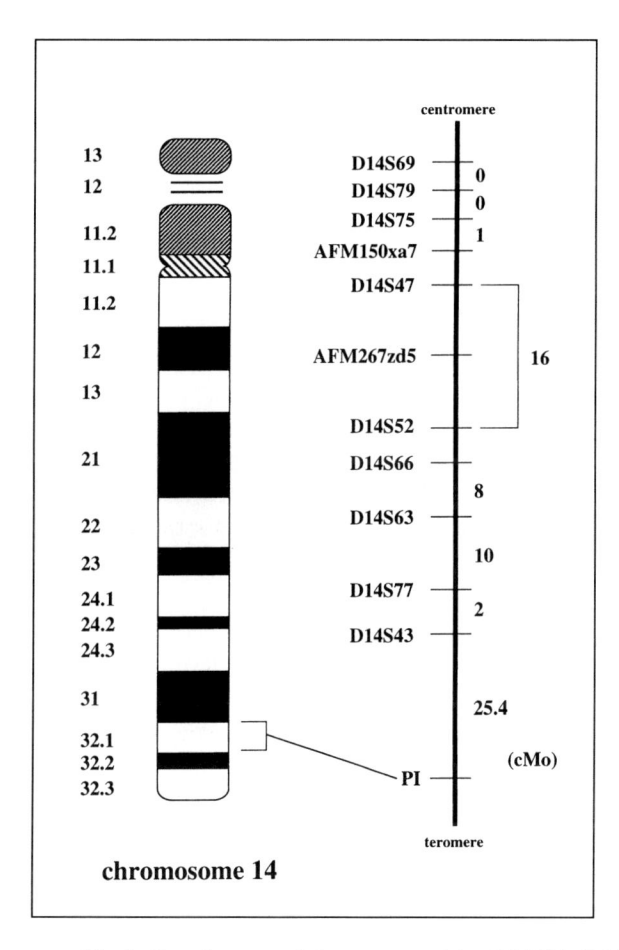

Fig. 2. Genetic map of chromosome 14q. D14S79, D14S47, D14S52, D14S66, and D14S77 were used in the present study.

Table 1. Pair-wise lod scores for HPD versus chromosome 14 markers

Locus	Recombination fraction									
	0.000	0.010	0.050	0.100	0.150	0.200	0.300	0.400	Z_{max}	(θ)
D14S79	1.464	1.442	1.353	1.236	1.113	0.668	0.341	0.983	1.464	0.000
D14S47	0.506	0.498	0.465	0.423	0.379	0.234	0.124	0.333	0.506	0.000
D14S52	2.681	2.638	2.459	2.226	1.979	1.146	0.524	1.717	2.681	0.000
D14S66	0.704	0.688	0.620	0.533	0.445	0.186	0.052	0.357	0.704	0.000
D14S77	0.506	0.498	0.465	0.423	0.379	0.234	0.112	0.333	0.506	0.000

The Gene for HPD

In the present study, the maximal lod core exceeded 2.0 in a single family, which strongly indicates that as far as the present HPD family is concerned, the gene for HPD is located on the chromosome 14q near D14S52. Nygaard et al. [13] described recombination events involving D14S47 and D14S63, which define a region of about 22 cM as containing the DRD gene [13]. As shown in figure 2, D14S52 is in the middle of the candidate region. Taken together, these data strongly indicate that the gene for HPD is located at the same locus as the gene for DRD.

The fact that the gene for HPD maps to the same locus indicates that HPD and DRD are most likely caused by allelic mutations on a gene located on chromosome 14q, despite some differences in their clinical manifestations as described above. To further narrow the candidate region, detailed linkage analyses using flanking microsatellite polymorphisms as well as the study of linkage disequilibrium will be required. Since most of HPD families are small, the search for loci showing linkage disequilibrium would be of particular help for the positional cloning of the gene for HPD (DRD).

Acknowledgment

This study was supported in part by a Grant in Aid for Scientific Research on Priority Areas, a Grant in Aid for Creative Basic Research (Human Genome Program), the Ministry of Education, Science and Culture, Japan, a Grant from Research Committee for Ataxic Diseases, the Ministry of Health and Welfare, Japan and special coordination funds of the Japanese Science and Technology Agency.

References

1 Segawa M, Hosaka A, Miyagawa F, Nomura Y, Imai H: Hereditary progressive dystonia with marked diurnal fluctuation; in Eldridge R, Fahn S (eds): Adv Neurol 1976;14:215–233.
2 Segawa M, Nomura Y, Kase M: Hereditary progressive dystonia with marked diurnal fluctuation – clinico-pathological identification in reference to juvenile Parkinson's disease. Adv Neurol 1986; 45:227–234.
3 Segawa M, Nomura Y, Kase M: Diurnal fluctuating hereditary progressive dystonia; in Vinken PJ, Bruyn GW, Klawans HL (eds): Handbook of Clinical Neurology. New York, Elsevier Science Publishers, 1986, pp 529–539.
5 Segawa M, Nomura Y: Hereditary progressive dystonia with marked diurnal fluctuation; in Segawa M (ed): Hereditary Progressive Dystonia with Marked Diurnal Fluctuation. Carnforth, Parthenon, 1993, pp 3–19.
6 Nygaard TG, Marsden CD, Duvoisin RC: Dopa-responsive dystonia. Adv Neurol 1988;50:377–384.
7 Nygaard TG, Snow BJ, Fahn S, Calne DB: Dopa-responsive dystonia: Clinical characteristics and definition; in Segawa M (ed): Hereditary Progressive Dystonia with Marked Diurnal Fluctuation. Carnforth, Parthenon, 1993, pp 21–35.
8 Ouvrier RA: Progressive dystonia with marked diurnal fluctuation. Ann Neurol 1978;4:412–417.

9   Kumamoto I, Nomoto M, Yoshidome M, Osame M, Igata A: Five cases of dystonia with marked diurnal fluctuation and special reference to homovanillic acid in CSF. Clin Neurol (Tokyo) 1984; 24:697–702.
10 Shimoyamada Y, Yoshikawa A, Kashii H, Kihara S, Koike M: Hereditary progressive dystonia – An observation of the catecholamine metabolism during L-dopa therapy in a 9-year-old girl. No-to-Hatatsu (Tokyo) 1986;18:505–509.
11 McGeer EG, McGeer PL: Some characteristics of brain tyrosine hydroxylase; in Mandel AJ (ed): New Concepts in Neurotransmitter Regulation. New York, Plenum Press, 1973, pp 53–68.
12 Tsuji S, Tanaka H, Miyatake T, Ginns EI, Nomura Y, Segawa M: Linkage analysis of hereditary progressive dystonia to the tyrosine hydroxylase gene locus; in Segawa M (ed): Hereditary Progressive Dystonia with Marked Diurnal Fluctuation. Carnforth, Parthenon, 1993, pp 107–114.
13 Nygaard TG, Wilhelmsen KC, Risch NJ, Brown DL, Trugman JM, Gilliam C, Fahn S, Weeks DE: Linkage mapping of dopa-responsive dystonia (DRD) to chromosome 14q. Nature Genet 1993;5: 386–391.
14 Sambrook JH, Fritsch EF, Maniatis T: Molecular Cloning: A Laboratory Manual, ed 2. Cold Spring Harbor, 1989.
15 Wang ZY, Weber JL: Continuous linkage map of human chromosome-14 short tandem repeat polymorphisms. Genomics 1992;13:532–536.
16 Weissenbach J, et al: A second-generation linkage map of the human genome. Nature 1992;359: 794–801.
17 Weber JL, May PE: Abundant class of human DNA polymorphisms which can be typed using the polymerase chain reaction. Am J Hum Genet 1989;44:388–396.
18 Lathrop GM, Lalouel JM, Jullier C, et al: Strategies for multilocus linkage analysis in humans. Proc Natl Acad Sci USA 1984;81:3443–3446.

Kotaro Endo, MD, Department of Neurology, Brain Research Institute, Niigata University,
1 Asahimachi, Niigata 951 (Japan)

Early-Onset, Generalized Dystonia Caused by DYT1 Gene on Chromosome 9q34

Laurie J. Ozelius[a,b], *Susan B. Bressman*[d], *Patricia L. Kramer*[e], *Neil Risch*[f], *Deborah de Leon*[d], *Stanley Fahn*[d], *Xandra O. Breakefield*[a,c,1]

[a] Molecular Neurogenetics Unit, Neurology Service, Massachusetts General Hospital, Boston, Mass.,
[b] Genetics Department and
[c] Neuroscience Program, Harvard Medical School, Boston, Mass.,
[d] Dystonia Clinical Research Center, Department of Neurology, Columbia Presbyterian Medical Center, New York, N.Y.,
[e] Department of Neurology, Oregon Health Sciences, Portland, Oreg., and
[f] Departments of Epidemiology and Public Health, and Human Genetics, Yale University School of Medicine, New Haven, Conn., USA

Dystonia refers to abnormal involuntary movements which are frequently twisting and repetitive in nature. These movements can be symptoms of a number of neurologic disease states (secondary dystonias) or can exist as distinct clinical entities without other symptoms (primary or idiopathic dystonia) [1]. A large proportion of the idiopathic types of dystonia are believed to be genetically determined. They can be subdivided into distinct subtypes based on the age of onset, nature of symptoms and drug responsiveness (table 1) [for review, see ref. 2, 3]. Subtypes include early onset with a tendency to generalize [4], late onset [5, 6]; myoclonic [7–9]; dopa-responsive [10, 11]; paroxysmal [12], rapid onset with Parkinsonian features [13], and an X-linked form with Parkinsonian features in the Philippines [14, 15]. All are inherited in an autosomal dominant mode with reduced penetrance except the Filipino form which is X-linked recessive. The early-onset form has been mapped to chromosome 9q34 in some families [4, 16].

---

[1] We thank the many fellow researchers and affected families who have helped us in this work. This work was supported by the Dystonia Medical Research Foundation, The Jack Fasciana Fund for Support of Dystonia Research and NINDS grant No. NS28384.

Table 1. Hereditary dystonias

Type	Mode of inheritance	Chromosomal location of gene
Early-onset, generalized	AD	9q34
Late-onset, focal	AD	?
Myoclonic, alcohol responsive	AD	?
Dopa-responsive	AD	14
Paroxysmal	AD	?
Rapid onset with Parkinson features	AD	?
X-linked with parkinsonism	XR	Xq21

This dystonia locus, DYT1, does not appear to be involved in, at least some, families with late-onset dystonia [Holmgren, pers. commun.; 6], alcohol-responsive myoclonic dystonia [17; Gasser, pers. commun.]; dopa-responsive dystonia [18]; whispering dysphonia [19] and rapid-onset dystonia [13]. The dopa-responsive locus has been mapped to chromosome 14 [20] while the X-linked locus has been mapped to Xq12 [21, 22]. It is now clear that at least three different genes are responsible for different forms of hereditary dystonia. It is still possible that some of the clinically distinct forms of hereditary dystonia are caused by different mutations in the same disease gene. Resolution of the genes causing dystonia will aid in clinical classification of this heterogeneous set of related syndromes.

'Classic' Early-Onset Dystonia

Classic, early-onset dystonia occurs in many different ethnic populations but has a particularly high frequency in the Ashkenazic Jewish group due to a founder mutation [23]. This form of hereditary dystonia can be distinguished by its tendency to manifest at an early age (mean 13 years, median 8–10 years, range 4–43 years) [4, 24]. Most cases have onset in a leg or arm, while only a few cases manifest first in the head or neck. The earlier the onset, the more likely the dystonia is to appear in a leg and dystonia almost always spreads to other limbs. In early-onset families with multiple affected members, all show this pattern of early limb onset dystonia and there are no other associated neurologic problems. However, it should be noted that in some other families having a predominance of members with late-onset dystonia, some early-onset members have been described [6]. It may be difficult to determine whether the DYT1 gene is responsible when there is only an isolated individual manifesting early-onset dystonia symptoms.

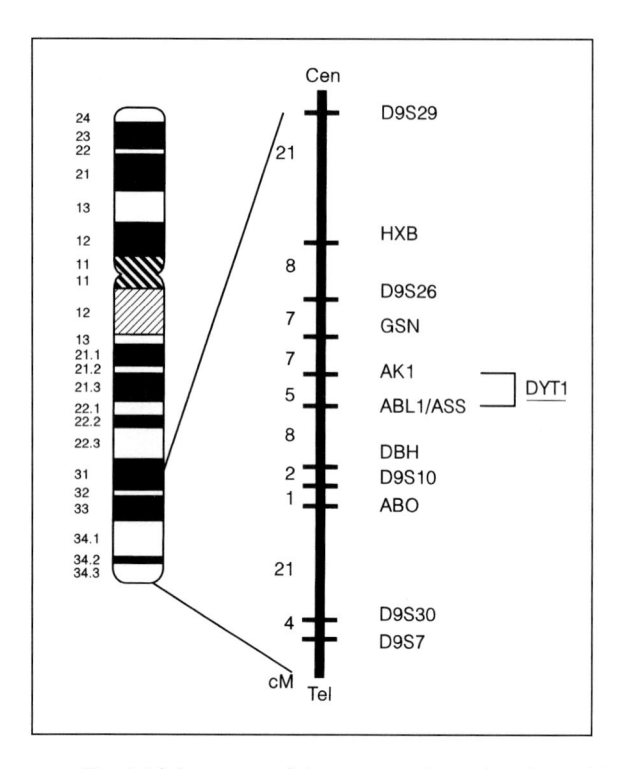

Fig. 1. Linkage map of chromosome 9q noting the position of the DYT1 gene between loci AK1 and ASS.

The classic form of early-onset dystonia does not consistently respond to drug therapy, although about half the cases benefit from high dose anticholinergics, and some individuals respond to dopamine agonists, dopamine antagonists or benzodiazepines [1]. In some cases, injury to a body part appears to precede onset of symptoms in that part, but it has been difficult to establish a cause-and-effect relationship to this phenomenon [25, 26].

Mutations in the DYT1 Gene

The involvement of the DYT1 gene in classic, early-onset dystonia was first demonstrated in a large, non-Jewish family [16] and in a pooled sample of 12 Ashkenazic Jewish families [4]. Initial studies mapped the locus to a 5 cM region between loci AK1 and ASS (fig. 1). Once the general location of the gene was known, a search was made for other polymorphic loci which could help to further define the

Table 2. Allele frequencies in Ashkenazic Jews near DYT1 gene

Locus:	D9S62	D9S62	D9S63	ASS	ABL	D9S64	
Allele:	2	8	16	12	4	2	10
Individuals affected (58)	0.88	0.90	0.95	0.81	0.93	0.33	0.26
Controls (306)	0.03	0.13	0.12	0.09	0.63	0.05	0.04

exact position of the gene. Three highly polymorphic dinucleotide repeat markers have been identified within the region [18, 27]. These markers have proven extremely useful in tracking the DYT1 gene in individuals, families and populations. Within the Ashkenazic Jewish population most (>95%) of the individuals with early-onset dystonia bear alleles 2/8 at the two D9S62 loci and 16 at D9S63 whereas these alleles are rarely seen in control Ashkenazic individuals (table 2). This haplotype then serves to mark the founder mutation which causes early-onset dystonia in this ethnic population, thus, allowing us to track the presence of this disease gene. Several important findings have been revealed in this analysis. First, within this Jewish population this mutant gene appears to be responsible for both hereditary and apparently some sporadic cases of early-onset dystonia [23]. This latter phenomenon is explained by the low penetrance of the disease gene (about 30%) and the small size of most families. Secondly, Ashkenazic individuals manifesting other forms of dystonia rarely have the same mutation as defined by the haplotype [24]. Thirdly, the mutation marked by these polymorphisms appears to have been introduced into the Ashkenazic Jewish populations only about 20–30 generations ago, probably in Russia [28]. These markers now allow molecular diagnosis of the early-onset dystonia gene in Ashkenazic Jewish families [de Leon, unpubl. data].

The syndrome of classic, early-onset dystonia occurs in many ethnic groups. We have identified six non-Jewish families with multiple affected members which manifest a syndrome essentially identical to that described for the Ashkenazic Jewish population, including the low penetrance [3]. Pairwise linkage analysis in these families with markers in the DYT1 region indicates that this gene is probably also responsible for dystonia in these families. In these non-Jewish families, different sets of alleles, as marked by the dinucleotide repeat polymorphisms, are found associated with the chromosome transmitting the disease (with the exception of two families of French-Canadian descent which share a common set of alleles and hence are probably related). These different sets of alleles appear to define five additional mutations in the DYT1 gene, in addition to the Ashkenazic Jewish one, which can yield autosomal dominant, early-onset dystonia with reduced penetrance.

Search for the DYT1 Gene

Positional cloning of a disease gene becomes feasible when the location of the disease gene is defined to an area of less than 1 megabase and when one or more mutant alleles have been defined in the disease gene. These criteria have now been met for the DYT1 gene. The first step is to clone all the genomic DNA from the region from control individuals. This has been undertaken by using fragments of unique sequence from the region to hybridize to and thus identify yeast artificial chromosomes (YACs) and cosmids which contain larger fragments of DNA from the region. These larger fragments are aligned by virtue of unique or repeat elements which they share. A megabase region of genomic DNA contains on average 10–30 genes. These genes can be found within the large stretches of noncoding sequences by using genomic clones to screen cDNA libraries or by 'trapping' exons from the genomic DNA contained in these clones [29]. Coding sequences of genes can be searched for mutations associated with disease state by screening for differences from normal sequences using SSCP analysis [30] and direct sequencing [31] of PCR-amplified fragments of genomic DNA or cDNA, prepared from mRNA by reverse transcriptase. In the case of the DYT1 gene, we believe we have DNA bearing at least six different mutations in the gene based on analysis of alleles in the region (see above). By having multiple mutations, the likelihood is increased of finding a mutation that will incriminate the gene. Genomic DNA is the most reliable in screening for mutations in a dominant disease gene, because the mutant and normal allele are present in equal amounts. However, genomic screening is very laborious because the exon structure of a gene must be determined first. Screening of mRNAs has potential limitations because the mutant message may not be expressed or may be unstable, and in the case of a neurologic disease, the message may not be expressed in tissues, such as blood cells and skin fibroblasts that can be obtained readily from patients. To date we have evaluated one gene in the critical region and have not found any mutations in it associated with dystonia.

Identification of the DYT1 gene should provide insight into the temporal and spatial organization of the basal ganglia, where the defective function is believed to originate [32]. The pattern of symptoms in the early-onset form of dystonia suggests that there is a progression of involvement from lower to higher body parts. In fact, within the basal ganglia themselves, neurons are laid down during development in a dorsal to ventral pattern that loosely reflects a foot-to-head control of movement [33]. It appears that in a person with a genetic susceptibility to the early-onset form of dystonia, some stimulus, usually early in life, can trigger a 'fixation' of neuronal activity that overwhelms other movement controls. If this fixation is possible only in neurons which retain some plasticity, then we can hypothesize that loss of plasticity also follows a foot-to-head pattern with age, and that neurons involved in control of lower body movement can exert a 'dominant' influence over those controlling upper body movements. In most gene carriers, however, this trigger is never pulled and they remain unaffected throughout life. Understanding the nature of the dystonia gene thus may help us gain insight into molecular controls of pattern formation and the neurons regulating motor activity in the basal ganglia.

References

1   Fahn S, Marsden CD, Caine DB: Classification and investigation of dystonia; in Marsden CD, Fahn S (eds): Movement Disorders London, Butterworth, 1987, vol 2, pp 332–358.
2   Gasser T, Fahn S, Breakefield XO: The autosomal dominant dystonias. Brain Pathol 1992;2:297–308.
3   Kramer P, Bressman S, Ozelius L, Fahn S, Breakefield XO: The Genetics of Dystonia; in Tsui J, Calne D (eds): Handbook of Dystonia. New York, Marcel Dekker, in press.
4   Kramer PL, de Leon D, Ozelius L, Risch M, Brin MF, Bressman SB, Burke RE, Kwiatkowski DJ, Schuback DE, Shale H, Gusella JF, Breakefield XO, Fahn S: Dystonia gene in Ashkenazi Jewish population located on chromosome 9q32–34. Ann Neurol 1990;27:114–120.
5   Forsgren L, Holmgren G, Almay BGL, Drugge U: Abnormal dominant torsion dystonia in a Swedish family. Adv Neurol 1988;50:83–92.
6   Bressman SB, Heiman MS, Nygaard TG, Ozelius LO, Hunt A, Brin MF, Gordon MF, Moskowitz CB, de Leon D, Burke RE, Fahn S, Risch NJ, Breakefield XO, Kramer PL: Genetic heterogeneity of idiopathic torsion dystonia in the non-Jewish population. Neurology, in press.
7   Kurlan R, Behr J, Medved L, Shoulson I: Myoclonus and dystonia: A family study. Adv Neurol 1988;50:385–390.
8   Quinn N, Rothwell J, Thompson P, Marsden C: Hereditary essential myoclonus: An area of confusion. Adv Neurol 1988;50:391–402.
9   Kyllerman M, Forsgren L, Sanner G, Holmgren G, Wahlstrom J, Drugge U: Alcohol-responsive myoclonic dystonia in a large family: Dominant inheritance and phenotypic variation. Mov Disord 1990;5:270–279.
10 Nygaard TG, Marsden CD, Duvoisin RC: Dopa-responsive dystonia. Adv Neurol 1988;50:377–384.
11 Segawa M, Hosaka A, Miyagawa F, Nomura Y, Imai H: Hereditary progressive dystonia with marked diurnal fluctuations. Adv Neurol 1976;14:215–233.
12 Kurlan R, Shoulson I: Familial paroxysmal dystonic choreoathetosis and response to alternate-day oxazepam therapy. Ann Neurol 1983;13:456–457.
13 Dobyns WB, Ozelius LJ, Kramer PL, Brashear A, Farlow MR, Perry TR, Walsh LE, Kasarskis EJ, Butler IJ, Breakefield XO: Rapid-onset dystonia-parkinsonism. Neurology 1993;43:2596–2602.
14 Kupke K, Lee L, Müller U: Assignment of the X-linked torsion dystonia gene to Xq21 by linkage analysis. Neurology 1990;40:1438–1442.
15 Fahn S, Moskowitz C: X-linked recessive dystonia and parkinsonism in Filipino males. Ann Neurol 1988;24:179.
16 Ozelius L, Kramer PL, Moskowitz CB, Kwiatkowski DJ, Brin MF, Bressman SB, Schuback DE, Falk CT, Risch N, de Leon D, Burke RE, Haines J, Gusella JF, Fahn S, Breakefield XO: Human gene for torsion dystonia located on chromosome 9q32-q34. Neuron 1989;2:1427–1434.
17 Wahlstrom J, Schuback D, Kramer P, Ozelius L, Kyllerman M, Forsgren L, Holmgren G, Drugge U, Sanner G, Fahn S, Breakefield XO: The gene for familial dystonia with myoclonic jerks responsive to alcohol is not located on the distal end of 9q. Clin Genet, in press.
18 Kwiatkowski DJ, Henske EP, Weimer K, Ozelius L, Gusella JF, Haines J: Construction of a GT polymorphism map of human 9q. Genomics 1992;12:229–240.

19　Ahmad F, Davis MB, Waddy HM, Oley CA, Marsden CD, Harding AE: Evidence for locus heterogeneity in autosomal dominant torsion dystonia. Genomics 1993;15:9–12.
20　Nygaard TG, Wilhelmsen KC, Risch NJ, Brown DL, Trugman JM, Gilliam C, Fahn S, Weeks DE: Linkage mapping of dopa-responsive dystonia (DRD) to chromosome 14q. Nat Genet 1993;5:386–391.
21　Kupke K, Graeber M, Müller U: Dystonia-parkinsonism syndrome (XPD) locus: Flanking markers in Xq12-q21.1. Am J Hum Genet 1992;50:808–815.
22　Wilhelmsen KC, Weeks DE, Nygaard TG, Moskowitz CB, Rosales RL, dela Paz DC, Sobrevega EE, Fahn S, Gilliam C: Genetic mapping of 'lubag' (X-linked dystonia-parkinsonism) in a Filipino kindred to the pericentromeric region of the X chromosome. Ann Neurol 1991;29:124–131.
23　Ozelius L, Kramer P, de Leon D, Risch N, Bressman SB, Schuback DE, Brin MF, Kwiatkowski DJ, Burke RE, Gusella JF, Fahn S, Breakefield XO: Strong allelic association between the torsion dystonia gene (DYT1) and loci on chromosome 9q34 in Ashkenazi Jews. Am J Hum Genet 1992;50:619–628.
24　Bressman SB, de Leon D, Ozelius L, Kramer P, Brin MF, Burke RE, Fahn S, Gusella JF, Breakefield XO, Risch N: Idiopathic torsion dystonia in Ashkenazim: A distinct phenotype of a single mutation (abstract 822). Am J Hum Genet 1992;51.
25　Koller WC, Wong GF, Lang A: Post-traumatic movement disorders: A review. Mov Disord 1989;4:20–36.
26　Fletcher NA, Harding AE, Marsden CD: The relationship between trauma and idiopathic torsion dystonia. J Neurol Neurosurg Psychiatry 1991;54:713–717.
27　Ozelius L, Kwiatkowski D, Schuback D, Breakefield XO, Wexler N, Gusella J, Haines J: A genetic map of human chromosome 9q. Genomics 1992;14:715–720.
28　Risch N, de Leon D, Ozelius LJ, Kramer PL, Bressman S, Kwiatkowski D, Brin M, Schuback D, Burke R, Gusella J, Fahn S, Breakefield XO: The genetics of idiopathic torsion dystonia in Ashkenazi Jews. Ministry of science and Technology, UK-Israel Binational Symposium on Molecular Genetics and Neuropsychiatric Disorders. Scientific Program and Abstracts, October 1991.
29　Buckler AJ, Chang DD, Graw SL, Brook JD, Haber DA, Sharp PA, Housmen DE: Exon amplification: A strategy to isolate mammalian genes based on RNA splicing. Proc Natl Acad Sci USA 1991;88:4005–5008.
30　Orita M, Suzuki Y, Sekiya T, Hayashi K: Rapid and sensitive detection of point mutations and DNA polymorphisms using the polymerase chain reaction. Genomics 1989;5:874–879.
31　Gorman KB, Steinberg RA: Simplified method for selective amplification and direct sequencing of cDNAs. Biotechniques 1989;7:326–328.
32　Marsden CD, Obeso JA, Zarranz JJ, Lang AE: The anatomical basis of symptomatic hemidystonia. Brain 1985;108:463–483.
33　Fishell G, Rossant J, van der Kooy D: Neuronal lineages in chimeric mouse forebrain segregated between compartments and in the rostrocaudal and radial planes. Dev Biol 1990;141:70–83.

Xandra O. Breakefield, PhD, Massachusetts General Hospital-East, Neuroscience Center, Building 149, 13th Street, Charlestown, MA 02129 (USA)

Neuronal Circuits and Compartments of the Basal Ganglia, and Their Clinical Manifestations

Some Aspects of Basal Ganglia-Thalamocortical Circuitry and Descending Outputs of the Basal Ganglia

Katsuma Nakano, Tetsuo Kayahara, Hiroshi Ushiro, Yasuo Hasegawa

Department of Anatomy, Faculty of Medicine, Mie University, Tsu, Japan

Recent methodological advances in neuroscience have provided substantial new information on the neural circuit of the basal ganglia. Lately, the dual innervation of the supplementary motor area (SMA), and motor and premotor areas by both the basal ganglia and cerebellar nuclei has been gradually elucidated [1–3]. Structurally and functionally distinct basal ganglia-thalamocortical circuits and neurochemically differentiated compartments in the striatum have also attracted renewed interest [4, 5]. Furthermore, dopaminergic influences in the striatum and its contribution to the motor, cognitive and emotional behaviors are coming to light [5–8]. This paper briefly reviews the neural circuits of the basal ganglia and focuses on new findings on the ascending and descending projections of the basal ganglia.

Neural Circuits of Basal Ganglia

The basal ganglia include the striatum, pallidum, substantia nigra (SN) and their related subthalamic nucleus (STN) and pedunculopontine nucleus (PPN). It is convenient to distinguish the basal ganglia into three categories, input, output and intrinsic nuclei, for description of its neural circuits. The input nuclei correspond to the caudate nucleus (CN), putamen, and nucleus accumbens; the output nuclei correspond to the internal segment of the globus pallidus (GPi), SN pars reticulata (SNr) and ventral pallidum, and the intrinsic nuclei correspond to the external segment of the GP (GPe), STN, and SN pars compacta (SNc) as well as the ventral tegmental area (VTA). The PPN, another key structure of the basal ganglia, is located in the midbrain tegmental area, laterally surrounding the decussation of the superior cerebellar peduncle. This nucleus is reciprocally connected with SN, GP, striatum, and the cerebral cortex, and sends descending fibers to the lower brainstem or spinal cord as described later [for reviews, see ref. 8–11].

The striatum (CN and putamen) is the major receptive and integrative center in the basal ganglia. It receives excitatory inputs from nearly all the cortical regions except for the primary visual cortex, and from the thalamus, SNc, dorsal raphe nucleus and STN. The minor striatal inputs originate from the GPe, locus ceruleus, parabrachial nucleus, and PPN [for reviews, see ref. 9–11]. The corticostriatal projections are glutamatergic fibers forming asymmetrical synapses on the heads of the dendritic spines of the medium-sized spiny neurons in the striatum, and are topographically organized [7, 12, 13]. The thalamostriatal projections receive inputs mainly from the intralaminar thalamic nuclei. In monkeys, the parafascicular nucleus (Pf) projects to the CN with a mediolateral topography, while the centromedian nucleus (CM) projects somatotopically to the putamen. The dorsomedial CM projects to the dorsolateral leg territory, the ventromedial CM to the ventromedial face territory, and the lateral CM to the intermediate arm territory of the putamen [14, 15]. The midline thalamic nuclei and nucleus reuniens, as well as its adjacent area ventromedial to the Pf, send fibers to the ventral striatum [15]. Neurotransmitters in the thalamostriatal projections are unknown to date, but glutamate may well be among them.

The nigrostriatal fibers are dopaminergic, and originate from the SNc. These fibers make synaptic contacts with the neck of the dendritic spine and dendritic shaft of the medium-sized spiny neurons [8, 13]. Dopaminergic nigrostriatal fibers exert an excitatory influence on the type II striatal spiny neurons containing GABA, substance P (SP), and dynorphin, all of which project direct pathways to the basal ganglia output nuclei (GPi and SNr), and exert inhibitory influence on type I striatal spiny neurons containing GABA and enkephalin (ENK), which project indirect pathways to the GPi and SNr via GPe and STN [7, 16]. The majority of striatonigral neurons express the D_1 dopamine receptor, whereas the striatopallidal neurons express the D_2 receptor [5]. The striatum also receives serotonergic projection fibers mainly from the dorsal raphe nucleus [13]. The minor projection was traced from the PPN to the striatum. Only large neurons in the PPN were labeled retrogradely, following HRP injections into the striatum [15].

The SNc, STN and GPe have internal connections of the basal ganglia. The STN receives GABAergic input from the GPe, and contributes excitatory output to both the GPi and GPe [7, 13]. The STN has reciprocal connections with SN and PPN, but fewer with the striatum [9, 10]. Thus, this nucleus exerts influence directly on the major output systems of the basal ganglia. In primates, separate neuronal populations in the STN project to the GPe and GPi [10]. Recent histochemical studies have shown STN neurons containing glutamate [17]. The STN is divided into the dorsolateral (motor loop related) and the ventromedial (associative loop related) portions [15].

Dopaminergic neurons in the midbrain, which are located in the VTA, SNc, islands in the SNr and in the retrorubral area, provide feedback to the striatum and to the prefrontal cortex [5]. Dopamine terminals in the prefrontal cortex have a role in memory-guided behavior [18]. The SNc sends dopaminergic fibers to the spiny neurons in the striatum bilaterally [8]. Less than 5% are crossed nigrostriatal fibers, and less than 10% are non-dopaminergic fibers [11]. The rostral two-thirds of the SNc are related to the head of the CN, while the nigral neurons projecting to the putamen are more caudally located. A mediolateral and inverse dorsoventral topographical relationship was demonstrated between the SN and striatum [19]. However, double-labeling studies in the monkey revealed that the nigro-CN and nigro-putamen neurons provide two different neuronal clusters, which distribute in a complex mosaic-like pattern in SNc [11]. More recently, Hedreen and DeLong [20] have reported that the midlateral and middle dopamine cell groups of the SN are related to the putamen ('motor striatum'), while the medial dopamine cell groups are related to the head CN and rostral putamen ('associative striatum'). Our findings in the macaque monkey also support these evidences [21].

The output nuclei of the basal ganglia (GPi, SNr and ventral pallidum) receive major input from the striatum, and send GABAergic major outputs to the thalamic nuclei, superior colliculus and PPN [7, 8]. The striatofugal pathways into the GPe, GPi and SN arise mostly from separate neuronal populations in the striatum [10, 22]. Only 2% of neurons project to both GPe and GPi [23]. There seems to be little overlap, if any, between the caudatonigral and putaminonigral projections in monkeys [24]. The SNc is locked in the nigro-striato-nigral loop. The GPi projects to the nucleus ventralis lateralis pars oralis (VLo), nucleus ventralis anterior pars principalis (VApc), VL pars medialis (VLm), centromedian nucleus (CM) and lateral habenular nucleus, and sends descending projections to the PPN [25]. Pallidal neurons projecting to the lateral habenular nucleus are few in number, and lie mainly in the periphery of the GPi. The pallido-intralaminar and pallido-VA/VL pathways arise largely from the same neurons; however, the former projection is less prominent than the latter. The pallidothalamic and pallidotegmental pathways in primates arise largely from the same neurons in the core of the GPi [11].

The SNr sends thalamic projections to the VA pars magnocellularis (VAmc) and nucleus medialis dorsalis pars paralamellaris (MDpl), the tectal projections to the intermediate layers of the superior colliculus, and the tegmental projections mainly to the PPN, bilaterally with ipsilateral predominance [26, 27]. These efferents from GPi and SNr are GABAergic fibers [13]. The striatonigral projection directly connects to the nigrothalamic and nigrotectal projection neurons [16, 28]. The pallidonigral projection, also, contacts nigral neurons projecting to the thalamus, superior colliculus and midbrain tegmentum [28, 29]. It makes synaptic contact mainly with perikarya and large dendrites of SN neurons [29], whereas most terminals of the striatonigral projection contact distal dendrites of SN neurons [30]. The SN receives strong inhibitory GABAergic as well as excitatory SP inputs from the striatum and GPe [31, 32]. Abundant GABAergic and SP synaptic inputs to dopaminergic neurons in the SNc and SNr were demonstrated and to GABAergic neurons in the SNr and SN pars lateralis [32].

The striatum has been divided into two compartments based upon staining patterns for many neurochemical, receptor, afferent and efferent systems: a small acetylcholinesterase (AchE)-poor compartment (patch or striosome) and an AchE-rich matrix compartment. The patches are round or ovoid areas 0.3–0.6 mm in diameter about the same diameter as columns in the cerebral cortex and embedded in AchE-rich matrix. They are most prominent in the head of the CN. The patches contain higher D_1 and opiate receptors. They receive afferents from limbic-related areas such as the amygdala, prelimbic cortex and insular cortex, and are reciprocally connected with SNc, whereas the matrix receives afferents from the thalamus, VTA (A10), retrorubral area (A8) and neocortex (i.e., sensorimotor, premotor and associative cortices), and sends efferents to the SNr, GPi and GPe. Somatostatin fibers and calbindin-immunoreactive neurons are located in the matrix. Neuropeptides do not reflect a strict patch-matrix organization. The cortical inputs to the matrix originate from the superficial layer V and supragranular layers, while the patches arise from the deep layer V and layer VI [see review by Gerfen, 5].

Kemel et al. [33] suggested that striatal cholinergic interneurons presynaptically control dopamine release from nerve terminals of the distinct dopamine cell groups innervating the patches and matrix. The matrix mediates information critical for motor and cognitive behavior. Glutamate markedly stimulates striatum dopamine release. Dopamine inhibits the striatal neurons projecting to the GPe (GABA, ENK, D_2 receptor), and activates those projecting to the SNr and GPi (GABA, SP, dynorphin, D_1 receptor). Acetylcholine tonically facilitates striatal dopamine release via the activation of both muscarinic and nicotinic receptors. On the other hand, dopamine tonically inhibits cholinergic interneurons [16].

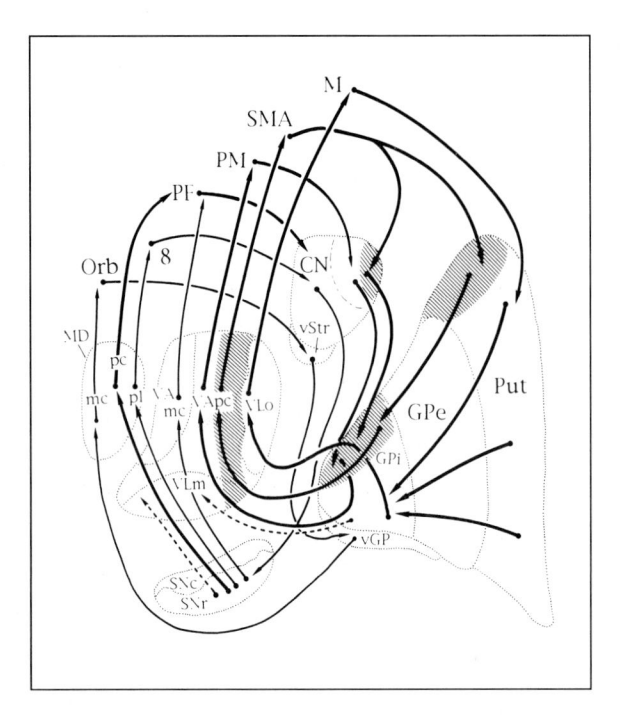

Fig. 1. Summary diagram showing the sensorimotor, supplementary motor and premotor subloops as well as the associative and limbic loops. Hatched portions indicate relay areas of the supplementary subloop. CN = Nucleus caudatus; GPe, GPi = globus pallidus externus et internus; vGP = ventral pallidum; MDmc, MDpc, MDpl = nucleus medialis dorsalis pars magnocellularis, pars parvicellularis et pars paralamellaris; M = motor area; Orb = orbital cortex; PM = premotor area; PF = prefrontal area; Put = putamen; SMA = supplementary motor area; SNc, SNr = substantia nigra pars compacta et reticulata; vStr = ventral striatum; VAmc, VApc = nucleus ventralis anterior pars magnocellularis et parvicellularis; VLm, VLo = nucleus ventralis lateralis pars medialis et oralis; 8 = area 8.

Basal Ganglia-Thalamocortical Circuits

An increasing number of studies have suggested segregate and parallel cortico-basal ganglia-thalamocortical loops [4]. These loops are divided into the motor, associative and limbic loops (fig. 1). The primary motor and sensory cortices project principally to the putamen in a somatotopic distribution: leg territory in the dorsolateral part, face territory in the ventro-medial part and arm in the intermediate part of the putamen in the form of obliquely arranged strips. The corticostriatal fibers arising from the premotor area terminate mainly in the later-

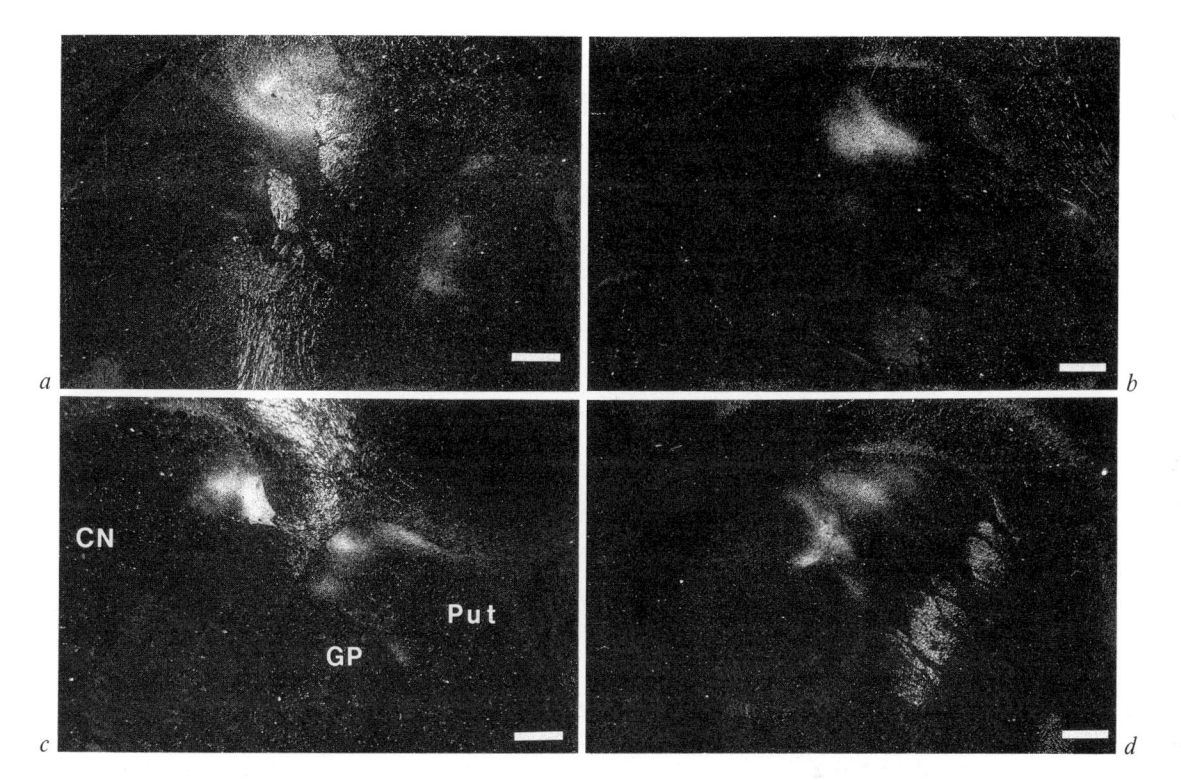

Fig. 2. Dark-field autoradiographs of isotope demonstrating labeled terminals in the lateral part of the caudate nucleus following ^{3}H-Leucine injection into the caudodorsal part of area 6 (area 6aα) (*a*), in the more medially located part of the caudate nucleus than in the former following injection into the rostrodorsal part of area 6 (area 6aβ) (*b*), in the lateral part of the caudate nucleus and dorsomedial part of the putamen following SMA injection (*c*), and in the dorsal to central parts of the caudate nucleus following injection into the frontal eye field (*d*). Scale bar = 1 mm.

al head of the CN (fig. 2a, b), whereas fibers arising from the SMA terminate in both areas of the lateral part of CN and dorsomedial zone of the putamen (fig. 2c). These striatal regions receiving fibers from motor-related cortical areas are called the 'motor striatum', and form the major relay station of the motor loop (fig. 1). The postcommissural putamen projects mainly to the ventral two-thirds of the pallidum. The CN and precommissural putamen innervate mainly the dorsal third of the pallidum. The ventral striatum is linked to the ventral GP complex and the medial tip of the GPi [21, 22]. A much smaller projection was observed from putamen to SNr [21].

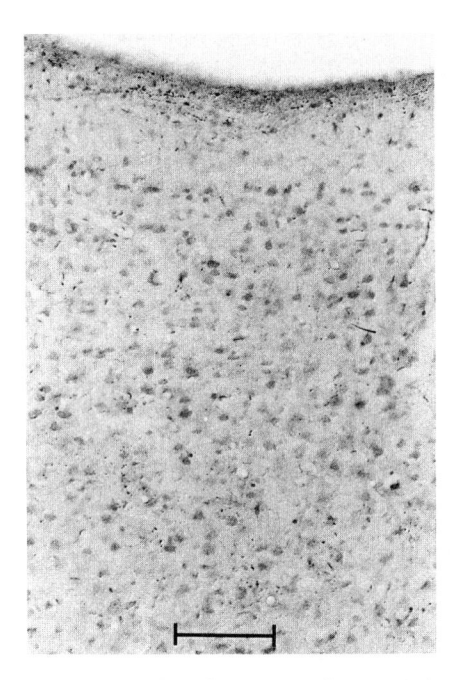

Fig. 3. Bright-field photomicrograph demonstrating anterogradely labeled terminals in the superficial part of lamina I of hand motor are following WGA-HRP injection into VLo. Scale bar = 100 µm.

The motor circuit arising from motor and sensory areas (sensorimotor sub-loop) projects to the ventrolateral part of caudal GPi via the motor striatum (putamen). The ventrolateral GPi in turn connects to the motor area and the caudal part of SMA via the VLo [2, 3, 21, 22, 24] (fig. 1). Following WGA-HRP injections into the VLo, retrograde-labeled neurons were seen in the ventrolateral part of the caudal two-thirds of GPi, and anterograde terminal labelings were detected in the superficial and deep layers of the motor cortex. The superficial thalamocortical projection, terminating in the superficial half of lamina I, was seen more clearly, especially around the hand motor area (fig. 3). These findings agreed well with our previous study using autoradiographic technique [3].

The motor circuit arising from the premotor area (premotor sub-loop) projects to the dorsomedial part of the rostral one-third of GPi via the lateral head of CN [22, 24]. This dorsomedial GPi projects back to the premotor area via the VApc (fig. 1). WGA-HRP injections within the VApc resulted in labeled neurons in the dorsomedial part of the rostral third of GPi, and in labeled terminals in the superficial and deep layers of the premotor area. This superficial terminal labeling was more dense in the regions around the arcuate spur and arcuate genu [3, 34].

The motor circuit originating from the SMA (SMA subloop) connects primarily to the dorsomedial GPi after relaying in the motor striatum, especially in the lateral CN and dorsomedial zone of the putamen. Then this pallidal region projects to the more lateral portion of the VApc that in turn connects with SMA (fig. 1). Superficial and deep thalamocortical terminations arising from the lateral VApc were seen in SMA [3, 34]. Since the reports by Shell and Strick [35], the VLo projection to SMA is now more widely accepted among researchers. However, our studies on the thalamocortical and corticothalamic connections using autoradiographic and HRP techniques demonstrate SMA connections with the VApc, especially with its lateral portion close to the VLo [2, 3, 34]. The SMA receives inputs mainly from the lateral VApc, but not from the VLo, if any, only the caudal SMA close to leg motor area seems to receive inputs from the VLo. The rostral part of SMA connects with the region of VApc located more medially [2].

The prefrontal association areas connect to the dorsal part of the head of the CN [36], and contribute to the 'prefrontal' basal ganglia-thalamocortical circuits (associative loop) (fig. 1). This caudate area is referred to as the 'associative striatum' and is related to cognitive and other high-level processes [4]. The rostralmost portion of the putamen and the ventromedial border zone of the postcommissural putamen are involved in the 'associative striatum'. These portions receive afferents also from the prefrontal association area and the Pf, but not from the CM. The associative striatum projects to the VAmc and lateral portion of the MD, mainly via the SNr, then back to the prefrontal association areas [4, 27, 34]. Following WGA-HRP injections within the VAmc, retrograde labeling neurons were seen in the lateral part of SNr, but these labeling neurons were scattered all over the mediolateral extent of the rostralmost part of the SNr; moreover, anterogradely transported labeling terminals were detected in the superficial layer of the prefrontal area [34]. The frontal eye field (area 8) sends fibers to the dorsal to central portions of the CN (fig. 2d), to relays in the lateral SNr, then to the MDpl, and finally back to area 8 [4, 6].

The prefrontal, temporal, parietal, and occipital association areas are divided into several subregions, cytoarchitectonically and functionally. Several subloops of the association loop might be anticipated, but the striatal regions related to these subregions are still inconsistent, and the related thalamic nuclei have not yet been studied in detail.

As the nucleus accumbens and its adjacent olfactory tubercular portion as well as extended amygdala have connections and histochemical features similar to those of the striatum, these regions are now referred to as the 'ventral' or limbic striatum. The ventral striatum receives afferent fibers from the allocortex (limbic and paralimbic cortical areas), anterior cingulate area (area 24), medial orbitofrontal cortex (Walker's area 13) and subcortical limbic structures such as the amygdala and hippocampus [4]. The limbic striatum projects to the subcommissural part of the pallidum (ventral pallidum), the rostral pole of the GPe, and the rostromedial portion of the GPi [37]. Then these regions project to the nucleus medialis dorsalis pars magnocellularis (MDmc). The limbic circuit is closed by thalamocortical projections from the MDmc to the anterior cingulate gyrus and medial orbitofrontal cortex [37]. The limbic striatum has a role in emotional and/or motivational behavior. The MD is involved in reward-related and memory processes. The ventral striatum is divided into the dorsolateral region, the so-called core zone (nucleus accumbens), and the ventromedial, neurotensin (NT)-rich region, that is the shell zone. The ventral GP also consists of a medial NT region and a lateral region. Different connections were indicated between these zones [37]. The ascending dopaminergic system from VTA connects to several circuit-related structures [6]. The nucleus accumbens represents a limbic-motor interface. Dopaminergic afferents to the striatum have a role in cognitive and emotional behavior and movement control, while the afferents to the prefrontal cortex are related to memory-guided behavior.

We demonstrated the superficial cortical projections of motor and associative loops in the hand motor area, arcuate premotor area, SMA, and prefrontal cortex. These areas receive polymodal sensory inputs from the posterior parietal lobule. Prominent corticocortical connections are present between these areas and adjacent cortical areas. The arcuate premotor area adjacent to the inferior limb of arcuate sulcus (F5) is related to the distal arm movement [38]. Dense dopaminergic terminals have been demonstrated in the superficial cortical layer of the frontal lobe in monkeys [39]. These superficial projections of the basal ganglia-thalamocortical loops via the VLo, VApc and VAmc appear to play a role in complex skilled movements.

Descending Fibers of Basal Ganglia

The information descending from the basal ganglia seems to be relayed to the PPN and SNr. The PPN receives afferents from the GPi, and connects reciprocally with the SN and STN [9]. The pallidotegmental fibers increase phylogenetically, and massive terminal labeling was indicated within the PPN after ^{3}H-amino acid injection in the GPi [25]. The nigropeduncular fibers appear to make synaptic contacts, containing pleomorphic vesicles, with the somata of noncholinergic large neurons and dendrites of cholinergic neurons in the PPN [40]. The PPN projects to the lower brainstem, and to the spinal cord directly or indirectly via the brainstem reticular formation [41]. The PPN relays extrapyramidal information to the lower brainstem or spinal cord generators. The nucleus appears to be an important integrative structure for locomotor activity.

The SNr is also one of the important relay nuclei of the descending projection of the basal ganglia. The lateral SNr is involved in orofacial and ocular movements [42, 43]. This part containing numerous GABA neurons receives input from the striatum, GPe, and minor input from the premotor and prefrontal areas [44]. The peripeduncular nucleus receives afferents from the subcommissural pallidum and dorsal amygdala, and is involved in the control of oral movements, especially those of the tongue [45]. In the squirrel monkey, the pallidonigral projection is modest and arises exclusively from the dorsal GPe [11]. In rats, both the SNr and SNc receive more dense pallidal afferents, especially the former [unpubl. data]. The striatopallidal fibers arborize twice in the GPi and GPe. Two subsets of pallidal neurons receiving the same striatal information are able to project to different brain centers [22]. The retrorubral nucleus and the caudolateral SNr project to the nucleus reticularis parvicellularis (Rtpc) [46]. With PHA-L injection in the lateral SNr or SN pars lateralis, the descending nigral fibers were clearly traced into the peritrigeminal area of motor trigeminal nucleus and into the Rtpc in the lower brainstem [47]. The nigrotegmental neurons receive convergent afferents from the striatum and GPe [46]. These nigral descending fibers seem to be involved in orofacial movements.

The nigrotectal neurons are located in the rostrolateral SNr. Striatal and pallidal afferents converge on the nigrotectal neurons [28]. The GABAergic nigral afferent fibers form symmetrical synapses with the somata or proximal dendrites of tectospinal neurons [48]. Visual cortical information seems to be relayed to the superior colliculus via the striatonigral projection [49]. The lateral part of the body of CN receives massive afferents from the visual association cortical areas [49]. With ^{3}H-leucine injection into the lateral part of the body CN in monkeys, labeled terminals were observed in the dorsomedial margin of the pallidum and in the dorsolateral SNr [unpubl. data]. The tectal afferents from the SNr, frontal eye-field (area 8), and PPN terminate preferentially in AchE-rich compartments of the intermediate layers of the superior colliculus [50]. Neurons in the superior colliculus give rise to tectobulbar (tectopreoculomotor) and tectospinal projections, and have a role in visuoocular behavior and postural control.

Conclusion

The thalamic VLo nucleus relays sensorimotor subloop to the superficial layer of motor area, especially the forelimb motor area. The lateral VApc links SMA subloop to the superficial layer of SMA, and the medial VApc mediates the premotor subloop to the superficial layer of the premotor area. The VAmc relays the associative loop to the superficial layer of the prefrontal area.

References

1 Wiesendanger R, Wiesendanger M: Cerebello-cortical linkage in the monkey as revealed by transcellular labeling with the lectin wheat germ agglutinin conjugated to the marker horseradish peroxidase. Exp Brain Res 1985;59:105–117.
2 Nakano K, Hasegawa Y, Kayahara T, Tokushige A, Kuga Y: Cortical connections of the motor thalamic nuclei in the Japanese monkey, *Macaca fuscata*. Stereotact Funct Neurosurg 1993;60: 42–61.
3 Nakano K, Tokushige A, Kohno M, Hasegawa Y, Kayahara T, Sasaki K: An autoradiographic study of cortical projections from motor thalamic nuclei in the macaque monkey. Neurosci Res 1992;13: 119–137.
4 Alexander GE, Crutcher MD, DeLong MR: Basal ganglia – Thalamocortical circuits: Parallel substrates for motor, oculomotor, 'prefrontal' and 'limbic' functions. Prog Brain Res 1990;85:119–146.
5 Gerfen CR: The neostriatal mosaic: Multiple levels of compartmental organization. Trends Neurosci 1990;15:133–139.
6 Cummings JL: Frontal-subcortical circuits and human behavior. Arch Neurol 1993;50:873–880.
7 Graybiel AM: Neurotransmitters and neuromodulators in the basal ganglia. Trends Neurosci 1990; 13:244–254.
8 Smith AD, Bolam JP: The neural network of the basal ganglia as revealed by the study of synaptic connections of identified neurones. Trends Neurosci 1990;13:259–265.
9 Carpenter MB: Interconnections between the corpus striatum and brain stem nuclei, in McKenzie JS, Kemm RE, Wilcock VN, (eds): The Basal Ganglia: Structure and Function. New York, Plenum Press, 1984, pp 1–68.
10 Parent A: Comparative connections of the basal ganglia. Trends Neurosci 1990;13:254–258.
11 Parent A: Comparative Neurobiology of the Basal Ganglia, New York, A Wiley-Interscience Publication, 1986, 335 pp.
12 Frotscher M, Rinne U, Hassler R, Wagner A: Termination of cortical afferents on identified neurons in the caudate nucleus of the cat. A combined Golgi-EM degeneration study. Exp Brain Res 1981;41:329–337.
13 McGeer PL, McGeer EG: Neurotransmitters and their receptors in the basal ganglia. Adv Neurol 1993;60:93–101.
14 Nakano K, Hasegawa Y, Kayahara T, Kuga Y: Topographical organization of the thalamostriatal projection in the Japanese monkey, *Macaca fuscata*, with special reference to the centromedian-parafascicular and motor thalamic nuclei, in Bernardi G et al. (ed): The Basal Ganglia III. New York, Plenum Press, 1991, pp 63–72.
15 Nakano K, Hasegawa Y, Tokushige A, Nakagawa S, Kayahara T, Mizuno N: Topographical projections from the thalamus, subthalamic nucleus and pedunculopontine tegmental nucleus to the striatum in the Japanese monkey, *Macaca fuscata*. Brain Res 1990;537:54–68.
16 Nieoullon A, Goff LK-L: Cellular interactions in the striatum involving neuronal systems using 'classical' neurotransmitters: Possible functional implications. Mov Disord 1992;7:311–325.
17 Smith Y, Parent A: Neurons of the subthalamic nucleus in primates display glutamate but not GABA immunoreactivity. Brain Res 1988;453:353–356.
18 Haber SN, Lynd-Balta E, Mitchell SJ: The organization of the descending ventral pallidal projections in the monkey. J Comp Neurol 1993;329:111–128.
19 Szabo J: Organization of the ascending striatal afferents in monkeys. J Comp Neurol 1980;189: 307–321.
20 Hedreen JC, DeLong MR: Organization of striatopallidal, striatonigral, and nigrostriatal projections in the macaque. J Comp Neurol 1991;304:569–595.
21 Hasegawa Y, Nakano K: Topographical projections from the putamen to the globus pallidus and substantia nigra in the Japanese monkey, *Macaca fuscata*. Fourth triennial meeting of the International Basal Ganglia Society, Giens, France, 1992 (abstract).
22 Hazrati L-N, Parent A: The striatopallidal projection displays a high degree of anatomical specificity in the primate. Brain Res 1992;592:213–227.

23   Flaherty AW, Graybiel AM: Output architecture of the primate putamen. J Neurosci 1993;13: 3222–3237.
24   Percheron G, Yelnik J, Fancois C: A Golgi analysis of the primate globus pallidus III. Spatial organization of the striato-pallidal complex. J Comp Neurol 1984;227:214–227.
25   Kim R, Nakano K, Jayaraman A, Carpenter MB: Projections of the globus pallidus and adjacent structures: An autoradiographic study in the monkey. J Comp Neurol 1976;169:263–289.
26   Beckstead RM, Frankfurter A: The distribution and some morphological features of substantia nigra neurons that project to the thalamus, superior colliculus and pedunculopontine nucleus in the monkey. Neuroscience 1982;7:2377–2388.
27   Carpenter MB, Nakano K, Kim R: Nigrothalamic projections in the monkey demonstrated by autoradiographic technics. J Comp Neurol 1976;165:401–416.
28   Smith Y, Bolam JP: Convergence of synaptic inputs from the striatum and the globus pallidus onto identified nigrocollicular cells in the rat: A double anterograde labelling study. Neuroscience 1991; 44:45–73.
29   Smith Y, Bolam JP: The output neurones and the dopaminergic neurones of the substantia nigra receive a GABA-containing input from the globus pallidus in the rat. J Comp Neurol 1990;296: 47–64.
30   Somogyi P, Hodgson AJ, Smith AD: An approach to tracing neuron networks in the cerebral cortex and basal ganglia. Combination of Golgi staining, retrograde transport of horseradish peroxidase and anterograde degeneration of synaptic boutons in the same material. Neuroscience 1979;4: 1805–1852.
31   Mendez I, Elisevich K, Flumerfelt BA: Gabaergic synaptic interactions in the substantia nigra. Brain Res 1993;617:274–284.
32   Mendez I, Elisevich K, Flumerfelt B: Substance P synaptic interactions with GABAergic and dopaminergic neurons in rat substantia nigra: An ultrastructural double-labeling immunocytochemical study. Brain Res Bull 1992;28:557–563.
33   Kemel ML, Desban M, Glowinski J, Gauchy C: Functional heterogeneity of the matrix compartment in the cat caudate nucleus as demonstrated by the cholinergic presynaptic regulation of dopamine release. Neuroscience 1992;50:597–610.
34   Nakano K, Kayahara T, Yasui Y, Hasegawa Y: Neural connections of the anterior motor thalamic nuclei in macaque monkeys (abstract). Neurosci Res 1993;suppl 18:S165.
35   Schell GR, Strick PL: The origin of thalamic inputs to the arcuate premotor and supplementary motor areas. J Neurosci 1984;4:539–560.
36   Yeterian EH, Pandya DN: Prefrontostriatal connections in relation to cortical architectonic organization in rhesus monkeys. J Comp Neurol 1991;312:43–67.
37   Haber SN, Lynd E, Klein C, Groenewegen HJ: Topographic organization of the ventral striatal efferent projections in the rhesus monkey: An anterograde tracing study. J Comp Neurol 1990;293: 282–298.
38   Rizzolatti G, Camarda R, Fogassi L, Gentilucci M, Luppino G, Matelli M: Functional organization of inferior area 6 in the macaque monkey II. Area F5 and the control of distal movements. Exp Brain Res 1988;71:491–507.
39   Gaspar P, Stepniewska I, Kaas JH: Topography and collateralization of the dopaminergic projections to motor and lateral prefrontal cortex in owl monkeys. J Comp Neurol 1992;325:1–21.
40   Spann BM, Grofova I: Cholinergic and non-cholinergic neurons in the rat pedunculopontine tegmental nucleus. Anat Embryol 1992;186:215–227.
41   Jackson A, Crossman AR: Nucleus tegmenti pedunculopontinus: Efferent connections with special reference to the basal ganglia, studied in the rat by anterograde and retrograde transport of horseradish peroxidase. Neuroscience 1983;10:725–765.
42   DeLong MR, Crutcher MD, Georgopoulos AP: Relations between movement and single cell discharge in the substantia nigra of the behaving monkey. J Neurosci 1983;3:1599–1606.
43   Hikosaka O, Wurtz RH: Visual and oculomotor functions of monkey substantia nigra pars reticulata. I. Relation of visual and auditory responses to saccades. J Neurophysiol 1983;49:1230–1253.
44   Künzle H: An autoradiographic analysis of the efferent connections from premotor and adjacent prefrontal regions (areas 6 and 9) in *Macaca fascicularis*. Brain Behav Evol 1978;15:185–234.

45 Spooren WPJM, Mulders WHAM, Veening JG, Cools AR: The substantia innominata complex and the peripeduncular nucleus in orofacial dyskinesia: A pharmacological and anatomical study in cats. Neuroscience 1993;52:17–25.
46 von Krosigk M, Smith Y, Bolam JP, Smith AD: Synaptic organization of GABAergic inputs from the striatum and the globus pallidus onto neurons in the substantia nigra and retrorubral field which project to the medullary reticular formation. Neuroscience 1992;50:531–549.
47 Yasui Y, Nakano K, Nakagawa Y, Kayahara T, Shiroyama T, Mizuno N: Non-dopaminergic neurons in the substantia nigra project to the reticular formation around the trigeminal motor nucleus in the rat. Brain Res 1992;585:361–366.
48 Bickford ME, Hall WC: The nigral projection to predorsal bundle cells in the superior colliculus of the rat. J Comp Neurol 1992;319:11–33.
49 Updyke BV: Organization of visual corticostriatal projections in the cat, with observations on visual projections to claustrum and amygdala. J Comp Neurol 1993;327:159–193.
50 Harting JK, Updyke BV, Van Lieshout DP: Corticotectal projections in the cat: Anterograde transport studies of twenty-five cortical areas. J Comp Neurol 1992;324:379–414.

Dr. Katsuma Nakano, Department of Anatomy, Faculty of Medicine, Mie University,
Tsu, Mie 514 (Japan)

Dopaminergic Innervation of Primate Cerebral Cortex

An Immunohistochemical Study in the Japanese Macaque[1]

Toshihiro Maeda[a], *Keiko Ikemoto*[b], *Keiji Satoh*[b], *Kunio Kitahama*[c], *Michel Geffard*[d]

Departments of
[a] Anatomy and
[b] Psychiatry, Shiga University of Medical Science, Otsu, Japan;
[c] Département de Médecine Expérimentale, Université Claude-Bernard, Lyon, et
[d] Laboratoire d'Immunologie, Université de Bordeaux II, Bordeaux, France

Since the aminergic neuronal system is phylogenetically very old, it has been assumed to be homologous in all mammalian species. This is the case for cortical dopaminergic innervation, which has thus long been thought to be restricted to a few cortical regions including the prefrontal cortex. This view was based on the results of studies on rodents. However, primates, including humans, manifest a striking development of the cerebral cortex, and cortical dopamine (DA) has been thought to be implicated in a wide range of mental processes. It has therefore been argued that a substantial difference exists between cortical dopaminergic innervation in rodents and in primates [1, 2]. This hypothesis was supported by studies examining the uptake of [^{3}H]DA [3] or ligand labeling of DA receptors [4]. Using a recently developed monoclonal antiserum directed against DA [5], dopaminergic innervation of the primate cerebral cortex, and particularly prefrontal cortex, was studied on the light- and electron-microscopic levels [6]. We used the same monoclonal antibody [5] to study the entire cerebral cortex of the Japanese macaque in an attempt to determine the organization of dopaminergic innervation in primate cortex. Extensive dopaminergic innervation and bilaminar patterning in primate cortex were confirmed, and important new findings were obtained.

[1] This work was supported by a grant-in-aid for scientific research (No. 05680656) to T.M. from the Ministry of Education, Science and Culture of Japan.

Materials and Methods

Three adult male Japanese macaques *(Macaca fuscata)* were deeply anesthetized with ketamine hydrochloride (Parke-Davis) (25 mg/kg, i.m.) and sodium pentobarbital (Pitman-Moore) (10 mg/kg, i.v.) and fixed by perfusion as previously described [7] using the fixative of Van Eden et al. [8] containing 5% glutaraldehyde (GA) and 1% sodium pyrophosphate (MBS). After perfusion for 30 min, the brain was removed and sliced at a thickness of 4 mm, followed by postfixation in solution containing 2% GA and 1% MBS for 3 h; the sections were then placed in a solution containing 15% sucrose and 1% MBS at 4°C. More than 2 months later, 30- and 60-μm-thick coronal sections were cut with a cryostat for light- and electron-microscopic study, respectively [9]. These sections were stored in the same sucrose solution. The sections for light microscopy were placed in 0.3% Triton solution for at least 3 days, and those for electron microscopy were treated with 0.01% trypsin for 5 min at 20°C prior to processing for DA immunohistochemistry. The brain sections were stained with monoclonal antiserum directed against DA [5] (diluted 1/10,000, incubated for 1 week), biotinylated goat antimouse IgG (Vector: diluted 1/1,000), an ABC kit (Vector) [10] and 3.3′-diaminobenzidine hydrogen peroxide solution containing 1% nickel ammonium sulfate [11]. After immunostaining, the sections for light microscopy were mounted onto gelatin-coated glass slides and counterstained with neutral red; the sections for electron microscopy were postfixed in 1% OsO_4, dehydrated and flat-embedded between silicon-coated slide glass and cover glass. Ultrathin sections were studied using a Hitachi H-500 electron microscope.

Results

General Observation

DA-immunoreactive (IR) varicose fibers were present in all cortical regions examined (fig. 1), including the prefrontal cortex (areas 46, 9 and 10), primary motor cortex (area 4), premotor cortex (area 6), the supplementary motor area (medial portion of area 6), parietal (areas 1, 2, 3, 5 and 7) and temporal (areas 21 and 22) cortices, anterior (area 24) and posterior (area 23) cingulate cortex and visual cortex (area 17). A general rostrocaudal gradient of decrease in regional density of DA-IR fibers was found. The fibers were most dense in the anterior cingulate cortex and least dense in the visual cortex. The central sulcus formed a rough boundary between densely innervated and lightly innervated regions of the cortex. A bilaminar pattern of DA innervation was also bounded caudally by the central sulcus. As shown in figure 1, a superficial lamina of dense DA-IR fibers in layer 1 was always found in every cortical region, whereas the lamina in the deep layers was less prominent or lost in portions of cortex caudal to the central sulcus. It should be emphasized that the deep laminar accumulation of DA-IR varicose fibers was not confined to one cytoarchitectonic layer; rather, the densely innervated deep layer(s) differed from area to area in the cortex.

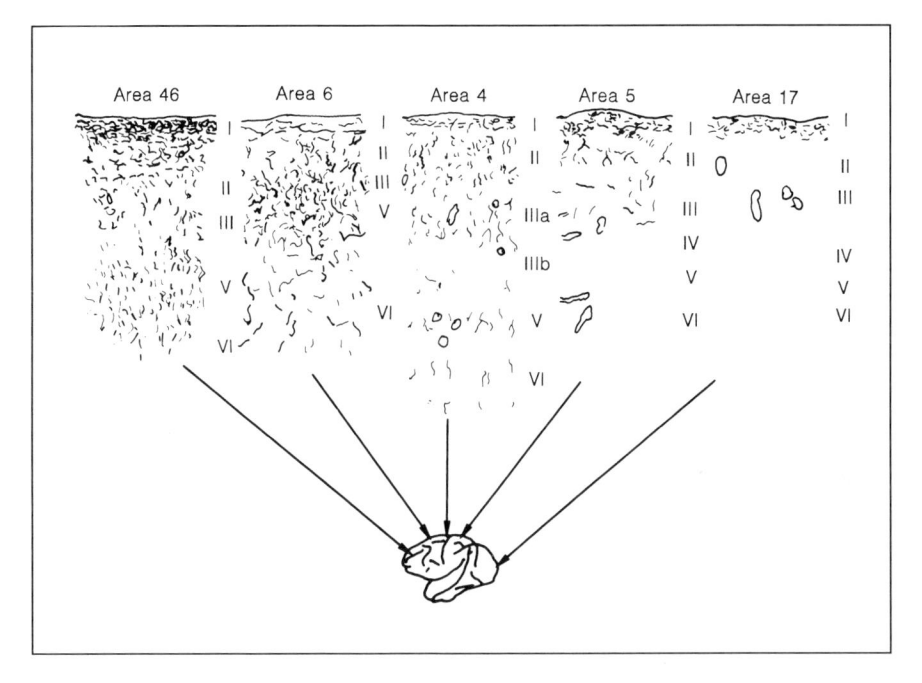

Fig. 1. Camera lucida drawings of DA-IR fibers observed on coronal sections in the prefrontal (area 46), frontal (area 4), parietal (area 5) and occipital (area 17) cortices of the macaque. Note that the bilaminar distribution pattern is conspicuous in agranular cortices (areas 46, 6 and 4) and also note the rostrocaudal gradient of decrease in density.

Medial Wall of Cerebral Hemisphere

DA afferents were very dense in the medial wall of the hemisphere in monkey cortex, including the medial part of area 6 and the cingulate cortex (areas 24 and 23), as shown in figure 2. The former region is the supplementary motor area. The cingulate cortex is on the ventral bank of the cingulate sulcus and is subdivided into a number of subfields, including areas 24 a–d and 23 [12]. Areas 24 and 23 differ in their cytoarchitectural features, including the absence of layer IV in area 24. However, a similar bilaminar distribution of DA-IR terminals was found in every subfield of the cingulate cortex (fig. 3), although area 24 was more densely innervated than area 23. DA-IR terminals are drawn by camera lucida in the frontal section of the ventral bank of the cingulate sulcus in figure 4. It clearly displayed continuous translocation of deeper lamina of dopaminergic innervation; from inside to outside in the sulcus (from left to right in fig. 4), the lamina of dense DA-IR terminals moved uphill, crossing the cortical layers from layer IV in area 24d to layer II/III in area 24b.

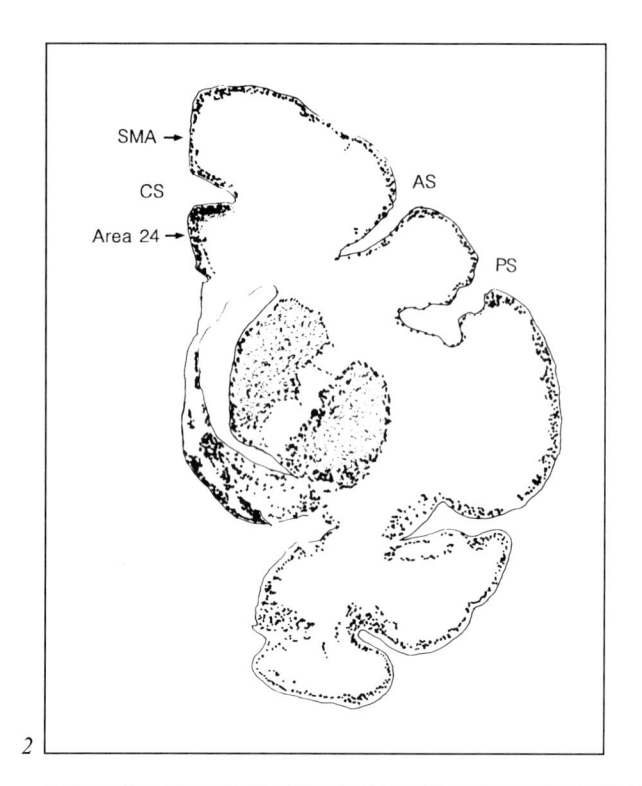

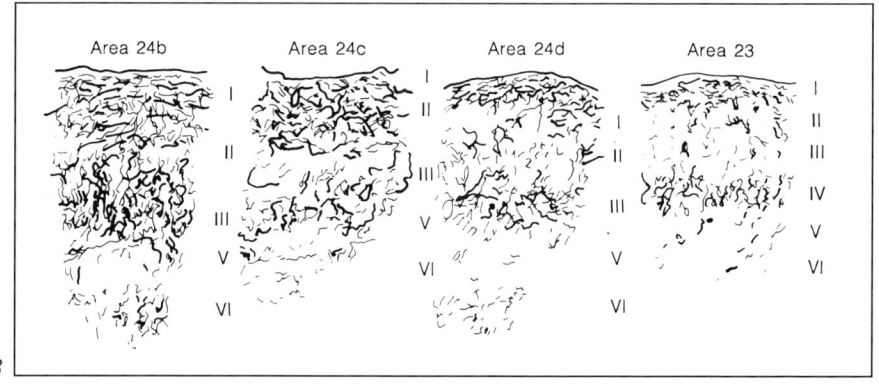

Fig. 2. Camera lucida drawing of DA-IR fibers in a coronal section through the striatum. Dense innervation is present in the medial wall of the cerebral hemisphere including the supplementary motor cortex (SMA) and cingulate cortex (area 24). AS = Arcuate sulcus; CS = cingulate sulcus; PS = principal sulcus.

Fig. 3. Camera lucida drawings of DA-IR fibers in coronal sections in subfields (areas 24b–d and 23) of the cingulate cortex. The deep laminar portion of the bilaminar pattern of innervation is localized to layer IV in area 23. A similar pattern is found in subfields of area 24 where layer IV is absent.

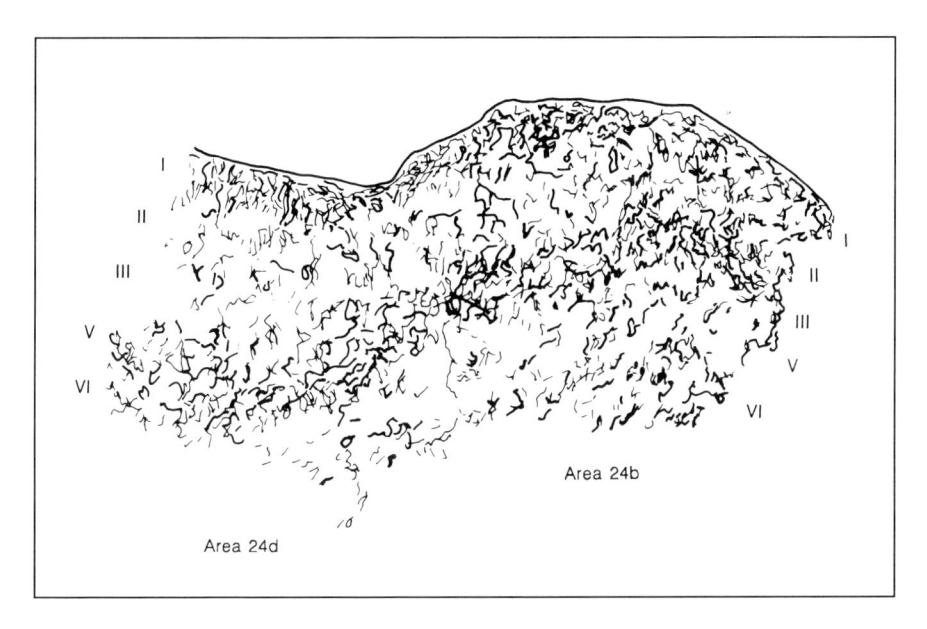

Fig. 4. Camera lucida drawing of DA-IR fibers on a coronal section of the ventral bank of the cingulate sulcus. The deep lamina of dopaminergic innervation rises from layer V/VI in area 24d to layer II/III in area 24b.

Photographs of immunohistochemical labelling (fig. 5) showed roughly two types of DA-IR terminal fibers of large size and small size. Most of the fibers had a varicose profile, but some of the large-sized DA-IR fibers were smooth and club-shaped. Large-sized fibers were abundant in the superficial lamina, although they were also present in the deep layers. The transversely running fibers were most conspicuous in the superficial layer, and the principal direction of the fibers in deep layers was nearly vertical.

Large-sized fibers usually formed large-sized terminal boutons. Giant terminal boutons of very large size were frequently found in deep layers in particular (fig. 6).

Electron Microscopic Observation

Small-sized DA-IR terminal fibers invaded synaptic complexes, in which many axon terminals synapsed with dendritic spines (58% of all DA-IR boutons examined) regardless of laminae. Many were included in the so-called triad of synaptic complex, in which an immunopositive bouton contacted both an immunonegative bouton and its target spine (fig. 7). However, there were significant numbers of DA-IR terminal fibers other than those forming simple triadic ele-

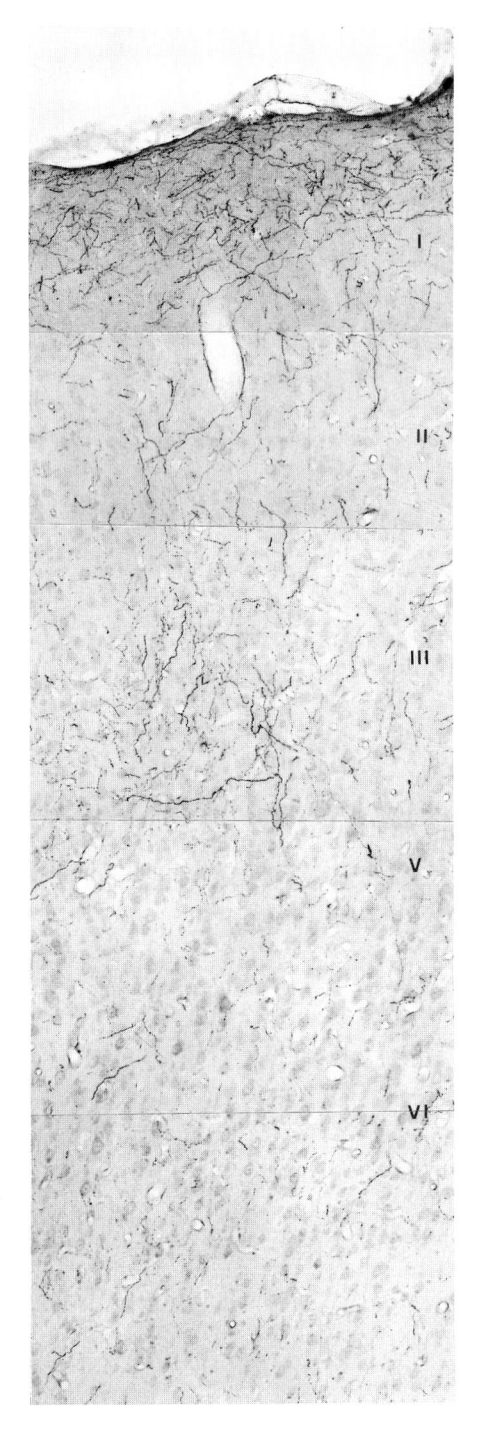

Fig. 5. Light-microscopic photomontage showing the distribution of DA-IR fibers across the layers of cortex in area 24b. Small- and large-sized DA-IR varicose fibers are present. × 15.

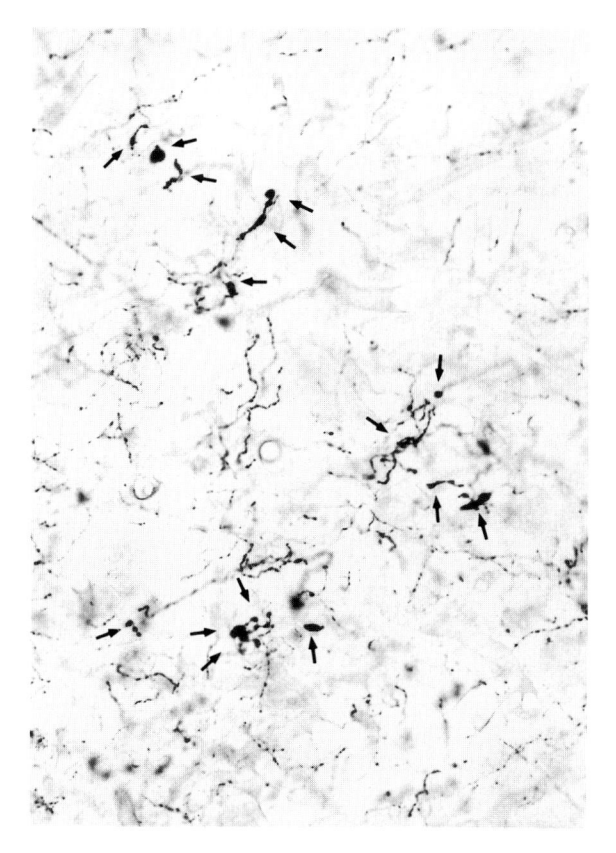

Fig. 6. Photomicrograph showing very large-sized boutons (arrows) observed in area 24b. × 260.

ments. Some fibers passed through the synaptic complex without contact with synaptic elements, while others enclosed other terminal elements constituting a triadic complex (fig. 7). These features were studied in serial ultrathin sections. A small but distinct number of asymmetric-type synapses on dendritic shafts or spines were found (17% of total boutons examined).

Large-sized DA-IR boutons were observed in the electron-microscopic as well as in the light-microscopic study. In addition to bouton profiles 1–2 μm in diameter, giant terminals 7 μm in length and a diameter of 3 μm were sometimes present (fig. 8a). No synaptic structure was observed in the giant terminals, although asymmetric synapses were occasionally observed on the DA-IR large-sized boutons of 1–2 μm in diameter (fig. 8b).

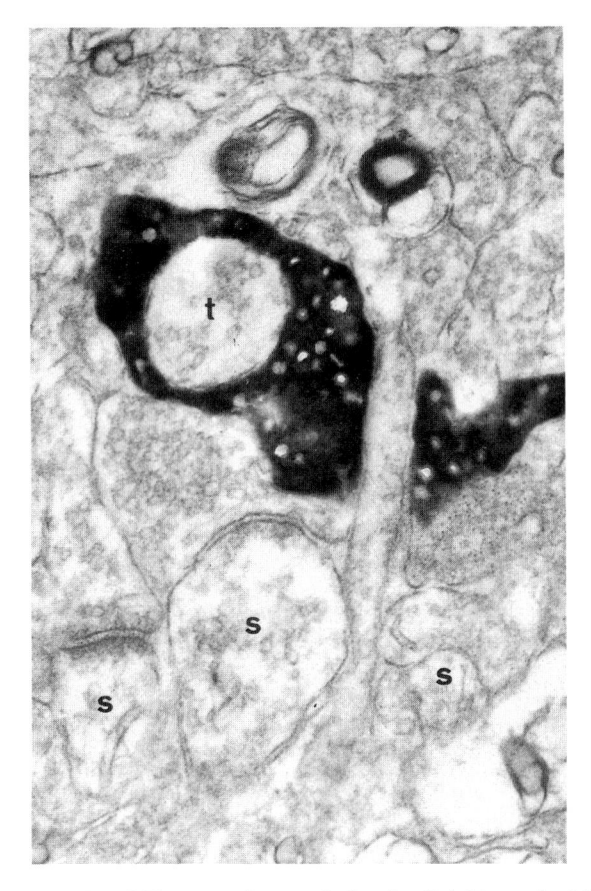

Fig. 7. Electron micrograph showing DA-IR terminal fibers invading the synaptic complex in which many input terminals synapse with dendritic spines (s) of pyramidal neurons. DA-IR boutons constitute not only a triadic complex but also enclose an immunonegative terminal element (t). × 60,000.

Discussion

The present study confirmed that, unlike that in lower animals, dopaminergic cortical innervation is very extensive and widespread in the primate, as has been suggested by recent studies using specific uptake of [^{3}H]DA [3] or specific ligand-labeling of DA receptors [4]. The general bilaminar pattern of dopaminergic innervation and its rostrocaudal decrease in density hypothesized by Berger et al. [1] were also confirmed. However, there is an obvious difference between their hypothesis and our results. They proposed that the expanded dopaminergic pro-

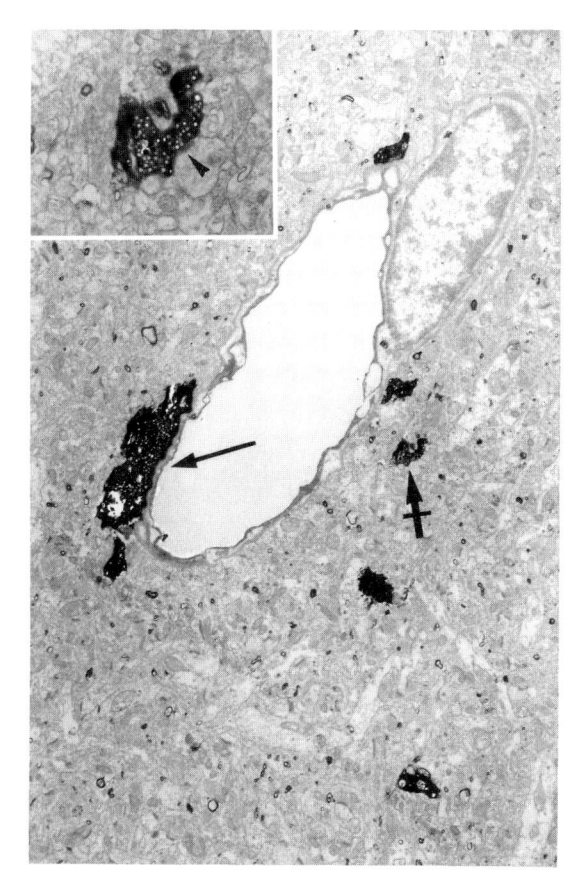

Fig. 8. Electron micrograph showing large-sized DA-IR terminal boutons. Note a giant terminal with a length of 7 µm and a diameter of 3 µm (arrow). On the large-sized bouton, asymmetric synapses (arrow head) are occasionally seen. × 4,300. *Inset:* High-power micrograph of the terminal indicated by the crossed arrow. × 14,000.

jections display a bilaminar pattern except in the agranular cortex, which is diffusely innervated. However, our findings indicated that this was not precisely the case. The frontal cortex, including the agranular regions, was densely innervated, but the bilaminar pattern of distribution was still evident. On the other hand, DA innervation in the deep lamina in the granular regions of the parietal, occipital and temporal cortex decreased in density roughly caudalwards, while DA innervation in the superficial lamina was dense in the entire region of the cortex. It should be emphasized that the bilaminar innervation pattern in the frontal cortex did not correspond to the cytoarchitectonic lamination of cortical neurons; rather, the layer(s) of cortex innervated by the deep lamina differed from area to area. Goldman-Rakic et al. [6], who studied the cortical dopaminergic innervation of the monkey *(Macaca mulatta)* using the same anti-DA monoclonal antiserum we used [5], found a bilaminar distribution pattern in DA-IR fibers across the thickness of the cortex in the prefrontal cortex (area 46). A high density of DA-IR fibers was present in layers I and II and in layers V and VI in area 46 in their study. A similar distribution of DA-IR fibers was observed in our study (fig. 1, area 46).

Perhaps the most significant finding of the present study was that dopaminergic innervation was dense in the medial wall of the cerebral hemisphere, including the supplementary motor area and cingulate cortex. Not only the dorsal but also the ventral banks of cingulate sulcus have been implicated in the so-called premotor cortex [13]. There is experimental evidence showing the direct origin of corticospinal neurons in the precingulate cortex [13]. It is clear, however, that cingulate cortex should be still considered a part of limbic system, since extensive connections are present between it and other portions of the limbic system including the midline and intralaminar thalamic nuclei [14]. The primate cingulate cortex therefore may play roles in the integration of emotional expression and sensorimotor function. In fact, there are many functions in which the cingulate cortex is thought to contribute, including attention, memory, somatic as well as autonomic motor control, representation of painful sensation, and others [see references in ref. 13 and 14]. The hypothesis that dopaminergic innervation of the cingulate cortex modulates the neuronal mechanisms underlying this integrative function by relating the activity of basal forebrain structures including the septum and nucleus accumbens which are densely innervated by the dopaminergic system in the primate [15] is intriguing. Whether or not cortical DA terminals are all provided by mesencephalic DA neurons and whether or not DA terminals in the medial wall of the cerebral hemisphere are included in the mesolimbic DA system, as is the case in rat brain, are important questions for future primate brain research.

Electron-microscopic study revealed two important features of DA-IR terminals in the monkey cortex. First, most DA terminal fibers invade the synaptic complex, where they form not only the triads of terminals described by Goldmon-Rakic et al. [6], but also a more complicated pattern of terminal associations. The synaptic complex, which is the main target of DA afferents, is a synaptic arrangement in which spines of pyramidal neurons receive corticocortical or thalamocortical inputs [6]. Accordingly, it appears that dopaminergic modulation in the primate cortex takes place primarily at the initial site of input to pyramidal neurons. In agranular cortex, pyramidal neurons are well developed and spines are distributed in greatest density on the surface of apical dendrites and their branches in pyramidal neurons. Thus, DA afferents innervated along with distribution of the

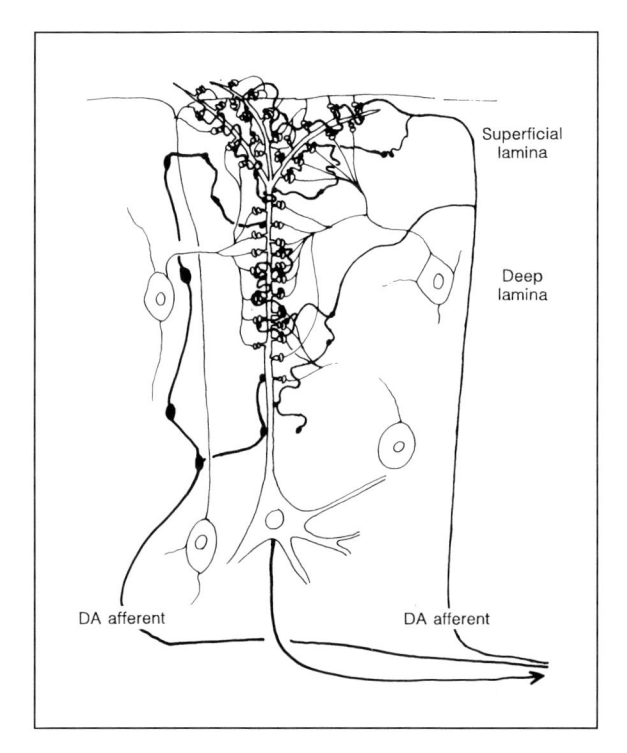

Fig. 9. Schematic representation of dopaminergic innervation of the macaque cortex. Dopaminergic fibers terminate chiefly in the synaptic complex on the apical dendrites of the pyramidal neuron, which may be the principal site of action of cortical DA. In addition, there are giant terminals without evidence of synaptic terminals; they may release DA diffusely as a neurohumoral agent.

synaptic complex on ascending apical dendrites. This may be the reason why layer(s) included in the deep lamina of dopaminergic innervation differ(s) from area to area, and why the superficial lamina is present in layer I, which features extensive ramification of apical dendrites in all regions of the cortex. In the granular cortex, the development of granule cells (layer IV) and/or regression of pyramidal cells (layer V) may correlate with a decrease in numbers of such synaptic complexes in the deep layers that results in loss of deep lamina of dopaminergic innervation.

Second, very large-sized DA-IR boutons are present in the primate cortex. Almost no synaptic structures were found around the boutons filled with synaptic vesicles even on careful examination of serial ultrathin sections. Although asymmetric synapses were very rarely observed between large-sized DA-IR boutons and dendritic shafts, the asynaptic morphology of DA-IR giant boutons does not support the hypothesis that monoamines in the cortex are released at specialized junctions [16] but does provide evidence for neurohumoral transmission of amines [17]. A hypothesis similar to the latter has been presented by Fuxe and Agnati [18] as that of volume transmission. The neurohumoral transmission of dopaminergic information seems to be compatible with the presence of an extensive bilaminar distribution of DA receptors in all cortical regions, demonstrated in ligand binding experiments [14], including granular cortex in which DA terminal fibers are present only in the superficial layer. The dopaminergic innervation of pyramidal neurons in the monkey cortex is diagramatically represented in figure 9.

References

1 Berger B, Gasper P, Verney C: Dopaminergic innervation of the cerebral cortex: Unexpected differences between rodents and primates. Trends Neurosci 1991;14:21–27.
2 Lewis DA, Foote SL, Goldstein M, Morrison JH: The dopaminergic innervation of monkey prefrontal cortex: A tyrosine hydroxylase immunohistochemical study. Brain Res 1988;449:225–243.
3 Berger B, Trottier S, Verney C, Gaspar P, Alvarez C: Regional and laminar distribution of the dopamine and serotonin innervation in the macaque cerebral cortex: A radioautographic study. J Comp Neurol 1988;273:99–119.
4 Richfield EK, Young AB, Penney JB: Comparative distribution of dopamine D-1 and D-2 receptors in the cerebral cortex of rats, cats, and monkeys. J Comp Neurol 1989;286:409–426.
5 Chagnaud JL, Mons N, Tuffet S, Grandier-Vazeilles S, Geffard M: Monoclonal antibodies against glutaraldehyde-conjugated dopamine. J Neurochem 1987;49:487–494.
6 Goldman-Rakic PS, Leranth C, Williams SM, Mons N, Geffard M: Dopamine synaptic complex with pyramidal neurons in primate cerebral cortex. Proc Natl Acad Sci USA 1989;86:9015–9019.
7 Iritani S, Fujii M, Satoh K: The distribution of substance P in the cerebral cortex and hippocampal formation: An immunohistochemical study in the monkey and rat. Brain Res Bull 1989;22:295–303.
8 Van Eden CG, Hoornemann EMD, Buijs RM, Matthussen MAH, Geffard M: Immunocytochemical localization of dopamine in the prefrontal cortex of the rat at light and electron microscopic level. Neuroscience 1987;22:849–862.
9 Wakabayashi Y, Tomoyoshi T, Fujimiya M, Arai R, Maeda T: Substance P-containing axon terminals in the mucosa of the human urinary bladder: Preembedding immunohistochemistry using cryostat sections for electron microscopy. Histochemistry 1993;100:401–407.
10 Hsu SM, Raine L, Fanger H: Use of avidin-biotin-peroxidase complex (ABC) in immunoperoxidase techniques: A comparison between ABC and unlabeled antibody (PAP) procedures. J Histochem Cytochem 1981;29:557–580.
11 Hancock MB: A serotonin-immunoreactive fiber system in the dorsal columns of the spinal cord. Neurosci Lett 1982;31:274–252.
12 Matelli M, Luppino G, Rizzolatti G: Architecture of superior and mesial area 6 and the adjacent cingulate cortex in the macaque monkey. J Comp Neurol 1991;311:445–462.
13 Dum RP, Strick PL: The origin of corticospinal projections from the premotor areas in the frontal lobe. J Neurosci 1991;11:667–689.
14 Vogt BA, Pandya DN, Rosene DL: Cingulate cortex of the rhesus monkey. I. Cytoarchitecture and thalamic afferents. J Comp Neurol 1987;262:256–270.

15 Ikemoto K, Satoh K, Maeda T: Immunohistochemistry of dopamine in the nucleus accumbens of macaque monkey: A light- and electron-microscopic study. Neurosci Res 1993(suppl 18):60.
16 Parnavelas JG, Papadopoulos GC: The monoaminergic innervation of the cerebral cortex is not diffuse and nonspecific. Trends Neurosci 1989;12:315–319.
17 Maeda T, Fujimiya M, Kitahama K, Imai H, Kimura H: Serotonin neurons and their physiological roles. Arch Histol Cytol 1989;52(suppl):113–120.
18 Fuxe K, Agnati LF: Two prinicpal modes of electrochemical communication in the brain: Volume versus wiring transmission; in Fuxe K, Agnati LF (eds): Volume transmission in the brain. Adv Neurosci. New York, Raven Press, 1991, vol 1, pp 1–9.

Dr. T. Maeda, Department of Anatomy, Shiga University of Medical Science,
Seta Tsukinowa-cho, Otsu, 520-21 (Japan)

Interactions between Glutamate and Dopamine in the Ventral Striatum: Evidence for a Dual Glutamatergic Function with Respect to Motor Control

Anders Svensson, Maria L. Carlsson, Arvid Carlsson

Department of Pharmacology, University of Göteborg, Sweden

Previous findings from our laboratory have shown that glutamate receptor blockade by means of systemic administration of the non-competitive NMDA antagonist MK-801 reverses the akinesia induced by reserpine and α-methyl-tyrosine in mice [1]. To investigate the target site(s) of the glutamate antagonist, a series of experiments based on intracerebral drug administration was performed in mice. These experiments pointed to a particularly important role of the nucleus accumbens, and suggested that glutamatergic neurotransmission in the nucleus accumbens exerts an inhibitory influence on behaviour [2]. In the present paper we report that glutamate in the ventral striatum can both stimulate and inhibit motor functions. The evidence for a dual function of glutamate is based on observations of the rotational behaviour induced by unilateral glutamate blockade in the nucleus accumbens.

Materials and Methods

Male albino mice of the NMRI strain weighing 18–20 g were purchased from ALAB, Sollentuna (Sweden).

Reserpine (Sigma) was dissolved in a few drops of glacial acetic acid and a 5.5% glucose solution. Ketamine hydrochloride (Sigma), α-methyl-para-tyrosine methylester hydrochloride (α-MT; Sigma, St. Louis, Mo., USA), quinpirole hydrochloride (Research Biochemicals,

Natick, Mass., USA), 2,3,4,5-tetrahydro-7,8-dihydroxy-1-phenyl-1H-3-benzazepine hydrochloride (SKF 38393; Research Biochemicals), raclopride tartrate (generously supplied by Prof. S. Ahlenius at Astra Läkemedel AB, Södertälje, Sweden) and (S)-(+)-8-chloro-2,3,4,5,-tetrahydro-3-methyl-5-phenyl-1H-3benzazepine-7-ol hydrochloride (SCH 23390; Research Biochemicals) were dissolved in physiological saline. The commercially obtained solution of xylazine chloride (Rompun vet.; Bayer, Leverkusen, Germany) was diluted to 0.75 mg/ml with physiological saline. The drugs were injected intraperitoneally (i.p.), with the exception of SCH 23390 and raclopride which were injected subcutaneously (s.c.). The injection volumes of systematically administered drugs were 10 ml/kg, except for reserpine which was given in a volume of 20 ml/kg.

In order to effectively reduce central dopaminergic neurotransmission, some mice in the present study were pretreated with the monoamine depleter reserpine and the catecholamine synthesis inhibitor α-MT. Reserpine (10 mg/kg) was administered 20 h and α-MT (500 mg/kg) 2 h before locomotor registration. One hour following reserpine administration and throughout the experiment, the ambient temperature was held at 26 °C. In addition, the animals were kept warm on electric pads until the locomotor registration commenced.

Stereotaxic surgery was performed under ketamine (approximately 150 mg/kg) and xylazine (approximately 7.5 mg/kg) anaesthesia. Guide cannules (diameter 0.60 mm, length 17 mm) were implanted unilaterally and fixed to the skull, the tips of the cannulas reaching just above the surface of the brain. The animals were allowed 3–4 days' recovery following surgery. DL-2-amino-5-phosphonopentanoic acid (AP-5; Sigma) was dissolved in an aqueous solution of NaOH, diluted with distilled water and adjusted to neutral pH and isotonicity by adding NaOH and NaCl, respectively. Methylene blue (20 µg/µl; Sigma) was added to the AP-5 solution to make it possible to locate the injection sites anatomically. Using injection cannulas (diameter 0.40 mm) AP-5 was injected in a volume of 0.1 µl into the nucleus accumbens of the freely moving animals.

After completion of the locomotor recordings, the brains were sectioned on a freezing sledge microtome and only animals with stained injection sites within the following coordinates, according to the atlas of Slotnick and Leonard [3] were included in the study: anterior/posterior 0.5–1.3 mm anterior to bregma, lateral 0.4–1.4 mm (on the right side) and vertical 4.0–4.8 mm.

Rotational behaviour was registered in circular open fields, surrounded by a plastic cylinder, diameter 25 cm, where the animals were videotaped for 20 min and the number of complete rotations was counted manually. Mann-Whitney U-test was used throughout for comparisons between groups.

Results

A unilateral injection of AP-5 (5 µg) into the nucleus accumbens caused the animals to rotate. The rotation was predominantly ipsilateral in animals with intact monoaminergic systems. In contrast, in monoamine-depleted mice, the rotation was exclusively contralateral (fig. 1a). This shift in the direction of rotation was apparently due to lack of dopamine receptor stimulation, because in monoamine-depleted animals treated with the mixed dopamine D_1/D_2 receptor

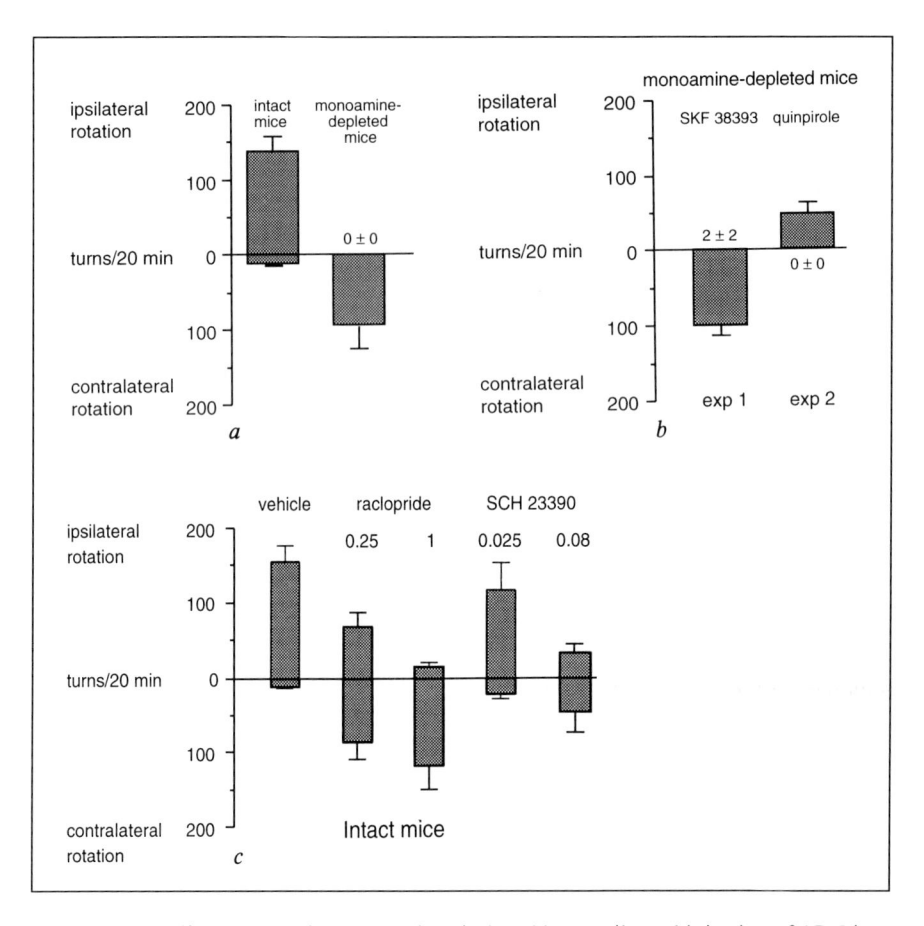

Fig. 1. a Effects on rotational behaviour induced by a unilateral injection of AP-5 into the nucleus accumbens of monoamine-depleted mice and monoaminergically intact mice, respectively. Immediately after the injection of AP-5 (5 µg), the animals were placed in the circular open fields and videotaped for 20 min. The number of complete rotations was counted manually. Means ± SEM; n = 7–8. b Effects on rotational behaviour induced by a unilateral injection of AP-5 into the nucleus accumbens of monoamine-depleted mice pretreated with either SKF 38393 or quinpirole. SKF 38393 (8 mg/kg) was administered 30 min and quinpirole (1 mg/kg) 10 min before recording of locomotor activity. Immediately (experiment 1) or 10 min (experiment 2) after the injection of AP-5 (2.5 µg), the animals were placed in the circular open fields and videotaped for 20 min. The number of complete rotations was counted manually. Means ± SEM; n = 6–8. c Effects on rotational behaviour induced by a unilateral injection of AP-5 into the nucleus accumbens of monoaminergically intact mice pretreated with either SCH 23390 or raclopride. SCH 23390 (experiment 1: 0.025, experiment 2: 0.08 mg/kg) was administered 5 min and raclopride (experiment 1: 0.25, experiment 2: 1 mg/kg) 20 min before recording of locomotor activity. Immediately after the injection of AP-5 (2.5 µg), the animals were placed in the circular open fields and videotaped for 20 min. The number of complete rotations was counted manually. Means ± SEM; n = 5–6.

agonist apomorphine (0.2 mg/kg s.c.), AP-5 caused the animals to rotate predominantly ipsilaterally (not shown).

In order to determine the consequence of dopamine D_1 and D_2 receptor stimulation, respectively, for the behavioural effects of NMDA receptor blockade in the nucleus accumbens, monoamine-depleted mice were pretreated with either the dopamine D_1 receptor agonist SKF 38393 or the D_2 receptor agonist quinpirole. Stimulation of D_2 receptors reversed the effects of monoamine depletion, since the rotation induced by AP-5 (2.5 µg) was consistently ipsilateral in animals pretreated with quinpirole. Stimulation of D_1 receptors by means of SKF 38393 did not influence the direction of the rotation (fig. 1b).

To confirm the crucial involvement of dopamine D_2 receptors, the rotation induced by AP-5 (2.5 µg) was studied in monoaminergically intact mice pretreated with the selective dopamine D_1 receptor antagonist SCH 23390 or the D_2 antagonist raclopride, each antagonist given in two different doses. The lower doses of SCH 23390 and raclopride were selected according to data from this laboratory showing that these doses reduced locomotor activity by approximately 50% in intact, non-habituated mice [4]. Given in these lower doses, SCH 23390 had no obvious effect on the direction of rotation induced by AP-5, whereas raclopride reduced the ipsilateral rotation and increased the contralateral rotation. Following a higher dose of raclopride, which did not reduce the total number of turns, AP-5 induced almost exclusively contralateral rotation. In animals pretreated with a higher dose of SCH 23390, which reduced the overall locomotor activity, AP-5 induced both ipsilateral and contralateral rotation (fig. 1c).

Discussion

It has been proposed that the basal ganglia are linked to the cerebral cortex and thalamus via a system of parallel feedback loops, exerting both motor and cognitive functions [5]. There are two major pathways through the basal ganglia to consider: (1) the direct pathway, via the nucleus entopeduncularis (rodent correspondent of the medial segment of the globus pallidus in primates) or the pars reticulata of the substantia nigra, encompassing two gabaergic neurons between the striatum and thalamus and (2) the indirect pathway, via the globus pallidus (lateral segment of the globus pallidus in primates) nucleus subthalamicus and nucleus entopeduncularis/pars reticulata of the substantia nigra, encompassing three gabaergic neurons between the striatum and the thalamus (fig. 2). The direct pathway appears to form part of a positive feedback loop, whereas the indirect pathway forms part of a negative feedback loop. The corticostriatal glutamatergic projections supply both pathways with excitatory inputs, enabling the cortex to facilitate as well as inhibit movement. The mesencephalostriatal dopaminergic

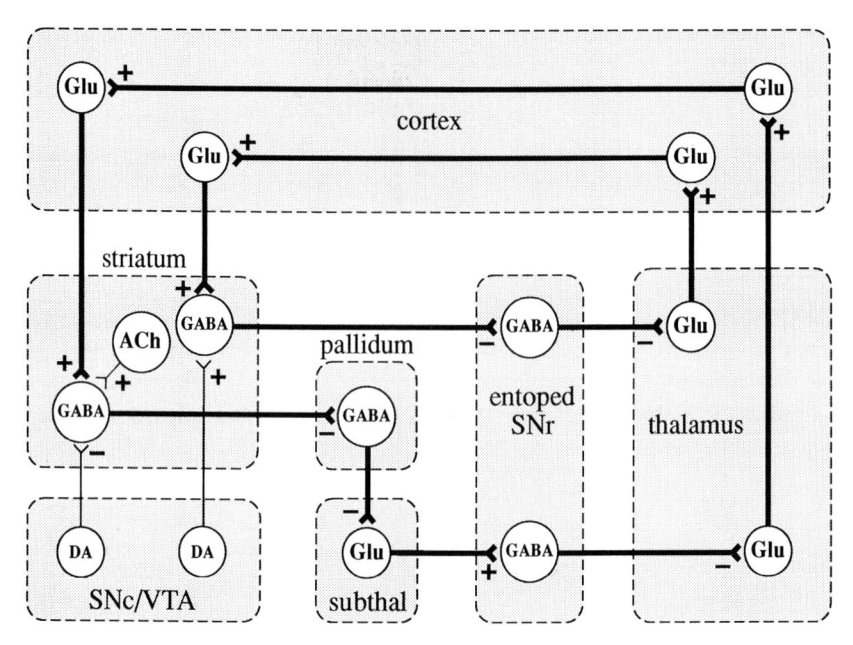

Fig. 2. The positive and negative feedback loops which connect the basal ganglia with the cerebral cortex.

projections facilitate behaviour acting on both pathways, since dopamine seems to excite the direct pathway and inhibit the indirect pathway [6–10]. However, also the dopaminergic system seems to have the possibility both to stimulate and inhibit psychomotor functions: recent findings suggest that postsynaptic dopamine D_3 receptors are inhibitory on behaviour, since the D_3 antagonist U 991194A stimulates locomotor activity in rats [11].

Although speculative, we suggest the following interpretation of the different rotation directions after glutamate blockade in the ventral striatum. The unilateral AP-5-injection into the nucleus accumbens of *monoamine-depleted* mice reduces the activity in the indirect inhibitory pathway on the AP-5-treated side. Hence, an asymmetrical stimulation of locomotor activity is induced, which gives rise to a turning behaviour directed away from the AP-5-injected side (contralateral turning). This finding is in line with the rotational behaviour induced by asymmetrical activation of the mesencephalostriatal dopaminergic system: in this case the animals turn away from the side where the dopaminergic activity is highest [12–15].

In contrast to monoamine-depleted mice, *monoaminergically* intact animals rotated predominantly towards the injected side (ipsilateral turning) following a unilateral injection of AP-5 into the nucleus accumbens. Consequently, this effect cannot be explained in terms of interference with the indirect pathway. Hence, we suggest that in animals with intact monoaminergic systems the balance between the two pathways is shifted in favour of the direct stimulant pathway, perhaps owing to the excitatory dopaminergic input to this pathway. In this case, the loss of glutamatergic tone in the positive feedback loop, following treatment with AP-5 in the striatum, should reduce the stimulant influence on the thalamus, and this should lead to a relatively higher activity in the positive feedback loop on the untreated side, and accordingly ipsilateral rotation. However, it might well be a postural rather than locomotor phenomenon that is responsible for the rotation in the monoaminergically intact mice: since these animals have spontaneous locomotor activity, the induction of a postural asymmetry would be sufficient to produce turning behaviour. In contrast, the turning behaviour in monoamine-depleted mice, whose baseline motor activity is practically zero, must be dependent on stimulation of locomotor activity.

This duality of the effect of intra-accumbens AP-5, which is dependent on the monoamingergic tone, was further investigated using selective dopamine D_1 and D_2 receptor agonists. The D_2 receptor agonsit quinpirole reversed the contralateral rotation induced by AP-5 in monoamine-depleted animals into ipsilateral rotation. In contrast, in monoamine-depleted mice treated with the dopamine D_1 receptor agonist SKF 38393, AP-5 still induced contralateral turning. These findings suggest that glutamate in the ventral striatum can both stimulate and inhibit motor functions and that interactions with dopamine D_2 receptors are crucial for the balance between glutamatergic stimulation and inhibition. The crucial involvement of the D_2 receptor is confirmed in the experiments with D_1 and D_2 antagonists.

The present findings and previous findings, showing that the motor stimulation induced by a dopamine D_1 receptor agonist is potentiated by the NMDA antagonist MK-801 whereas the effects of a D_2 agonist are antagonized [16], suggest that the responsiveness of the D_1 receptors is suppressed by glutamate, whereas D_2 receptors seem to operate in concert with the glutamatergic system, with respect to the control of psychomotor functions. It might thus be hypothesized that the striatal gabaergic neurons pertaining to the direct pathway (fig. 2) are principally equipped with D_2 receptors, whereas the D_1 receptors are principally involved in the indirect pathway. Experimental support for this hypothesis is also provided by work of Herrera-Marschitz and Ungerstedt [17]. Against this interpretation of the striatal distribution of the dopamine D_1 and D_2 receptors stand data of Gerfen et al. [10] showing that the majority of rat striatonigral neurons (the direct pathway) express mRNAs encoding the D_1 receptor but not the D_2

receptor, whereas striatopallidal neurons (the indirect pathway) express mRNAs encoding the D_2 receptor but not the D_1 receptor. However, a considerable degree of colocalization of D_1 and D_2 receptor mRNA in rat striatal cells has also been reported [18].

Acknowledgments

The excellent technical assistance of Malin Lundgren and Marie Nilsson is gratefully acknowledged. This study has been supported by grants from the Swedish Medical Research Council (155 and 9067), Marion Merrell Dow Research Institute (Cincinnati, Ohio, USA) and Scottish Rite Schizophrenia Research Program (NMJ, USA).

References

1   Carlsson M, Carlsson A: The NMDA antagonist MK-801 causes marked locomotor stimulation in monoamine-depleted mice. J Neural Transm 1989;75:221–226.
2   Svensson A, Carlsson ML: Injection of the competitive NMDA receptor antagonist AP-5 into the nucleus accumbens of monoamine-depleted mice induces pronounced locomotor stimulation. Neuropharmacology 1992;31:513–518.
3   Slotnick BM, Leonard CM: A stereotaxic atlas of the albino mouse forebrain. US Department of Health, Education and Welfare, 1975.
4   Martin P, Svensson A, Carlsson A, Carlsson ML: On the roles of dopamine D_1 vs. D_2 receptors for the hyperactivity response elicited by MK-801. J Neural Transm [Gen Sect], in press.
5   Alexander GE, DeLong MR, Strick PL: Parallel organization of functionally segregated circuits linking basal ganglia and cortex. Annu Rev Neurosci 1986;9:357–381.
6   Penney JB, Young AB: Striatal inhomogeneities and basal ganglia function. Mov Disord 1986;1:3–15.
7   Alexander GE, Crutcher MD: Functional architecture of basal ganglia circuits: Neural substrates of parallel processing. Trends Neurosci 1990;13:266–271.
8   Bernath S, Zigmond MJ: Dopamine may influence striatal GABA release via three separate mechanisms. Brain Res 1989;476:373–376.
9   Girault JA, Spampinato U, Glowinski J, Besson MJ: In vivo release of [^{3}H]γ-aminobutyric acid in the rat neostriatum. II. Opposing effects of D_1 and D_2 dopamine receptor stimulation in the dorsal caudate putamen. Neuroscience 1986;19:1109–1117.
10 Gerfen CR, Engber TM, Mahan LC, Susel Z, Chase TN, Monsma FJ Jr, Sibley DR: D_1 and D_2-dopamine receptor-regulated gene expression of striatonigral and striatopallidal neurons. Science 1990;250:1429–1432.
11 Waters N, Svensson K, Haadsma-Svensson SR, Smith W, Carlsson A: The dopamine D_3 receptor: A postsynaptic receptor inhibitory on rat locomotor activity. J Neural Transm [Gen Sect] 1993;94:11–19.
12 Ungerstedt U: Striatal dopamine release after amphetamine or nerve degeneration revealed by rotational behaviour. Acta Physiol Scand Suppl 1971;367:49–68.
13 Ungerstedt U: Postsynaptic supersensitivity after 6-hydroxy-dopamine induced degeneration of the nigro-striatal dopamine system. Acta Physiol Scand Suppl 1971;367:69–93.
14 Colle M, Wise RA: Circling induced by intra-accumbens amphetamine injections. Psychopharmacology 1991;105:157–161.
15 Miller R, Beninger RJ: On the interpretations of asymmetries of posture and locomotion produced with dopamine agonists in animals with unilateral depletion of striatal dopamine. Prog Neurobiol 1991;36:229–256.

16 Svensson A, Carlsson A, Carlsson ML: Differential locomotor interactions between dopamine D_1/D_2 receptor agonists and the NMDA antagonist dizocilpine in monoamine-depleted mice. J Neural Transm [Gen Sect] 1992;90:199–217.
17 Herrera-Marschitz M, Ungerstedt U: The dopamine-γ-aminobutyric acid interaction in the striatum of the rat is differently regulated by dopamine D_1 and D_2 types of receptor: Evidence obtained with rotational behavioural experiments. Acta Physiol Scand 1987;127:371–380.
18 Meador-Woodruff JH, Mansour A, Healy DJ, Kuehn R, Zhou Q-Y, Bunzow JR, Akil H, Civelli O, Watson SJ: Comparison of the distributions of D_1 and D_2 receptor mRNAs in rat brain. Neuropsychopharmacology 1991;5:231–242.

Anders Svensson, Department of Pharmacology, Medicinaregatan 7,
S–413 90 Göteborg (Sweden)

Locomotion and Posture: Modulation by Upper Neurons

Shigemi Mori[a], *Yutaka Homma*[a], *Katsumi Nakajima*[b]

[a] National Institute for Physiological Sciences, Okazaki, and
[b] Department of Physiology, Asahikawa Medical College, Nishikagura, Japan

    A central pattern generator related to the generation of locomotor rhythm exists in the spinal cord, and is under the control of supraspinal structures [1, 2]. Locomotor-drive signals arising from the subthalamic locomotor region in the posterior hypothalamus and the mesencephalic locomotor region in the caudal midbrain are relayed by the reticulospinal neurons to the spinal cord [3]. These locomotor-related reticulospinal neurons seem to be under the control of the sensorimotor cortex by way of the corticoreticular fibers and corticospinal collaterals, and of the cerebellar cortex by way of the fastigioreticular fibers. Studies in both acute decerebrate [2, 4] and intact, awake cats [5] suggested that satisfactory locomotor movements can be executed by activating neural networks related to reflex-dependent and context-dependent adaptation. Functional integration of the locomotor-related signals routing through these two neural networks seems to be a prerequisite for execution of purposive locomotor movements with incessant interactions with the environment [6, 7]. It is highly possible that such integration occurs at the level of the pontomedullary reticulospinal neurons. In this chapter, activities of locomotor-related reticulospinal neurons and higher-order neurons are first introduced and then new aspects of the reticulospinal fiber system are described. Finally, higher-order control of reticulospinal neurons is discussed in relation to reflex-dependent and context-dependent adaptation of locomotor movement.

Contribution of Reticulospinal Neurons and Higher-Order Neurons to the Control of Locomotor Movements

Single-unit recordings from high-decerebrate cats, often walking on a treadmill with only the hindlimbs, showed that reticulospinal neurons within the medullary reticular formation (MRF) are phasically modulated and their discharge rate is increased once or twice in each step cycle, primarily in relation to the activity of ipsilateral hindlimb flexor muscles [4]. Shimamura et al. [8] also recorded the activity for reticulospinal neurons during locomotion in thalamic cats, and showed a positive correlation between the rhythmically discharging neurons and flexor muscle activity, principally of the ipsilateral side. Drew et al. [9], examined the temporal relationships between neuronal and muscle activity in chronically prepared unrestrained cats walking on a treadmill. Their results showed that the discharge of many of the MRF reticulospinal neurons was frequency modulated during locomotion, and was indeed related to the activity of extensor as well as flexor muscles of both sides of the body and of fore- and hindlimbs. Some cells even appeared to be related to muscles of more than one limb, indicating they are involved in the control of both posture and locomotion [9].

Neurons in the cerebral cortex, cerebellar cortex and cerebellar nuclei exhibited either patterned (rhythmic) or unpatterned (tonic) discharges during locomotor movements [5]. Most cells in the motor cortex exhibited one discrete burst of impulses per step or displayed one period of peak activity superimposed on a tonic discharge, and the proportion of rhythmically active cells was approximately the same among corticospinal (pyramidal tract) neurons (PTNs) and non-PTNs. Rhythmic activity was present in both fast- and slow-axon PTNs, but the former tended to achieve higher discharge rates and were more likely to discharge in discrete bursts [10]. The cerebellum is, by way of the spinocerebellar tracts, constantly informed of the activity of the spinal locomotor automatism, and is also informed of the cerebral activity via the cerebrocerebellar pathways arising from the pericruciate (sensorimotor) cortex. The cerebellum seems to select only essential parts of the detailed information related to the current state of locomotor movements in relation to the environment [11, 12]. According to the study by Armstrong and Edgley [13], paravermal Purkinje cells showed some degree of frequency modulation time-locked to the walking movements. Some cells had two activity peaks per step, but most had one peak. The cells varied greatly in their discharge timing relative to the step cycle in the ipsilateral forelimbs but, in contrast to the findings of Orlovsky [14] in decerebrate cats, the 'average cell' was more active during swing than stance phase.

The efferents from the interpositus nucleus terminate mainly in the contralateral red nucleus producing monosynaptic excitation of rubrospinal neurons.

Part of the efferents extend further into the ventrolateral nucleus of the thalamus which, in turn, project to the sensorimotor cortex forming the interposito-thalamo-cortical path. Armstrong and Edgley [15] found that most interpositus neurons discharged tonically in the absense of overt movements, but discharged rhythmically in time with the stepping movements, and their population activity was almost precisely opposite in phase to that in PTNs. These interpositus neurons had elbow-related receptive fields, and elbow flexions were evoked by intracortical microstimulation of the sensorimotor cortex from which the PTNs were recorded. Although the question as to what inputs do initiate the rhythmic modulation of the PTNs remains unanswered, Armstrong and Edgley [15] proposed that the aforementioned reciprocal relationship between the activities of PTNs and interpositus neurons would be in excellent accord with the conclusion of Volsi et al. [16] that the 'interpositus nucleus efferents which activate a particular muscle via the rubrospinal path inhibit (via the interposito-thalamo-cortical path) the discharge of PTNs controlling that muscle and excite PTNs controlling the antagonistic muscles'.

It is also generally recognized that the medial zone of the cerebellum, the vermis, is specially concerned with maintained postures, such as standing, and with the automatic movements leading up, to and following on such postures, e.g., walking and the associated body and neck movements [17, 18]. One efferent cerebellar pathway is via the lateral vestibular nucleus (Deiter's nucleus) and the vestibulospinal tract, and the other is via the fastigial nucleus and the reticulospinal tract [19, 20]. In acute decerebrate cats, recording of the activity of Purkinje cells, cells in the fastigial nucleus, interpositus nucleus, and Deiter's nucleus were made also during walking with hindlimbs and even fictitious scratching movements of the hindlimbs [14, 21]. Most these cells discharged rhythmically in a particular phase of a step cycle. In general, the rhythmical activity of cells of cerebellar nuclei was considerably more regular than that of Purkinje cells [12]. It should be mentioned here that the decerebrate cats are provided with the capability of adjusting their locomotor movements reflexively, but are deprived of the capability of adjusting them to the changes in the environment [6].

Termination Mode and Branching Patterns of Reticulospinal Neurons, and Convergence of Corticoreticular and Fastigioreticular Fibers on Them

Reticulospinal systems originate from the medial two thirds of the pontomedullary reticular formation where large-sized cells are located. Most of the pontine fibers originate from the caudal part of the nucleus reticularis pontis oralis (NRPo) and the entire nucleus reticularis pontis caudalis (NRPc), while most of the medullary reticulospinal fibers come from the nuclei reticularis gigantocellularis (NRGc) and magnocellularis (NRMc) [19, 22]. Reticulospinal fibers give off a considerable number of axon collaterals along their trajectory through the spinal cord and characteristically terminate on 'long' propriospinal neurons, some of them making connections with motorneurons of axial and proximal muscles. Many reticulospinal axons projecting to the thoracic and lumbar spinal cord are also shown to send axon branches into the gray matter of the cervical enlargement [23]. Peterson [24] postulated as follows 'the widespread branching of individual reticulospinal neurons can be thought of as imposing a type of "hard-wired coordination" within the descending reticulospinal system in that activation of a single reticulospinal neuron leads to direct synaptic input to neurons located at multiple levels along the spinal neuraxis'. However, information related to the trajectory, collateral distribution and termination mode of individual reticulospinal axons has not been provided yet.

Termination Mode of Reticulospinal Fibers to the Spinal Interneurons and Motor Neurons

In view of the recent new findings on the contribution of reticulospinal neurons to the control of posture and locomotion, we have tried to study the trajectory, branching patterns and termination mode of single reticulospinal fibers by utilizing an anterograde neural tracer, *Phaseolus vulgaris* leukoagglutinin (PHA-L) [25, 26]. Corticoreticular and fastigioreticular innervation of the reticular neurons have also been studied by utilizing the same anterograde tracing method [27]. To do this, PHA-L was focally microinjected into the NRPo, NRGc, pericruciate cortex and fastigial nucleus in a separate group of cats. After an appropriate survival period, serial frontal sections (50 μm) of the brain stem and spinal cord were made and immunoreacted according to the procedure described previously. Branching patterns of single axon collaterals were reconstructed as much as possible from 5–15 serial sections of the spinal cord by a camera lucida drawing method [25, 26].

A great number of PHA-L-labeled thin fibers from the NRPo descended bilaterally coursing through the medial part of the pontine and medullary reticular formation with an ipsilateral predominance. Labeled terminal boutons were closely apposed to somata of various sized pontomedullary reticular neurons, possibly including reticulospinal neurons. Labeled thick fibers descended ipsilaterally, coursing through the ventral half of the medial longitudinal fasciculus, and further descended through the ventral funiculus of the spinal cord. At the levels of the cervical and upper thoracic cord, these reticulospinal fibers gave off axon collaterals sending terminal fibers to small- to large-sized neurons in Rexed's laminae VII and VIII. Some of the axon collaterals innervated not only ipsilateral but also contralateral gray matter. Descending axons from the NRGc coursed through the ventral and ventrolateral funiculi bilaterally. Single axon collaterals arising from the descending axons gave off terminal fibers to the left or the right gray matter. Their terminals were located in laminae V–IX, mainly in laminae VII and VIII. In lamina IX, they were distributed in the medial portion where motor neurons innervating proximal muscles are located. As has been well demonstrated, long-propriospinal neurons are located in Rexed's laminae VII and VIII, while short-propriospinal neurons are located in laminae V–VII [22].

The thick pontine reticulospinal fibers gave off at least 2–3 axon collaterals at each segmental level of the cervical and upper thoracic cord [26]. Each axon collateral innervated spinal neurons located in a disc-like spinal segment with a width less than 1.0 mm. A few of the axon collaterals innervated ventral gray bilaterally. The thicker the diameter of the stem axons, the greater tended to be the number of axon collaterals. At the level of the lumbosacral enlargement, the number of axon collaterals was much fewer than that at the cervical enlargement along with a decrease in the number of branching preterminal and terminal fibers. One of the interesting findings was that the distribution patterns of preterminal and terminal fibers of the axon collaterals, which originated from a single-stem axon, were similar at each segmental level but were dissimilar depending on the stem axons. All these findings indicated that each reticulospinal axon has preference in the innervation area of the ventral gray. It should be mentioned here that with scratch-related rhythmical movements of the hindlimbs, relative number of rhythmically active cells was maximum at lamina VII of the fifth lumbar cord [28], where both short- and long-propriospinal neurons are located.

Termination Mode of Corticoreticular and Fastigioreticular Fibers in the Pontomedullary Reticular Formation

Throughout the brain stem, PHA-L-labeled fibers coursed through the pyramidal tract ipsilateral to the PHA-L injection site. These fibers possibly included both corticospinal and corticoreticular fibers. At the level of the rostral pons, PHA-L-labeled fibers run dorsally to the ipsilateral tegmental reticular nucleus and paraleminiscal tegmental field (FTP). Contralaterally, a small number of PHA-L-labeled fibers also run dorsally crossing the midline to the FTP. At the level of the caudal pons, labeled fibers run dorsally to the gigantocellular tegmental field (FTG) bilaterally with an ipsilateral predominance. At the level of the medulla, PHA-L-labeled fibers run dorsally to the FTG and the magnocellular tegmental field (FTM) bilaterally. The fibers projecting to the contralateral FTG and FTM crossed at the level of nucleus raphe magnus. With regard to the rostrocaudal distribution of PHA-L-labeled fibers in the brain stem, several characteristics emerge from the present study. In the rostral pons, distribution of labeled fibers was dominant in the FTP ispilateral to the PHA-L injection site. In the caudal pons, the distribution of labeled fibers in the FTG bilaterally became denser than that of the rostral pons, although ispilateral predominance remained. In the medulla, distribution of labeled fibers became much denser in both FTG and FTM bilaterally than that in the caudal pons. At this level, the ispilateral predominance disappeared, and the density of labeled fibers in both FTG and FTM became almost the same bilaterally.

PHA-L-labeled fastigioreticular fibers projected bilaterally to the brain stem. Ipsilaterally to the injection side, labeled fibers were densely distributed in the vestibular complex, especially in the lateral vestibular nucleus. Some of these fibers were found dorsomedially to the medial longitudinal fasciculus, the FTG and FTM of the caudal pons and medulla. Contralaterally to the injection side, labeled fibers were found to cross the midline in the cerebellum, in part passing through the opposite fastigial nucleus and forming the hook bundle of Russell [19]. A few fibers were distributed around the dorsal part of the brachium conjunctivum. Labeled fibers were densely distributed also in the vestibular complex. At the level of the rostral pons, labeled fibers were distributed in the FTP coinciding the NRPo. At the level of the caudal pons, labeled fibers coursed through the facial nerve, and some of these fibers were distributed in the lateral part of the medial longitudinal fasciculus, and the medial part of the FTG and FTM. At the level of the medulla, labeled fibers were distributed much more densely in the FTM than in the FTG.

Terminals of corticoreticular fibers, including those of the corticospinal collaterals, were distributed with an ipsilateral predominance in the rostral pontine reticular formation (PRF), while they were distributed bilaterally in the caudal PRF and MRF. The distribution of terminals tended to be denser in the ventral region including the NRMc than in the dorsal region including the NRGc. Labeled terminals were observed in close apposition to the somata of small-sized reticular neurons, and possibly to the distal dendrites of the large-sized reticular neurons. Terminals of the fastigioreticular fibers distributed with a contralateral predominance in the rostral PRF, and were distributed bilaterally in the caudal PRF. At the level of the medulla, the terminals were distributed much more densely in the contralateral than in the ipsilateral MRF to the PHA-L injection side. There was no clear difference in the distribution patterns of the terminals between the dorsal and ventral regions. These terminals were again observed in close apposition to the somata of large- and small-sized reticular neurons. Comparison of the plots of the terminal distribution regions suggests that reticular neurons located at the level of the caudal PRF and the ventral region of the MRF are possibly under the dual innervation of the corticoreticular and fastigioreticular fibers.

Higher-Order Control of Reticulospinal Neurons in Relation to Reflex- and Context-Dependent Adaptation of Locomotor Movements

To initiate and terminate locomotor movements both in the biped and in the quadruped, a smooth transition from and to standing is necessary. It is our current understanding that postural and locomotor control systems do no reside clustered in particular sites, instead, they are formed by subprograms that are distributed within the central nervous system [6]. Postural and locomotor control systems have been shown to share their subprograms at the levels of the brain stem and the spinal cord [29, 30]. As mentioned already, reticulospinal neurons receive locomotor-drive signals from the subthalamic and mesencephalic locomotor regions, and locomotor-control signals from the cerebellum by way of the cerebellar nucleus [3], and even from the cerebral cortex [15]. It has recently been demonstrated that the pedunculopontine nucleus (PPN), which is under the control of the substantia nigra, project their descending fibers to the cells in the pontomedullary reticular formation. Garcia-Rill et al. [31] proposed that part of the PPN belongs to the mesencephalic locomotor region, and demonstrated that removal of GABAergic inhibitory inputs to the mesencephalic locomotor region/PPN complex from the substantia nigra resulted in production of spontaneous locomotor movements in acute decerebrate cats.

In the cat, there are two chief terminal regions of the corticoreticular fibers, a caudal one which coincides approximately with the NRGc, and a rostral one covering the NRPc and the caudal part of the NRPo, from which reticulospinal tracts originate [19]. Keizer and Kuypers [32] demonstrated that up to 30% of the corticospinal neurons in the medial and anterior parts of the pericruciate motor area give off axon collaterals. With regard to the functional role played by the corticospinal collaterals, there is an interesting suggestion that spinal interneurons and motoneurons are under the dual control of cortico-reticulo-spinal and corticospinal pathways [22]. The indirect cortico-reticulo-spinal activity reaches the spinal interneurons and motoneurons earlier than the direct corticospinal activity, because of the relatively large diameter of the reticulospinal axons. Such a dual control can be made by the same cortical neurons or by different group of cortical neurons. It seems that slower direct corticospinal connections superimpose a finer regulation upon the general regulation of the limb and body position by the brain stem, adjusting the locomotor movements to the required 'context'. It has been well demonstrated that loss of the sensorimotor cortex does not impair the locomotion process, but disturbs the context in which these movements are performed [5].

For automatic coordination and regulation of locomotor movements, the cerebellum seems to occupy a position of reflex center, its afferent and efferent limbs are the spinocerebellar pathways and the plural pathways descending from the brain stem such as the rubro-, vestibulo- and reticulospinal tracts. The nucleus ruber is under the control of the interpositus nucleus. Major termination sites of the fastigioreticular fibers are the Deiters nucleus and the medial reticular formation, from which vestibulospinal and reticulospinal tracts originate, respectively. The terminal regions of the fastigioreticular fibers appear to cover parts of the regions, which give off both descending and ascending fibers. The fastigial nucleus, via the reticular formation, could influence not only to the spinal cord but also to 'higher levels' of the brain [19]. The descending influence will contribute to the reflex-dependent adaptation of locomotor movements in the decerebrate cats, while the ascending influence will contribute, along with the activity relayed by the interposito-thalamo-cerebral pathway, to their context-dependent adaptation in intact cats. Removal of the cerebellum deprives the animal from its automatic control capability of locomotor movements, because it interrupts the transmission of posture and locomotor-related signals routed through the spinocerebello-spinal pathways, resulting in abnormally variable coupling between the step cycles in the different limbs, which leads to frequent loss of equilibrium and unevenness in stride length [12].

Finally, contribution of basal ganglia need to be discussed. Reticulospinal neurons are subject to the 'feedforward' control signals arising from the cerebral cortex, cerebellum and the basal ganglia, and the sensory feedback signals arising from the spinal cord and locomotor apparatus. The basal ganglia circuits are composed of a sequence of fibers connecting the cerebral cortex, striatum, globus pallidus, substantia nigra and thalamus [33]. The striatum receives inputs from almost entire cerebral cortex, and in turn, the basal ganglia output is transmitted mostly back to the frontal cortex (mainly premotor cortex) via the thalamus. Nakamura et al. [34] and Nakamura and Sato [35] have provided evidence that there exist monosynaptic striatal inputs to the nigrotegmental neurons, and monosynaptic nigral inputs to the PPN neurons that project their descending axons to the medioventral part of the MRF including the NRMc. Terminals of the striatonigral and nigro-PPN projections showed only symmetric (inhibitory) synapses, while the PPN-MRF and reticulospinal terminals showed both asymmetric (excitatory) and symmetric synapses. However, possible functional roles played by the aforementioned multisynaptic neuronal networks have not been elucidated clearly.

Dystonia, which often accompanies the basal ganglia lesions, can be partly explained by the abnormal integration or disintegration of posture- and locomotor-related signals at multiple sites within the central nervous system. As to the output pathways from the basal ganglia, there are at least two main routes: the one is through the cerebral cortex to the spinal cord by ways of the corticospinal tract and cortico-reticulo-spinal tract [22, 24, 33], and the other is through the SN-PPN-reticulospinal tract to the spinal cord [33–35]. Some of the reticulospinal neurons are under the convergent and possibly even divergent control of cortico-reticular, fastigioreticular and PPN-reticulospinal pathways. It is possible that disintegration of all these synaptic inputs at the levels of individual pontomedullary reticular neurons results in the abnormal setting of postural muscle tone [30]. At this stage of our study, we do not yet know the relative contribution of each pathway or the relative combination of these pathways to the control of pontomedullary reticular neurons, nor do we know the resultant execution of purposive locomotor movements. Elucidation of putative neurotransmitters acting at each relay station of the multisynaptic pathways, along with identification of detailed synaptic mechanisms, will contribute greatly to our understanding of the neuronal mechanisms related to the genesis of abnormal movements.

References

1 Grillner S: Locomotion in bipeds – Central mechanisms and reflex interactions. Physiol Rev 1975; 55:274–304.
2 Shik ML, Orlovsky GN: Neurophysiology of locomotor automatism. Physiol Rev 1976;55:465–501.
3 Orlovsky GN: Connections of the reticulo-spinal neurons with the 'locomotor sections' of the brain stem. Biophysics 1970;15:178–186.
4 Orlovsky GN, Shik ML: Control of locomotion: A neurophysiological analysis of the cat locomotor system. Int Rev Physiol Neurophysiol 1976;2:282–317.
5 Armstrong DM: Supraspinal contributions to the initiation and control of locomotion in the cat. Prog Neurobiol 1986;26:273–361.
6 Mori S: Integration of posture and locomotion in acute decerebrate cats in awake, freely moving cats. Prog Neurobiol 1987;28:161–195.
7 Mori S, Takakusaki K: Integration of posture and locomotion; in Amblard B, Bertohoz A, Clarac F (eds): Posture and Gait. Amsterdam, Excerpta Medica, 1988, pp 341–354.
8 Shimamura M, Kogure I: Discharge patterns of reticulospinal neurons corresponding with quadrupedal leg movements in thalamic cats. Brain Res 1983;230:27–34.
9 Drew T, Dubuc R, Rossignol S: Discharge patterns of reticulospinal and other reticular neurons in chronic, unrestrained cats walking on a treadmill. J Neurophysiol 1986;55:375–401.
10 Armstrong DM, Drew T: Discharges of pyramidal tract and other motor cortical neurones during locomotion in the cat. J Physiol 1984;346:471–495.
11 Arshavsky Yu I, Gelfand IM, Orlovsky GN: The cerebellum and control of rhythmical movements. Trends Neurosci 1983;6:417–422.
12 Arshavsky Yu I, Gelfand IM, Orlovsky GN: Cerebellum and Rhythmical Movements. Berlin, Springer 1986.
13 Armstrong DM, Edgley SA: Discharges of Purkinje cells in the paravermal part of the cerebellar anterior lobe during locomotion in the cat. J Physiol 1984;352:403–424.
14 Orlovsky GN: Work of the Purkinje cells during locomotion. Biophysics 1972;17:935–941.
15 Armstrong DM, Edgley SA: Discharges of nucleus interpositus neurones during locomotion in the cat. J Physiol 1984;351:411–432.
16 Volsi GL, Pacitti C, Perciavalle V, Sapienza S, Urbano A: Interpositus nucleus influences on pyramidal tract neurons in the cat. Neuroscience 1982;7:1929–1936.
17 Chambers WW, Sprague JM: Functional localization in the cerebellum. I. Organization in longitudinal cortico-nuclear zones and their contributions to the control of posture both extrapyramidal and pyramidal. J Comp Neurol 1955;103:105–112.

18 Chambers WW, Sprague JM: Functional localization in the cerebellum. II. Somatotopic organization in cortex and nuclei. Arch Neurol Psychiatry 1955;74:653–680.
19 Brodal A: Neurological Anatomy in Relation to Clinical Medicine. Oxford, Oxford University Press, 1981.
20 Walberg F, Pompeiano O, Westrum LE, Hauglie-Hanssen E: Fastigioreticular fibers in the cat. An experimental study with silver method. J. Comp Neurol 1962;119:187–199.
21 Orlovsky GN: Work of the neurons of the cerebellar nuclei during locomotion. Biophysics 1972;17:1117–1185.
22 Kuypers HGJM: Anatomy of the descending pathways; in Handbook of Physiology, section 1: The Nervous System, vol 2, Brookhart JM, Mountcastle VB (sect eds). Bethesda, American Physiological Society, 1981, pp 597–666.
23 Peterson BW, Maunz RA, Pitts, NG, Filon, M: Patterns of projection and branching of reticulospinal neurons. Exp Brain Res 1975;23:333–351.
24 Peterson BW: The reticulospinal system and its role in the control of movement; in Barnes CD (ed): Brainstem Control of Spinal Cord Function. Orlando, Academic Press, 1964, pp 28–87.
25 Matsuyama K, Ohta Y, Mori S: Ascending and descending projections of the nucleus reticularis gigantocellularis in the cat demonstrated by the anterograde neural tracer, *Phaseolus vulgaris* leucoagglutinin (PHA-L). Brain Res 1988;460:124–141.
26 Matsuyama K, Kobayashi Y, Takakusaki K, Mori S, Kimura H: Termination mode and branching patterns of reticuloreticular and reticulospinal fibers of the nucleus reticularis pontis oralis in the cat: An anterograde PHA-L tracing study. Neurosci Res 1993;17:9–21.
27 Homma Y, Yokoyama T, Matsuyama K, Kobayashi Y, Mori S: Projections from the fastigial nucleus to brainstem and cerebellar cortex in the cat. Jap J Physiol 1993;43(suppl 2):s243.
28 Berkinblit MB, Deligiana TG, Feldman AG, Gelfand IM, Orlovsky GN: Generation of scratching. I. Activity of spinal interneurons during scratching. J Neurophysiol 1978;41:1040–1047.
29 Mori S, Nishimura H, Kurakami C, Yamamura T, Aoki M: Controlled locomotion in the mesencephalic cat: Distribution of facilitatory and inhibitory regions within pontine tegmentum. J Neurophysiol 1978;41:1580–1591.
30 Mori S, Kawahara K, Sakamoto T, Aoki M, Tomiyama T: Setting and resetting of postural muscle tone in the decerebrate cat by stimulation of the brain stem. J Neurophysiol 1982;48:737–748.
31 Garcia-Rill E, Skinner RD, Fitzgerald JA: Chemical activation of the mesencephalic locomotor region. Brain Res 1985;330:43–54.
32 Keizer K, Kuypers HGJM: Distribution of corticospinal neurons with collaterals to lower brain stem reticular formation in cat. Exp Brain Res 1984;54:107–120.
33 Garcia-Rill E: The basal ganglia and the locomotor regions. Brain Res Rev 1986;11:47–63.
34 Nakamura Y, Kudo M, Tokuno H: Monosynaptic projection from the pedunculopontine tegmental nuclear region to the reticulospinal neurons of the medulla oblongata. An electron microscopic study in the cat. Brain Res 1990;524:353–356.
35 Nakamura Y, Sato F: Multisynaptic connections from the striatum to the spinal motoneuron in the cat; in Mano N, Hamada I, Delong, MR (eds): Role of the Cerebellum and Basal Ganglia in Voluntary Movement. Amsterdam, Excerpta Medica, 1993, pp 123–130.

Shigemi Mori, National Institute for Physiological Sciences, Myodaji-cho, Okazaki 444 (Japan)

Function of the Indirect Pathway in the Basal Ganglia Oculomotor System: Visuo-Oculomotor Activities of External Pallidum Neurons

Makoto Kato, Okihide Hikosaka

Laboratory of Neural Control, National Institute for Physiological Sciences, Okazaki, Japan

A recent model of the basal ganglia has suggested that there are two main pathways, the direct and indirect pathways [1] (fig. 1). The direct pathway is thought to facilitate the target structures through double inhibitions, as shown for the oculomotor part of the basal ganglia [2, 3]. In a resting state, caudate neurons discharge with very low frequencies, whereas substantia nigra pars reticulata (SNr) neurons discharge with very high frequencies. The tonic firing of SNr neurons inhibits the superior colliculus (SC), thus suppressing saccadic eye movements. Activation of caudate neurons by cerebral cortical inputs leads to inhibition of SNr neurons, and consequently disinhibition of SC neurons, and therefore facilitates saccadic eye movements.

Activation of the indirect pathway would have effects opposite to the direct pathway, as judged by the signs of anatomical connections. Activation of caudate neurons would inhibit external pallidum (GPe) neurons, which leads to disinhibition of subthalamic nucleus (STN) neurons, facilitation of SNr neurons, thus increasing the tonic inhibition upon the SC. However, it is still unclear how these two pathways work in natural behavioral conditions. Do they compete with each other? Or do they work in different behavioral contexts? To answer these questions, we recorded single cell activities from the GPe in the monkey performing visuo-oculomotor tasks.

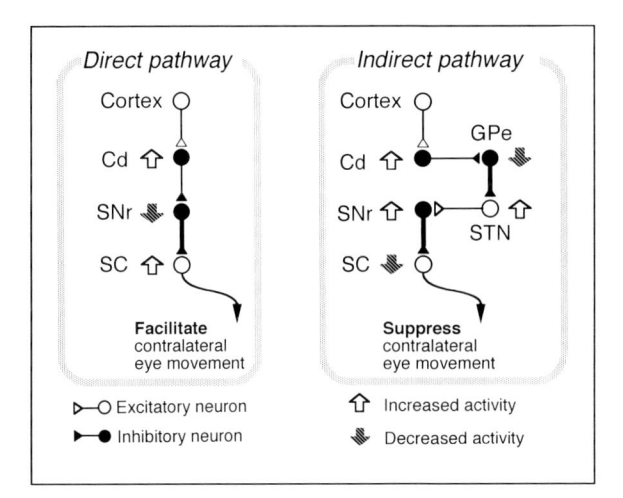

Fig. 1. Connections and hypothetical activity changes of oculomotor areas through the direct (left) and indirect (right) pathways in the basal ganglia. Open and filled neurons indicate inhibitory and excitatory neurons, respectively. Open upward arrows indicate increases in activity changes; hatched downward arrows indicate decrease in activity changes. Cd = Caudate nucleus; GPe = external segment of globus pallidus; SC = superior colliculus; SNr = substantia nigra pars reticularis; STN = subthalamic nucleus.

Materials and Methods

We used two Japanese monkeys trained to perform mainly two tasks, visually guided and memory-guided saccade tasks [4] (fig. 2). The tasks were controlled by a conventional microcomputer. The animal sat on a monkey chair in a quiet, dimly lit shielded room facing a dome-shaped black panel implanted with an array of small yellow LEDs (8 direction and 4 eccentric amplitudes plus the center). After sufficient training using these tasks, the animal was anesthetized with pentobarbital sodium (5 mg/kg/h), and underwent an operation for mounting a head holder and a recording chamber stereotaxically on its head and for implanting an eye coil. At the time of experiment, the head of the animal was fixed to the chair by the head holder and eye position was monitored with the search coil method [5]. A glass-insulated elgiloy electrode was introduced into the target area through the chamber for recording extracellular spikes with a conventional amplifier. The spikes were converted into pulses by a discriminator with spike amplitude and duration windows. The pulses, task events, and AD-converted eye position signals were stored in the computer for later offline analysis. Small electrolytic lesions were made at several selected points in and around the GPe by passing weak DC currents (+5 µA, 100 s) through the recording electrode. After the series of experiments, the animal was deeply anesthetized with pentobarbital sodium (50 mg/kg) and perfused with saline containing heparin, followed with 10% formalin. The brain was cut into several blocks, sectioned at 50 µm thickness, and stained with cresyl violet for reconstruction of recording sites.

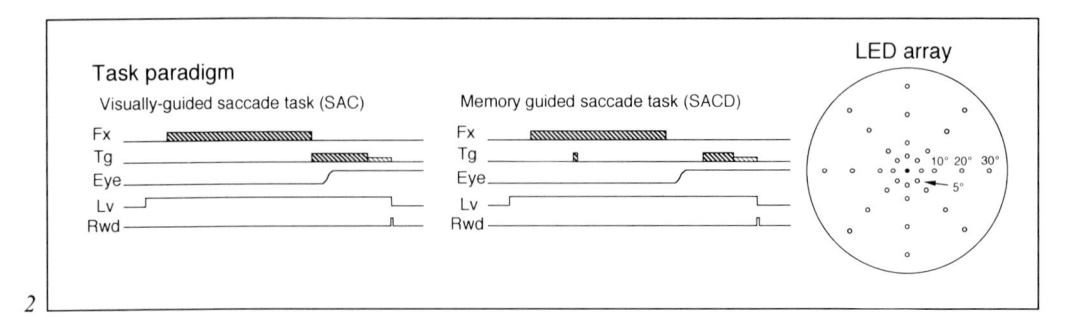

Task paradigm

Visually-guided saccade task (SAC)

Fx
Tg
Eye
Lv
Rwd

Memory guided saccade task (SACD)

Fx
Tg
Eye
Lv
Rwd

LED array

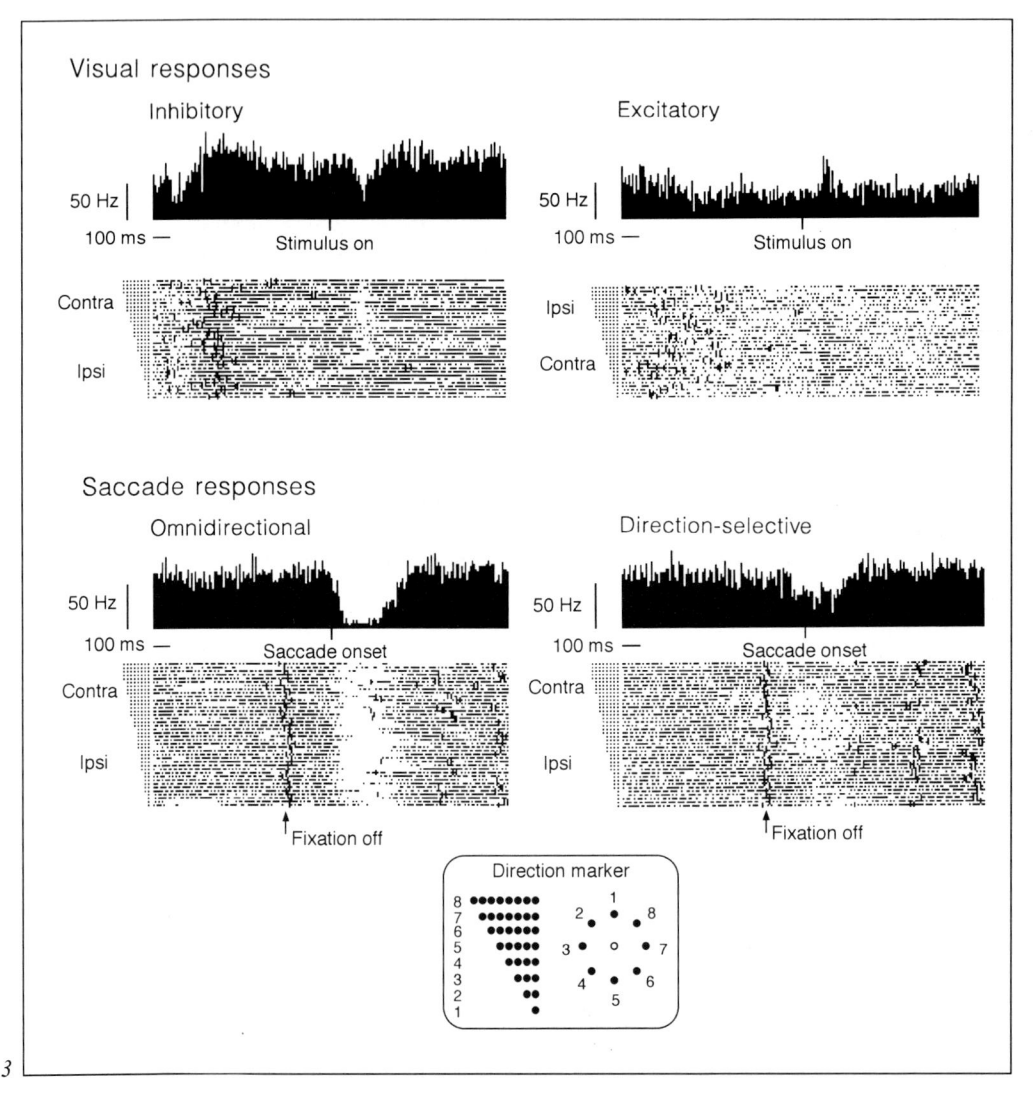

Visual responses

Inhibitory

50 Hz
100 ms — Stimulus on

Contra

Ipsi

Excitatory

50 Hz
100 ms — Stimulus on

Ipsi

Contra

Saccade responses

Omnidirectional

50 Hz
100 ms — Saccade onset

Contra

Ipsi

Fixation off

Direction-selective

50 Hz
100 ms — Saccade onset

Contra

Ipsi

Fixation off

Direction marker

Kato/Hikosaka

Results

Out of 400 GPe neurons, 100 were responsive to the tasks. Sixty-four neurons showed responses time-locked to saccades and 34 neurons showed responses time-locked to visual events. Many saccadic neurons also responded to other events such as visual stimulus, reward or lever release, and in addition showed tonic responses during eye fixation.

Visual responses were observed in response to a brief flash of the LED light (on-response) or to the offset of fixation light (off-response). Figure 3 shows representative visual on-responses of 2 cells with a decrease and an increase in discharge rate. On-responses were obtained in 23 neurons with decreased (n = 13) or increased (n = 10) activity. These neurons had large receptive fields, including,

Fig. 2. Schematic diagrams of a visually guided saccade (left) and memory-guided saccade (center) task. Fx = Fixation light; Tg = target light; Eye = eye position; Lv = lever switch. Right: schematic illustration of distribution of an array of small yellow LEDs (the center, and 8 direction and 4 eccentric amplitudes) implanted on a dome-like black panel. The task was initiated with a lever press by the monkey. For the visually guided saccade task, a LED light of a fixation point turned on at the center of the panel. The monkey had to fixate the fixation point. And after 1.5–3 s, the fixation point turned off, and simultaneously, another LED light (target point) turned on. The animal had to make a saccade to the target point. After 0.5–1.5 s, the target point dimmed for 0.5 s. The animal had to release the lever during the dim. If successful, a drop of water was delivered. For memory guided saccade task, a LED light turned on briefly at 1 s after the onset of the fixation light, as the cue to indicate where the target point would appear. After the fixation point turned off, the monkey must make a saccade according to his memory to where the cue had appeared. The target point turned on at 0.6 s after the fixation point turned off, and thereafter dimmed.

Fig. 3. Raster-histograms of representative visual-on and saccade responses. For visual-on responses (top), inhibitory (left) and excitatory (right) responses were obtained during the memory guided saccade task. While the monkey was fixating the center spot light, another spot light (visual stimulus) was presented at one of 8 directions with a 20° eccentric amplitude. These raster-histograms were aligned at the onset of the stimulus; they have been rearranged according to the direction of the stimulus. Direction markers are indicated at the left to each raster line. The bottom inset shows what direction the marker indicates. 'Contra' and 'Ipsi' indicate that the stimulus was presented in the visual field contralateral and ipsilateral to the recording site of the neuronal activity, respectively. For saccade responses (bottom), omnidirectional (left) and direction-selective (right) saccade responses were obtained during the visually guided saccade task. Saccade amplitudes were 20°. These raster-histograms were aligned at the onset of saccades; they have been rearranged according to the direction of the saccade. 'Contra' and 'Ipsi' indicate that the saccade was directed to the visual field contralateral and ipsilateral to the recording site of the neuronal activity, respectively.

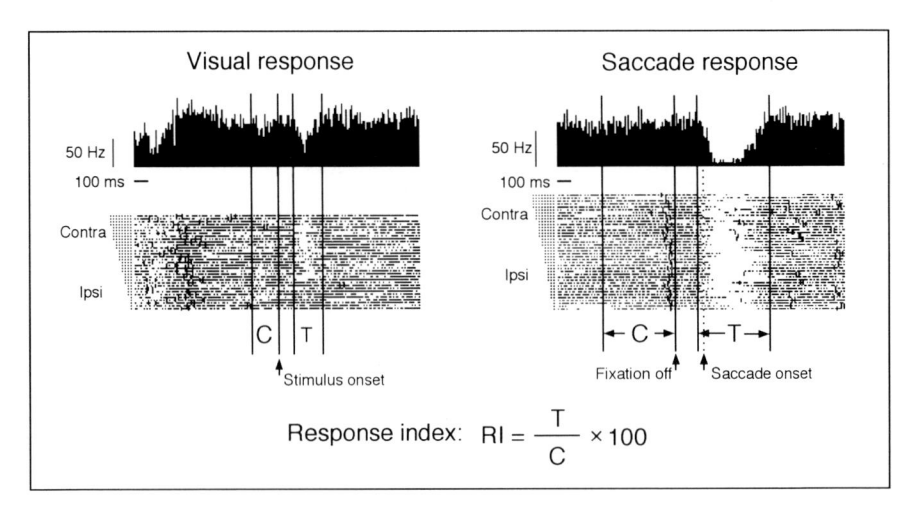

Fig. 4. For quantitative analysis of direction selectivity, a response index (RI) was defined as the ratio of spike activity within a test period (T) to spike activity within a control period (C) (represented in percentage). For visual responses (left), the control period was set just before the visual stimulus onset; for saccade responses, it was set just before the offset of the fixation light. The test period was set during the response phase. Durations of the control and test periods were made the same. Thus the excitatory and inhibition responses are indicated by the response index more than and less than 100, respectively.

the contralateral hemifield. Off-responses were obtained in 11 neurons with decreased (n = 5) or increased (n = 6) activity.

For quantitative analysis of the direction selectivity of visual on-responses, we calculated the response index for each of 8 directions and made polar diagrams (fig. 4). Figure 5 (upper) shows examples of visual on-responses. Broadly tuned neurons tended to have multiple peaks of response.

For saccade responses, the dominant response was a decrease in activity (40 out of 64 saccadic neurons; fig. 3). We found neurons showing omnidirectional responses (n = 32) and direction-selective responses with a contralateral (n = 11), ipsilateral (n = 5), or up-down (n = 16) preference. These direction selectivities were broadly turned as well as those for visual responses.

For quantitative analysis of the direction selectivity, we calculated the response index as well as visual responses (fig. 4). Figure 5 (lower) shows representative polar diagrams of saccade responses. Saccade responses were broadly tuned, including more than 4 responsive directions. Some neurons showed different saccade responses between visually guided and memory-guided saccades. Saccade responses were often associated with responses to other task events such as reward.

Visual responses

Saccade responses

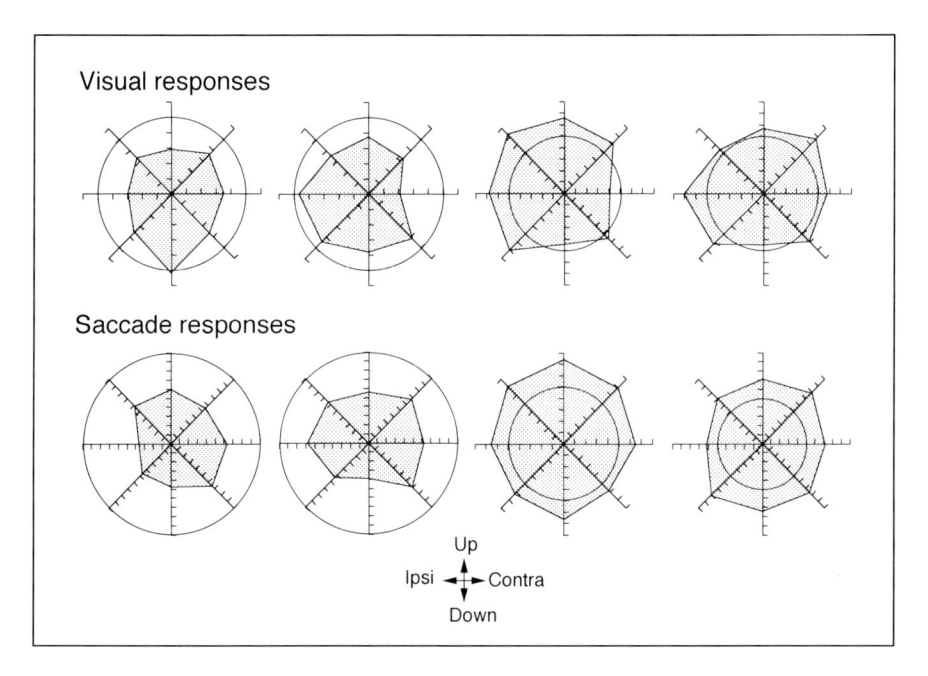

Up
Ipsi ←→ Contra
Down

Fig. 5. Polar diagrams of representative visual (upper, 4 neurons) and saccade (lower, 4 neurons) responses in each direction. The right side of each polar diagram indicates the response to the stimulus presented contralaterally to the recording site of the neurons. Circles indicate the control level in spike activity.

In addition, we found neurons showing tonic responses. Figure 6 shows representative tonic responses of 2 cells. Unit 1 showed tonic decreases in activity during fixation and after target light onset. Unit 2 showed a slight tonic increase during fixation followed by an abrupt depression of activity after a saccade, and gradual recovery after target light onset. The decreased activities during fixation to the center and visual target, and waiting for the appearance of the target light may contribute to maintain eye position by preventing external signals from triggering unnecessary eye movements. Tonic responses were often associated with phasic saccade or visual responses.

Response latencies of visual and saccade responses were measured from stimulus and saccade onset, respectively. The latency for a saccadic activity preceding the saccade onset was represented by a minus value. The mean latencies of visual and saccade responses were 128 (n = 21) and 4 ms (n = 59), respectively. In both visual and saccade responses, the mean latency of the excitatory response (113 and −20 ms, respectively) was slightly shorter than that for the inhibitory response (139 and 4 ms, respectively). Since the signals from the striatum (the

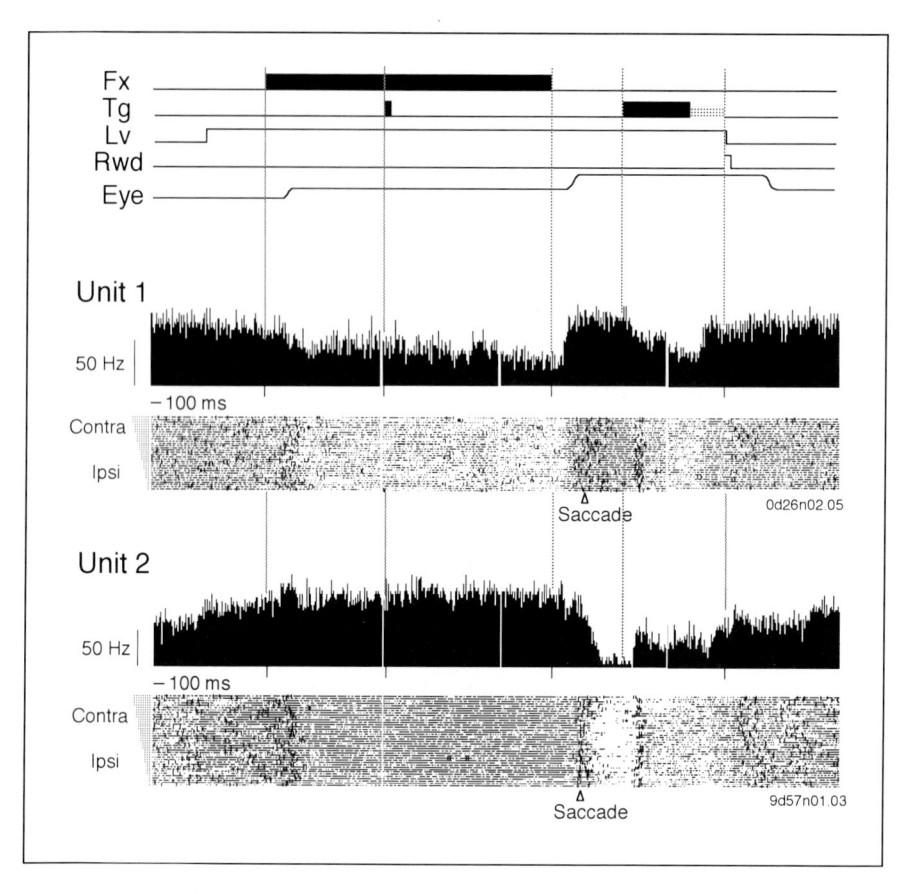

Fig. 6. Raster-histograms of representative tonic responses during the memory guided saccade task obtained from two neurons (Unit 1 and 2). The upper diagram indicates schematic task events. These raster-histograms were aligned at the onset of fixation light, the onset of cue light, the offset of fixation light, the onset of target light, and the lever release, each indicated by a vertical line. They have been rearranged according to the direction of the saccade. Pairs of short lines on each raster indicate the onset and offset of detected saccades. An arrow head under each raster-histograms indicates the approximate time of the occurrence of the memory guided saccades. Other conventions are the same as in figures 2 and 3.

caudate nucleus and putamen) would lead to inhibitions of GPe neurons, the excitatory responses of GPe neurons are probably evoked by the STN inputs. The shorter latency of the excitatory response is consistent with the fact that the excitatory cortico-subthalamo-pallidal pathway is faster than the inhibitory cortico-striato-pallidal pathway [6].

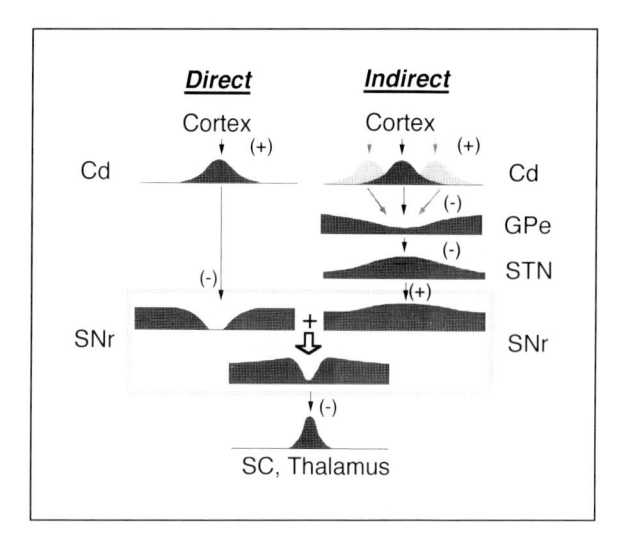

Fig. 7. Schematic diagrams of response fields in each levels of structures involved in the direct and indirect pathways in the basal ganglia. Abbreviations are the same as in figures 1. (+): Excitatory transmission; (–): inhibitory transmission. Details are explained in the Discussion.

Histological analysis for the locations of recorded cells revealed that visuo-oculomotor cells were distributed mainly in the dorsal portion of the GPe. In a recent study [7], these areas received afferents from the caudate. We could not find visuo-oculomotor neurons in the internal pallidum.

Discussion

The present study revealed that GPe neurons have large receptive and movement fields including more than a half of the visual field. Their responses consist of various combinations of phasic and tonic components. Recent studies [3, 8] have suggested that caudate neurons have more specific visuo-oculomotor information. Eccentricities of visual-receptive or movement fields of caudate neurons are less than those of GPe neurons, and the centers of these fields of caudate neurons are distributed in the contralateral visual fields. These results, taken together, indicate that there is a great degree of convergence from caudate neurons onto single GPe neurons (fig. 7). SNr neurons also have relatively specific visuo-oculomotor information [9]. In our laboratory, visuo-oculomotor activity was recorded from STN neurons [10]. Although excitatory responses were dominant, they also had relatively large receptive and movement fields. These results suggest that the indirect pathway provides the SNr with surrounding effects in contrast with specific signals provided through the direct pathway.

Our results are at variance with some previous studies. Georgopoulos et al. [11] reported that the discharge of 41% of internal pallidum (GPi) neurons and 48% of GPe neurons during movement was related to the direction of step-tracking elbow movements. Mitchell et al. [12], using a task with torque loads for dissociating movement direction from EMG pattern, found that 30% of GPi neurons and 30% of GPe neurons had directional responses and only a few neurons (3%) had bidirectional responses. These studies suggested that neuronal signals in the GPe and GPi are not differentiated.

Other studies, however, seem consistent with our results. Mink and Thach [13] reported differences in the population of directional cells between the GPe and GPi during step-tracking wrist movements. They found that 41% of GPi neurons and 28% of GPe neurons were directional, whereas 18% of GPi neurons and 40% of GPe neurons were bidirectional. Using a task with tracking wrist movements, Hamada et al. [14] found that all wrist-related neurons in the GPe were bidirectional. These studies, in agreement with our studies, suggest that neuronal signals conveyed by the GPe are less specific than those in the GPi.

Why are there the discrepancies in the results between the different studies? One possibility is the difference in the degree of the freedom of movement. The elbow movement occurs only in one-dimensional direction, flexion and extension. In contrast, the wrist movement occurs in two-dimensional space, including flexion, extension, abduction and adduction. The eye movement also occurs in two-dimensional space. When wrist and eye movements occur, it is often necessary that they are limited to one dimension by suppressing dimension. In such a condition, the indirect pathway could provide the surrounding effects, in contrast with specific signals provided through the direct pathway. Such a mechanism would be suitable for selecting neuronal signals to facilitate or release a proper movement.

References

1 Bergman H, Wichmann T, DeLong MR: Reversal of experimental parkinsonism by lesions of the subthalamic nucleus. Science 1990;249:1436–1438.
2 Hikosaka O, Wurtz RH: The basal ganglia; in Wurtz RH, Goldberg ME (eds): The Neurobiology of Saccadic Eye Movements. Amsterdam, Elsevier, 1989, pp 257–281.
3 Hikosaka O, Sakamoto M, Usui S: Functional properties of monkey caudate neurons. I. Activities related to saccadic eye movements. J Neurophysiol 1989;61:780–798.
4 Hikosaka O, Wurtz RH: Visual and oculomotor functions of monkey substantia nigra pars reticulata. III. Memory-contingent visual and saccade responses. J Neurophysiol 1983;49:1268–1284.
5 Robinson DA: A method of measuring eye movement using a scleral search coil in a magnetic field. IEEE Trans Biomed Eng 1963;10:137–145.

6   Nambu A, Yoshida SI, Jinnai K: Discharge patterns of pallidal neurons with input from various cortical areas during movement in the monkey. Brain Res 1990;519:183–191.
7   Parent A, Bouchard C, Smith Y: The striopallidal and strionigral projections: Two distinct fiber systems in primate. Brain Res 1984;303:385–390.
8   Hikosaka O, Sakamoto M, Usui S: Functional properties of monkey caudate neurons. II. Visual and auditory responses. J Neurophysiol 1989;61:799–813.
9   Hikosaka O, Wurtz RH: Visual and oculomotor functions of monkey substantia nigra pars reticulata. I. Relation of visual and auditory responses to saccades. J Neurophysiol 1983;49:1230–1253.
10 Matsumura M, Kojima J, Gardiner TW, Hikosaka O: Visual and oculomotor functions of monkey subthalamic nucleus. J Neurophysiol 1992;67:1615–1632.
11 Georgopoulos AP, DeLong MR, Crutcher MD: Relations between parameters of step-tracking movements and single cell discharge in the globus pallidus and subthalamic nucleus of the behaving monkey. J Neurosci 1983;3:1586–1598.
12 Mitchell SJ, Richardson RT, Baker FH, DeLong MR: The primate globus pallidus: Neuronal activity related to direction movement. Exp Brain Res 1987;68:491–505.
13 Mink JW, Thach WT: Basal ganglia motor control. II. Late pallidal timing relative to movement onset and inconsistent pallidal coding of movement parameters. J Neurophysiol 1991;65:301–329.
14 Hamada I, DeLong MR, Mano N: Activity of identified wrist-related pallidal neurons during step and ramp wrist movements in the monkey. J Neurophysiol 1990;64:1892–1906.

Makoto Kato, MD, Lab. of Neural Control, Department of Biological Control System,
National Institute of Physiological Sciences, 38 Nishigonaka, Myodaiji, Okazaki 444 (Japan)

Role of the Nigrostriatal Dopamine System in Corticostriatal Signal Transmission in Alert Animals

Minoru Kimura, Toshihiko Aosaki

Faculty of Health and Sport Sciences, Osaka University, Toyonaka, Osaka,
Laboratory for Neural Circuits, Bio-Mimetic Control Research Center,
The Institute of Physical and Chemical Research (RIKEN), Atsuta, Nagoya, Japan

The nigrostriatal dopamine system plays an essential role in the basal ganglia function and mechanisms for the initiation and control of centrally programmed movement, since dysfunction of this system results in severe motor disturbances, typically known as Parkinson's disease. In Parkinson's disease, there are not only severe motor disturbances including akinesia, rigidity and tremor but also sensory neglect and deterioration in cognitive functions [1, 2]. On the other hand, unilateral destruction of the nigrostriatal dopamine system in experimental animals results in turning movements toward the dopamine-depleted side [3] and in neglect of sensory stimuli in the contralateral hemifield [4, 5].

Abnormal neuronal activity in dopamine-depleted striatum has been described in animal models of Parkinson's disease. The sensory responsiveness of striatal neurons decreased [6, 7], while the threshold of somatosensory responsiveness of globus pallidus neurons lowered and the receptive field became larger [8].

In the present study, we aimed at clarifying the mechanism of nigrostriatal dopamine effects on striatal neuron responses to afferent inputs. Specifically, afferent projections from the cerebral motor cortex to the striatum were electrophysiologically examined in animals in which the nigrostriatal dopamine system was selectively destructed on one side of the brain. We selectively inactivated the nigrostriatal system by injecting a dopaminergic neurotoxin 1-methyl-4-phenylpyridium (MPP$^+$) in one hemisphere, and the ipsilateral and controlateral corticostriatal projections were electrophysiologically examined by recording evoked field potentials in the striatum to electrical stimulation of the cerebral cortex.

Methods

General

Behavioral observation and neurophysiological recordings were made in 5 cats (3.2 – 4.3 kg). A recording chamber was stereotaxically implanted over an opening in the skull and fixed with dental acrylic. The axis of the chamber was aimed at the caudate nucleus vertically. Cortically evoked field potentials in the striatum were recorded through a glass-coated Elgiloy microelectrode (0.5–2 MΩ). The microelectrode was lowered by a Narishige microdrive. The head of the animal was rigidly held in the stereotaxic frame by inserting two pairs of bars into a transverse tube mounted on the skull. During the experiment, the cat was comfortably supported without restraint by a hammock and was fed with milk. Some cats tended to fall asleep after they were mounted in the apparatus. Sixty to 90 days elapsed between the implantation of the recording chamber on the skull and completion of striatal recordings.

MPP$^+$ Treatment

MPP$^+$ (5–7.5 mg) was dissolved in 200 µl 0.9% NaCl and infused into the left striatum of each animal under deep pentobarbital sodium anesthesia by means of an Alzet osmotic minipump (0.5 µl/h) [9]. The infusion needle 0.8 mm OD 0.5 mm ID) was implanted at the central part of the caudate nucleus at A13–A19 (fig. 1). It took 7 days for the pump to infuse one side of the striatum with 200 µl of the toxin.

Stimulation of Motor Cortex and Recording in the Striatum

Electrical Stimulation of Motor Cortex. Two insulated acupuncture silver needles arrayed mediolaterally with a 1.5-mm separation were inserted into the precruciate motor cortex on both sides. The needle electrodes were chronically implanted by fixating them to the skull with dental acrylic. Current pulses (20–500µA, 0.1–0.2 ms duration) were applied bipolarly between neighboring needles at a rate of 1/s. Evoked field potential was recorded in the ipsilateral and contralateral caudate nucleus by inserting the Elgiloy microelectrode into the striatum. The evoked field potential was amplified and digitized by computer (PC9801vx) and recorded at 0.5 mm depth intervals during single electrode penetrations through the caudate nucleus by averaging data obtained during 100–200 trials of stimulation.

Histology

After the recording session was terminated, the animals were deeply anesthetized with a pentobarbital sodium injection (60–90 mg/kg), and transcardially perfused with 4% paraformaldehyde phosphate buffer solution. The brain was frozen-sectioned at the frontal plane at 50 µm thickness and stained with cresyl violet. The motor cortex was also frozen-sectioned to identify stimulating sites histologically.

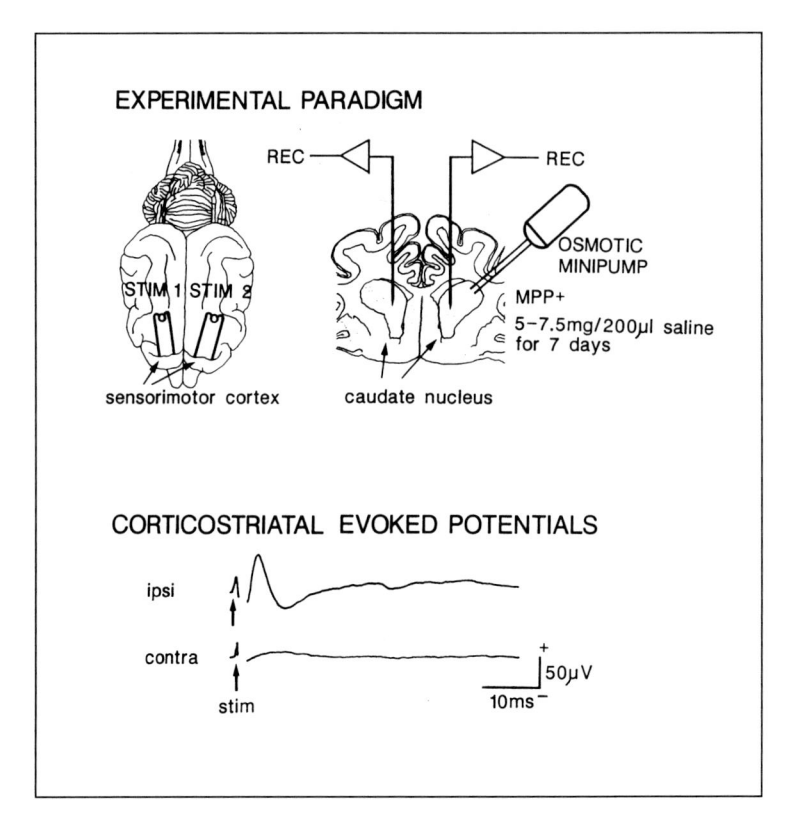

Fig. 1. Experimental paradigms and corticostriatal evoked potential waveforms. The upper panel shows the positioning of the bipolar stimulating electrode in both sides of the precruciate motor cortices (left), the positioning of the osmotic minipump containing MPP$^+$ in the left caudate nucleus and the recording from the striatum on both sides (right). The lower panel shows field potentials recorded in the caudate nucleus after stimulation of the ipsilateral and contralateral motor cortex. Upward arrows indicate the onset of stimulation.

Results

Behavioral Observations

Behavioral abnormalities began to appear 2–3 days after implantation of the minipump. The most remarkable motor sign was head and whole body turning and circling movement toward the MPP$^+$-injected (dopamine-depleted) side. The circling movement consistently started on the 3rd day after implantation of the minipump, and lasted for about 3 weeks (stage I). Then circling gradually disap-

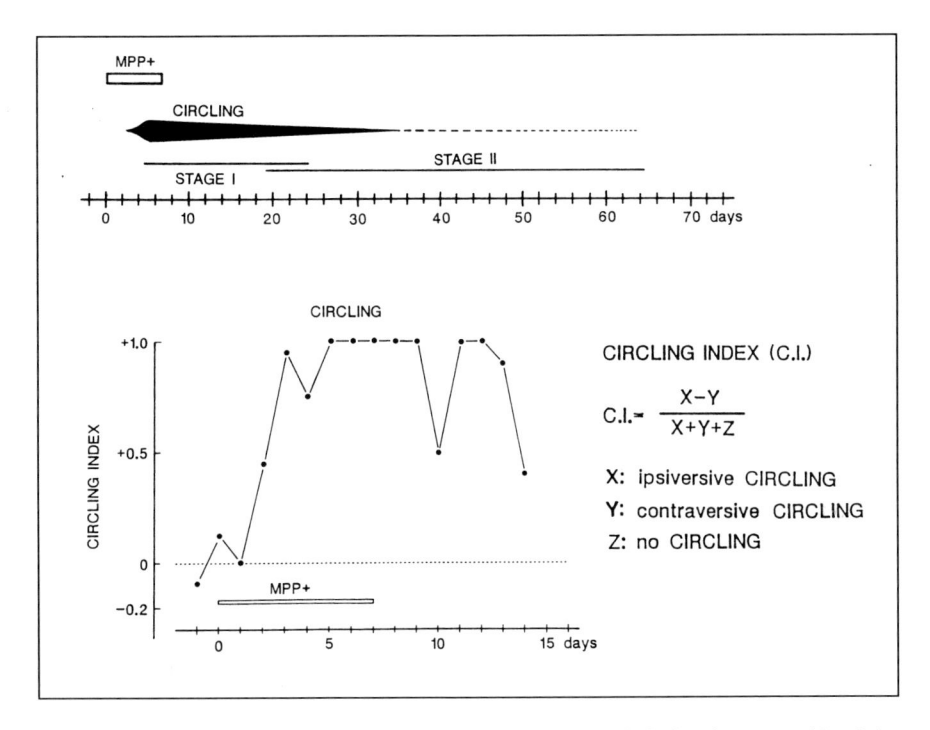

Fig. 2. Development of circling movement after MPP⁺ infusion into one side of the caudate nucleus. The upper panel shows the time course of the circling induced by MPP⁺ infusion. The lower panel shows a quantitative measurement of circling movements. The cats were held by both hands of the experimenter at knee height and dropped on the floor. In most cases, intact cats turned their heads to either the right or left side and circled. But after MPP⁺ infusion, the probability of circling toward the MPP⁺ infusion side (ipsiversive) increased abruptly.

peared, (stage II; upper panel of fig. 2). In 3 cats, we measured quantitatively the occurrence and development of the circling movement before, during and after MPP⁺ infusion. Circling movements toward MPP⁺ infusion side increased abruptly after implantation of the minipump (lower panel of fig. 2). We tried a treatment with a dopamine agonist, apomorphine, when the animal was manifesting severe ipsiversive circling movement after MPP⁺ infusion. Following treatment with apomorphine (0.1 mg/kg, i.p.), the direction of circling movement changed to the contralateral side to MPP⁺ infusion in about 5 min, attained a maximum after 10–20 min and lasted for approximately 1 h. This was consistent with the observation in a primate model of hemiparkinsonism induced by MPTP infusion [9].

Histology

After about 2 months of recording experiments, the animals were perfused with 0.9% NaCl solution and 4% paraformaldehyde. Tissue damage was observed along the track of the infusion needle of the osmotic minipump. Except for this, there was no detectable difference between the striatal tissue on the MPP^+ injected side and the tissue on the intact side. Almost all microelectrode tracks (118 tracks) were histologically reconstructed; tracks (9 tracks) which did not pass through the caudate nucleus were omitted in the present analysis.

Corticostriatal Signal Transmission

In normal cats, a single pulse of sufficient intensity delivered to the motor cortex (precruciate gyrus) evoked a sequence of potential waves in the caudate nucleus. The largest and constant components of this evoked potential were short latency (3–5 ms) positive potentials and following slow negative potentials in the dorsolateral part of the caudate nucleus on the side of cortical stimulation (fig. 1, lower panel). This pattern of response was also observed by Liles [10] and Blake et al. [11].

Figure 3 shows the spatial distribution of potentials evoked by precruciate cortex stimulation in the ipsilateral caudate nucleus. In three microelectrode penetration tracks, evoked potential traces obtained in the caudate nucleus are marked by vertical bars on the left side of each trace. Short-latency positive-negative potentials were obtained in the caudate nucleus through 22 out of 57 tracks. These potentials were located in the lateral and rostral part of the caudate nucleus. In single tracks, significant potentials were evoked in the location of the striatum which extended to 0.5–1.0 mm. This was consistent with previous tract tracing [12, 13] and electrophysiological [14] studies of cortico-caudate projections.

Projections from the precruciate cortex to the caudate nucleus showed drastic changes after dopamine depletion in the caudate nucleus by MPP^+ infusions. As illustrated in the lower panel of figure 3, precruciate cortex stimulation evoked almost no significant potentials in the ipsilateral caudate nucleus. Microelectrode penetrations in the caudate nucleus covered the same rostrocaudal and mediolateral coordinate levels of the nucleus through the recording chamber (see Methods) as those made before MPP^+ application. Depth profile of recording, such as an average frequency and amplitude of neuron discharges in the cerebral cortex and caudate nucleus did not show drastic changes after MPP^+ injection. But a spontaneous discharge frequency of striatal neurons increased slightly after dopamine depletion (3.5 ± 3.6 spikes/s, mean ± SD before MPP^+; 2.3 ± 2.4 spikes/s after MPP^+). There was no increase in spontaneous discharge frequency in the intact striatum. Figure 4 shows evoked potentials recorded in the caudate nucleus after electrical stimulation of the contralateral precruciate cortex. Cortical stimulation evoked small amplitude potentials in the contralateral caudate nucleus

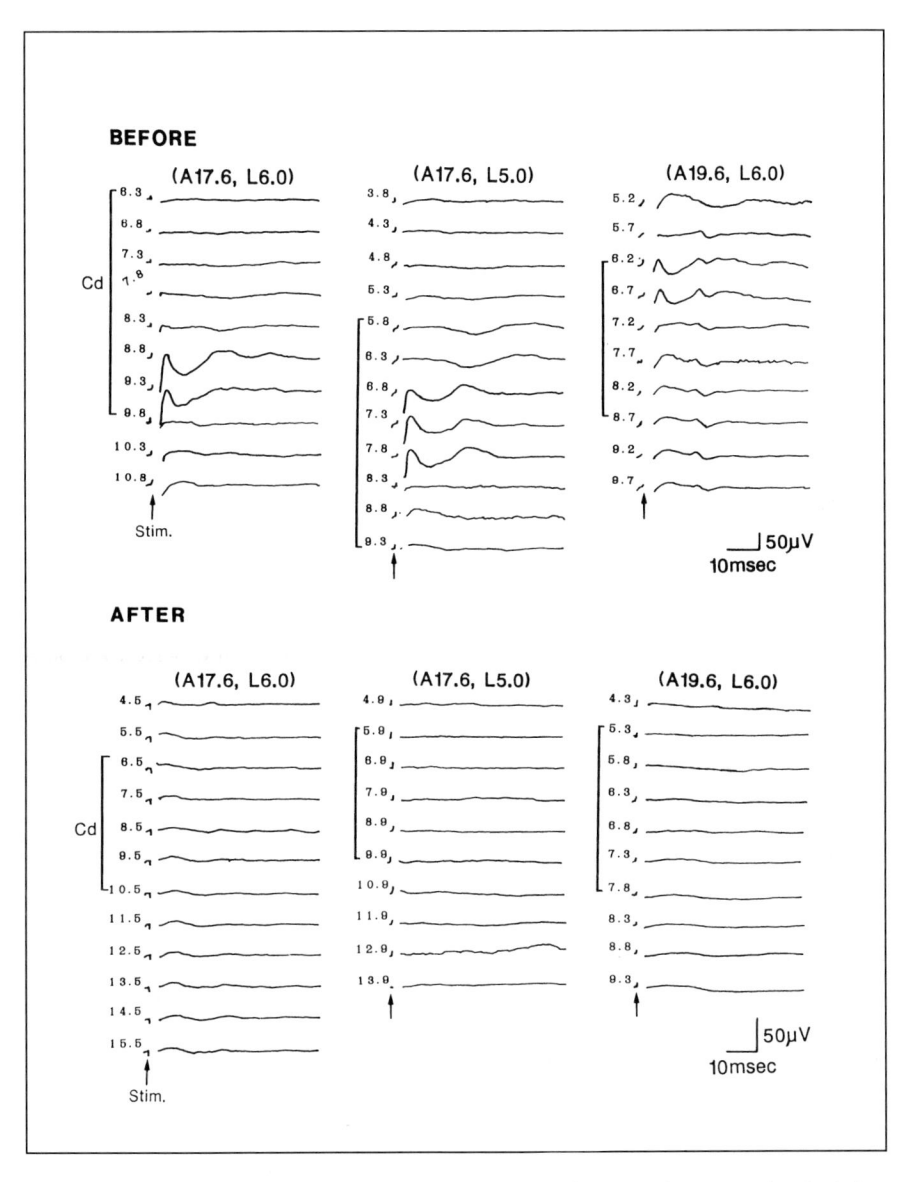

Fig. 3. Ipsilateral corticostriatal evoked potentials before and after dopamine depletion in the striatum. Depth profile of corticostriatal evoked potentials recorded through 3 microelectrode penetrations are shown vertically through the caudate nucleus. Figures in bracket indicate Horsley-Clark stereotaxic coordinates. Figures on the left side of each potential trace are depth readings (mm) of microelectrode. Upward deflection in the potential traces indicates positivity.

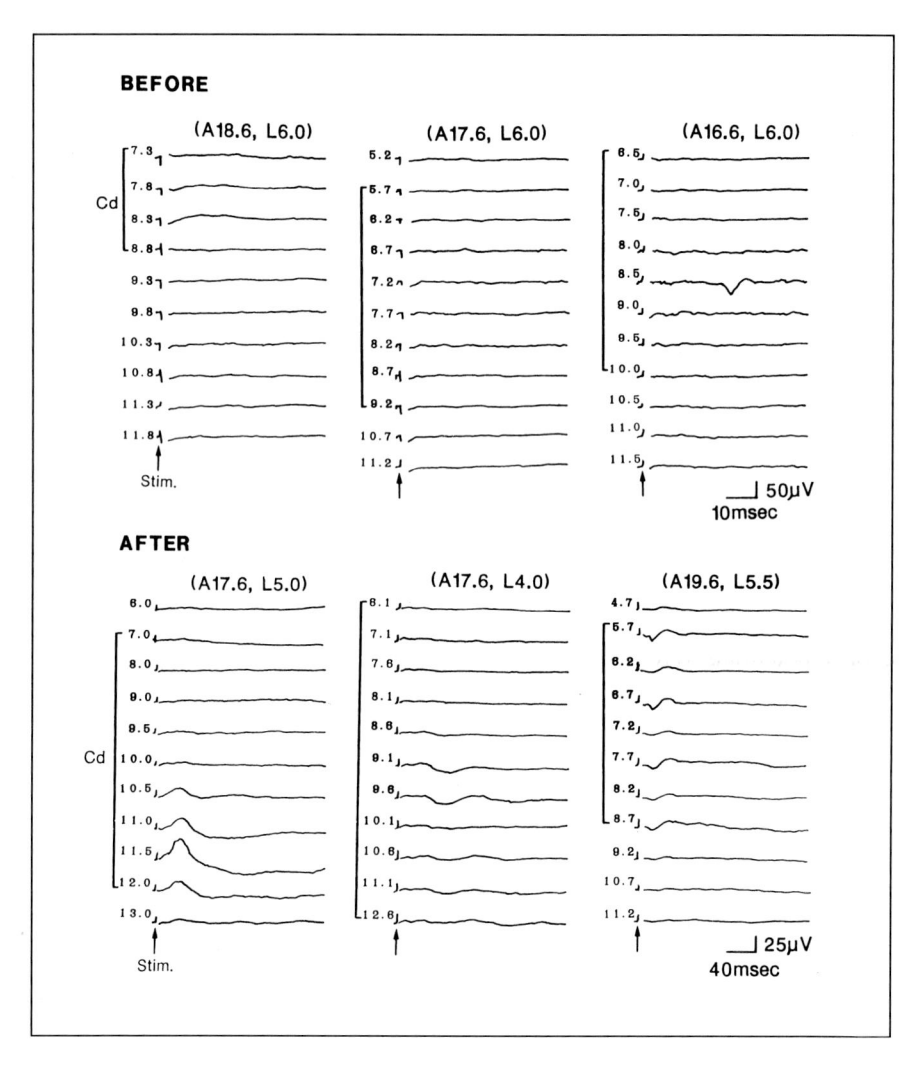

Fig. 4. Contralateral corticostriatal evoked potentials before and after dopamine depletion in the striatum.

before MPP+ application (fig. 4, upper panel). After MPP+ application, however, short-latency potentials were consistently obtained at particular locations in the caudate nucleus (fig. 4, lower panel). The potential wave was similar to that obtained on the ipsilateral side, but in some of the tracks, not only positive-negative waves but also negative-positive waves were observed (right track in lower panel of fig. 4). These drastic changes of corticostriatal projections were observed in all the animals examined (n = 5). The change was observed at the 3rd week after completion of injection of MPP$^+$ and could still be observed 4–5 weeks after the injection.

Discussion

The most striking observation in this study is the drastic reorganization of corticostriatal projections after unilateral depletion of nigrostriatal dopamine. The vast majority of the striatal neurons are medium-size spiny neurons which have axons projecting to the globus pallidus and substantia nigra, and a few percent of interneurons [15]. This characteristic cytoarchitecture is not regionally different. Therefore, a reversal of the polarity of the potential wave after cortical stimulation does not seem to occur during electrode penetrations through the caudate nucleus [10, 11]. Actually, in the present study, cortical stimulation evoked positive potentials with 4–7 ms latency and the polarity of the potentials did not change depending on the depth of recording in the striatum. When current pulses were delivered through recording electrodes in the caudate nucleus where positive evoked potentials were recorded, antidromically activated short-latency (3–5 ms) potentials were obtained through the electrode used for the precruciate cortical stimulation. It has been demonstrated that cortical stimulation evoked monosynaptic EPSPs in cat striatal neurons at a latency of 6–12 ms [14]. Therefore, it would be strongly suggested that at least an initial component of the positive field potentials obtained in the striatum after stimulation of the cruciate cortex was evoked monosynaptically.

Recently, Rothblat and Schneider [16] showed that in the cat rendered parkinsonian by 1-methyl-4-phenyl-1,2,3,6-tetrahydropyridine (MPTP) injections, the percentage of cells responding to cortical stimulation significantly increased above that observed in normal animals. They also showed that the number of striatal neurons responsive to bilateral peripheral somatosensory stimuli and of neurons with large receptive field increased after MPTP application. Fillion et al. [8] also showed that globus pallidus neurons of MPTP-induced parkinsonism in monkeys responded to somatosensory stimulation applied not only to contralateral but also to ipsilateral limbs. It seemed that input selectivity of basal ganglia neurons became poor. In the present study, nigrostriatal dopamine was depleted unilaterally by MPP$^+$ and the dominance of ipsilateral and contralateral corticostriatal projection was dramatically modified. All these observations indicate that the nigrostriatal dopamine system may play an essential role in a signal transmission in the striatum, especially corticostriatal transformations of neural information may be under the strong control of the nigrostriatal dopamine system.

References

1. Denny-Brown D: The Basal Ganglia and Their Relation to Disorders of Movement. Oxford, Oxford University Press, 1962.
2. Marsden CD: The mysterious motor function of the basal ganglia: The Robert Wartenberg lecture. Neurology 1982;32:514–539.
3. Ungerstedt U: Postsynaptic supersensitivity after 6-hydroxy-dopamine induced degeneration of nigrostriatal dopamine system in the rat brain. Acta Physiol Scand 1971;82:69–93.
4. Apicella P, Legallet E, Nieoullon A, Trouche E: Neglect of contralateral visual stimuli in monkeys with unilateral striatal dopamine depletion. Behav Brain Res 1991;46:187–195.
5. Heilman KM, Bowers D, Valenstein E, Watson RT: Hemispace and hemispatial neglect; in Jeannerod M (ed): Neurophysiological and Neuropsychological Aspects of Spatial Neglect. Amsterdam, Elsevier, 1987, pp 115–150.
6. Schneider JS: Responses of striatal neurons to peripheral sensory stimulation in symptomatic MPTP-exposed cats. Brain Res 1991;544:297–302.
7. Aosaki T, Ishida A, Watanabe K, Imai H, Graybiel AM, Kimura M: Effects of dopaminergic agents on the tonically active neurons of the striatum in hemi-parkinsonian monkeys. Soc Neurosci Abstr 1992;18:693.
8. Filion M, Tremblay L, Bedard PJ: Abnormal influences of passive limb movement on the activity of globuls pallidus neurons in parkinsonian monkeys. Brain Res 1988;444:165–176.
9. Imai H, Nakamura T, Endo K, Narabayashi H: Hemiparkinsonism in monkeys after unilateral caudate nucleus infusion of 1-methyl-4-phenyl-1,2,3,6-tetrahydropyridine (MPTP): Behavior and histology. Brain Res 1988;474:327–332.
10. Blake DJ, Zarzecki P, Somjen GG: Electrophysiological study of corticocaudate projections in cats. J Neurobiol 1976;7:143–156.
11. Liles SL: Cortico-striatal evoked potentials in cats. Electroenceph Clin Neurophysiol 1973;35:277–285.
12. Kemp JM, Powell TPS: The site of termination of afferent fibers in the caudate nucleus. Phil Trans Roy Soc London 1971;262:413–427.
13. Malach R, Graybiel AM: Mosaic architecture of the somatic sensory-recipient sector of the cat's striatum. J Neurosci 1986;6:3436–3458.
14. Kitai, ST, Kocsis, JD, Preston RJ; Sugimori M: Monosynaptic input to caudate neurons identified by intracellular injection of horseradish peroxidase. Brain Res 1976;109:601–606.
15. Graveland G, Williams RS, DeFiglia M: A Golgi study of the human neostriatum: Neurons and afferent fibers. J Comp Neurol 1985;234:317–333.
16. Rothblat DS, Schneider JS: Responses of caudate neurons to stimulation of intrinsic and peripheral afferents in normal, symptomatic, and recovered MPTP-treated cats. J Neurosci 1993;13:4372–4378.

Dr. Minoru Kimura, Faculty of Health and Sport Sciences, Osaka University, Toyonaka, Osaka 560 (Japan)

The Role of the Descending Pallido-Reticular Pathway in Movement Disorders

F. Shima[a], S. Sakata[a], S.-J. Sun[a], M. Kato[a], M. Fukui[b], R.P. Iacono[c,1]

Departments of
[a] Clinical Neurophysiology and
[b] Neurosurgery, Neurological Institute, Kyushu University, Fukuoka, Japan, and
[c] Division of Neurosurgery, Loma Linda University Medical Center, Loma Linda, Calif., USA

Stereotactic neurosurgery traditionally dismissed the akinesia and postural problems of Parkinson's disease (PD) as unamenable to therapeutic lesioning of the brain [1]. Responsive symptoms, particularly tremor, attracted the efforts of stereotactic surgery, and indeed, surgical interventions in PD such as thalamotomy and classical anterodorsal pallidotomy aimed at eliminating tremor or rigidity [2–5]. Leksell's posteroventral pallidotomy (PVP) [6–8] demonstrates significant and encouraging therapeutic effects for the hypokinetic symptoms resistant to classical pallidotomy or thalamotomy. Moreover, PVP improves idiopathic dystonia which did not respond to previous ventrolateral nucleus (VL) thalamotomy [9]. These therapeutic effects suggest that the main mechanism of PVP does not participate in the pallido-thalamic circuit ordinarily considered as the anatomical substrate of involuntary movements [1].

Neuronal organization of the basal ganglia became evident over the past decade, due to significant advances in the anatomy, physiology and pharmacology of these structures [1, 10–15]. Most of the internal pallidal neurons in the monkey were described to project efferents to the thalamus and also to the midbrain pedunculopontine nucleus (PPN) [16–18], known to exert locomotor and postural

---

[1] We are grateful to Professor emeritus Hirotaro Narabayashi, Juntendo University Hospital, Tokyo, for his valuable suggestions and encouragement for this paper.

regulating functions [19–21] by connecting to the reticulospinal pathway [21, 22]. Microelectrode-recorded neuronal hyperactivity in the medial segment of the globus pallidus (P.m.) [23, 24] correlates with the proposed mechanism of akinesia, which explains the success of posterior pallidal lesioning [18, 25–27].

The purpose of this paper is to demonstrate a role for the descending pallidoreticular pathway in movement disorders by describing the characteristics of PVP and altered neuronal activities in the basal ganglia recorded during surgery.

Materials and Methods

Patients
PVP was carried out in 18 patients with PD (11 men and 7 women aged 44–74, mean = 61). The neurological findings included marked bradykinesia, freezing of gait and a defect of postural equilibrium associated with rigidity and tremor in 13 patients (bradykinetic type), similar gait and postural abnormalities with minimal rigidity and tremor in 4 (akinetic type), and severe flexed posture without gait disturbance in 1. There was moderate or severe hand tremor in 8 patients, dyskinesia of the extremities in 4 and intolerable pain in both legs in 2. When 'off', 13 patients demonstrated grade 4 Hoehn and Yahr and 4 patients grade 5. One patient with a marked flexed posture had grade 3. Symptoms were optimally controlled by medical treatment prior to surgery. Most patients had received L-threo-dops [28] (daily dose range 600–1,200 mg), but improvement only lasted for the first 1–2 years. Two patients had undergone unilateral ventral intermediate nucleus thalamotomy 3 and 5 years before pallidotomy, respectively.

PVP was also performed in 7 patients with idiopathic dystonia (6 men and 1 woman aged 12-71, mean = 40). The motor signs were generalized in 4 patients and restricted to the neck muscles in 2 and to the buccolingual region in 1. They were refractory to medical treatments or botulinum injection. Five patients had undergone VL thalamotomy (unilaterally in 2 and bilaterally in 3), which improved them only temporarily or not at all. The mean interval between thalamotomy and pallidotomy was 3.7 years (0.5–9 years).

Clinical Assessment
Clinical signs were assessed by neurological examination and surface electromyography, and were recorded with videotape. Parkinsonian akinesia in particular was evaluated by examining freezing of gait, arising from chair, turning body, postural stability and bradykinesia. The patients were followed up until 4–17 months after surgery.

Surgical Procedures
All patients received preoperative T1-weighted magnetic resonance imaging (MRI) with a 1.5-tesla system (Sigma Advantage, GE Medical system) providing slices of 5 mm in the midsagittal plane, oblique coronal plane through the bregma and posterior aspect of the mamillary body, and axial plane 4–6 mm below the intercommissural baseline (AC–PC line). The stereotactic target on MRI was located at the ventral border of the internal pallidum at the level of the posterior aspect of the mamillary body. The three-dimensional coordinates were calculated referring to the midpoint of the AC-PC line.

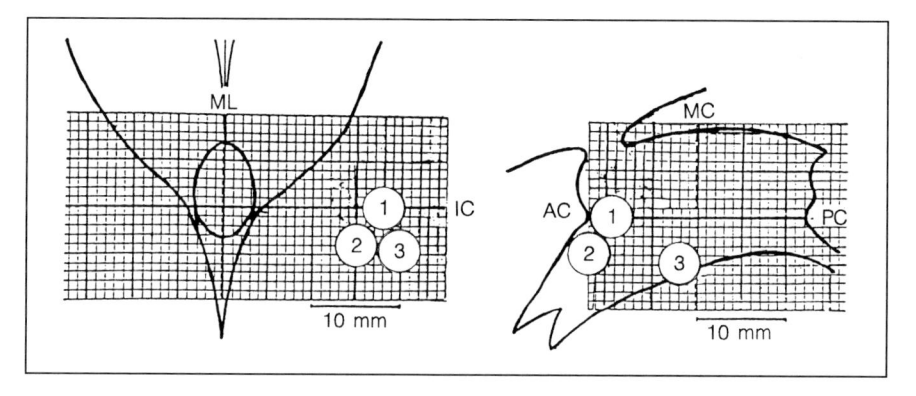

Fig. 1. Illustrations showing the targets of classical anterodorsal pallidotomy (1), Spiegel's pallidoansotomy (2) and Leksell's posteroventral pallidotomy in relation to the anterior commissure (AC), the posterior commissure (PC), the midcommissural point (MC) and the midline of the third ventricle (ML).

Surgery was performed on fully conscious patients under local anesthesia with 2% xylocaine used for bone fixation of a stereotactic frame (Tokairika, BSD-02) and a scalp incision. Through a burr-hole made at 20 mm lateral to the bregma, third ventriculography was performed by stereotactics using isotonic iohexol (2–3 ml). The stereotactic coordinates measured on MRI were transferred to the conventional ventriculograms. The pallidal target lay 1–2 mm rostral to the midcommissural point, 4–6 mm below the AC–PC line and 19–22 mm to the midline of the third ventricle (fig. 1). The final target was defined based on the following physiological findings.

Neuronal activity in the basal ganglia was recorded using a tungsten microelectrode, 100 nm in diameter, insulated except for the tapered tip, 10 nm in length, with an impedance of 1–2 MΩ. Except for the distal 15 mm, the microelectrode was enforced with stainless steel tubing insulated except for 1 mm at the distal end. An electric microdrive was used to introduce the electrode into the target structure for extracellular recording. The microelectrode signal was monitored by listening to an audio monitor and observing the spike train displayed on the oscilloscope screen. The neuronal activity was monitored for evidence of alteration of the firing pattern during voluntary or involuntary and passive movements. The microelectrode signal, electromyographic signal and descriptive voice channel were recorded on tape for postoperative off-line analysis. The activity of a single unit was identified by feeding the microelectrode signal into a window discriminator, of which pulses were then fed into a digital data analyzer (Mac Lab system) for sampling the firing rate of cellular activities. The same computer system was used to integrate the spontaneous background activities of the neurons in the basal ganglia.

Electrical stimulation of the target area was done by connecting the microelectrode to the negative output of an optically isolated stimulator and the tubing to the positive output. The parameters consisted of rectangular pulses cf 0.1 ms, 60 Hz and 1–5 mA.

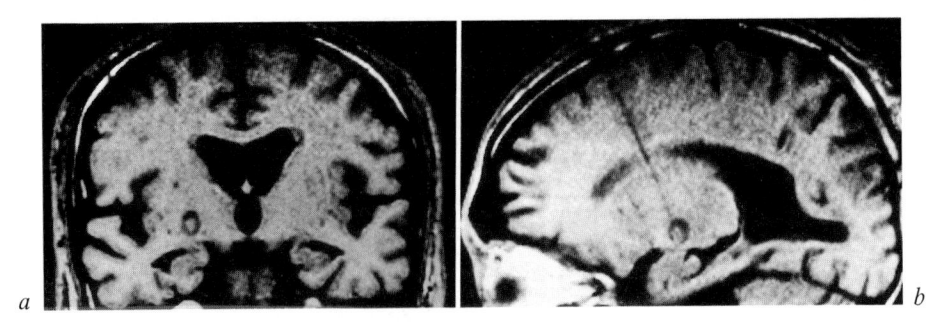

Fig. 2. Postoperative coronal (*a*) and sagittal (*b*) T1-weighted MRIs showing an electrolesion located in the right posteroventral pallidum (arrow) in a patient with PD. The MRIs were recorded 2 weeks after surgery.

Radiofrequency thermocoagulation was carried out using a 2-mm-thick electrode with a 3-mm uninsulated tip, introduced into the target point at which the microelectrodes signal rapidly ceased. An electrolesion was made at 70°C for 40 s so that the dorsoventral extent of the lesion was 5–6 mm and the transverse diameter 4–5 mm, which was verified by postoperative MRI (fig. 2).

Results

Clinical Findings
Parkinson's Disease. Seventeen patients underwent unilateral and 1 bilateral pallidotomy within 3 months. None of them suffered any visual field deficits or exacerbation of symptoms, including worsened bradykinesia or motor weakness. Three patients experienced transient exacerbation of low-voice speech and one of dysphasia; however, they fully recovered within a couple of months.

In all patients with bradykinetic type, PD, rigidity and disturbed reciprocal movements in the contralateral extremities were improved immediately after pallidotomy. Bradykinesia, freezing of gait and postural instability were markedly improved postoperatively. Tremor in the contralateral extremities was usually attenuated, but did not cease completely. Two patients underwent additional VOP thalamotomy simultaneously with pallidotomy to alleviate residual tremor. In some patients, tremor worsened temporarily and then subsided at 2–3 months after surgery. Dyskinesia completely resolved contralaterally. Pain in both legs occurring in 2 patients was relieved bilaterally. Recurrent akinetic symptoms after unilateral surgery in 1 patient were improved by a second pallidotomy performed contralaterally. This effect persisted without reduction for a long period of time.

In 4 patients with akinetic type, PD, gait freezing and postural instability were significantly improved at 1–3 months after surgery, however, all of these symptoms recurred.

Dystonia. All patients benefited from pallidotomy without any surgical complications. The motor signs almost completely resolved in 2 of 4 patients with generalized dystonia; however, in 1 of them there was a 50% recurrence of the symptoms. One patient had a moderate benefit from surgery and the other had minimal benefit. In a patient with hemidystonia, spasmodic torticollis or buccolingual dystonia, the symptoms ceased, and their daily life became normal. The surgical improvement of dystonia usually began 2–3 days after surgery and continued for 1–2 months.

Microrecording

The volume of the spontaneous electrode signal changed along the needle track corresponding to each structure of the basal ganglia (fig. 3a). The striatal roof was encountered at a considerable depth (about 30 mm above the target) and the lower border of the P.m. (target) was located easily by a sudden decrease of the neural background activity.

The spontaneous activity was moderate in the lateral segment of the globus pallidus (P.l.) and it increased in the P.m. Both segments of the globus pallidus were easily distinguishable by recording the silent zone located 8–9 mm above the target, corresponding to the medial pallidal lamina separating the two segments. In patients with bradykinetic type, PD, the background activity in the P.m. was significantly increased compared with that in patients with other disease types or dystonia (fig. 3). Most neurons in the P.m. showed tonic activity, exhibiting high discharge rates of 80–150 spikes per second. In the PD patients of the akinetic type, the background activity in both segments of the globus pallidus was significantly lower than that of patients with bradykinetic type PD, however, the discharge rate of single neurons was similarly high, especially in the P.m. (fig. 3b).

Background activity in dystonia patients was moderate in the striatum and globus pallidus (fig. 3c). Some pallidal neurons relating to involuntary movements were activated. Some pallidal neurons were potentiated similarly by voluntary movements of the different joints in different direction losing somatotopic selectivity, which was preserved in PD patients.

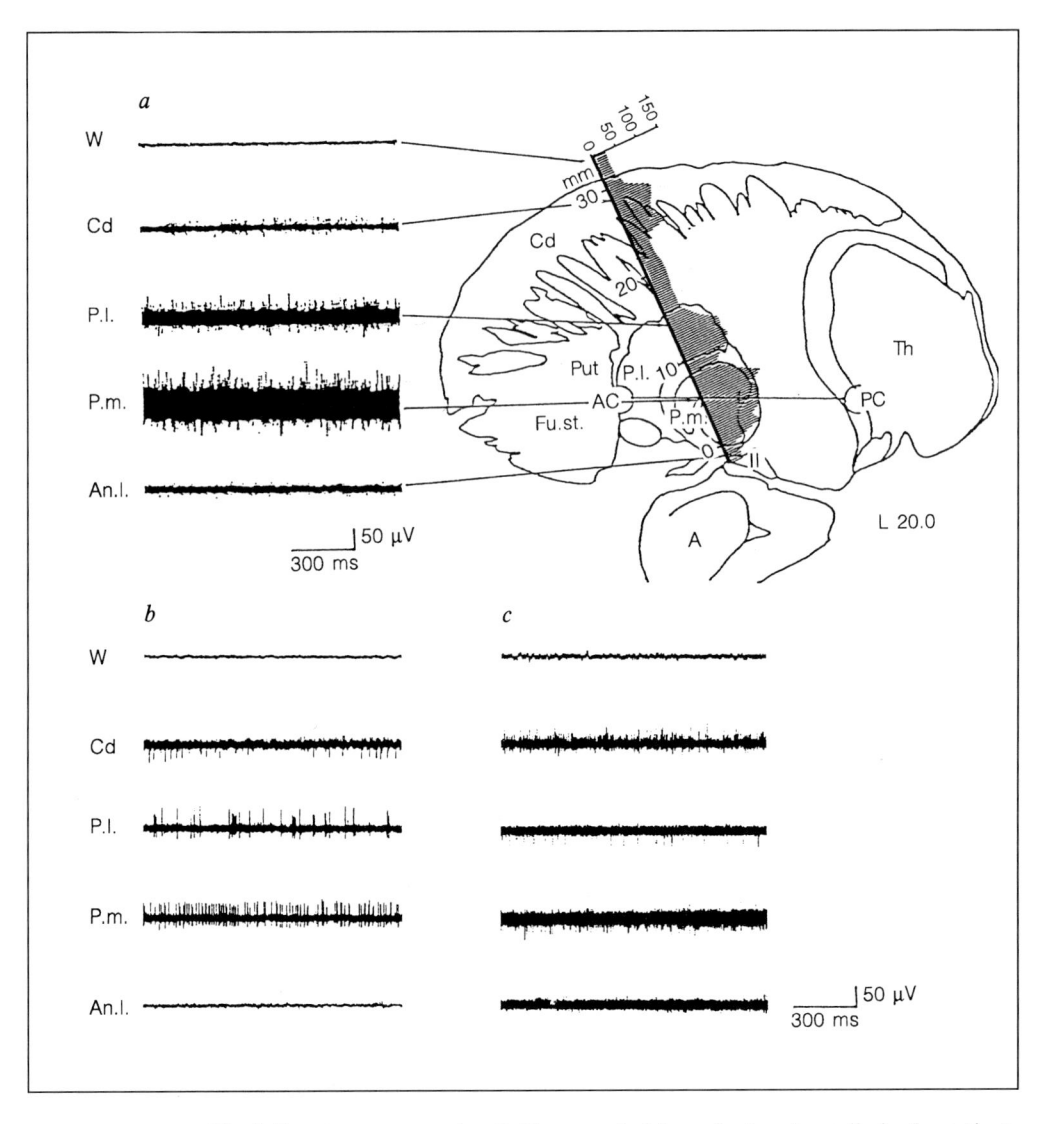

Fig. 3. Spontaneous neural activities recorded from the basal ganglia in the patients with Parkinson's disease of bradykinetic type (*a*), with Parkinson's disease of akinetic type (*b*) and with idiopathic dystonia (*c*). Illustrations show five different patterns of background activities recorded from respective subcortical structures in the course of lowering the electrode and in *a* the sequential alteration of the background activity along the needle track. The integrated values for 1 s are plotted along the needle track superimposed on the corresponding profile of the basal ganglia through the parasagittal plane 20 cm lateral to the midline. W = White matter; Cd = caudate nucleus; P.l. = lateral pallidum; P.m. = medial pallidum; An.l. = ansa lenticularis; Put = putamen; Fu.st. = fundus striati; A = amygdala; II = optic tract; Th = thalamus.

Table 1. Characteristics of stereotactic therapeutic effects on PD

	Thalamotomy (VOP or VIM)	Classical anterodorsal pallidotomy	Posteroventral pallidotomy
Tremor	+++	+	++*
Rigidity	+++	+++	+++
Akinesia	–	–	+++
Postural problems	–	–	+++
Dyskinesia	+++	+++	+++
Dysarthria	–	–	+*
Pain	–	–	+++

+++ = Markedly effective; ++ = moderately effective; + = slightly effective; – = not effective; * = delayed effect.

Discussion

The effects of PVP on PD are characterized by improvement of bilateral negative motor signs such as bradykinesia, gait freezing and postural problems, while incompletely reducing tremor (table 1). Although hypokinetic symptoms cannot be improved by VL or VIM thalamotomy, they can be exacerbated, especially by bilateral surgery. Furthermore, PVP eliminated dystonic symptoms never improved by previous VL thalamotomy [9]. Therapeutic lesions in the present study were localized in the posteroventral part of the P.m. and ansa lenticularis, and avoided the target area of the classical anterodorsal pallidotomy [3] or of Spiegel's pallidoansotomy [4, 5] (fig. 1), which was abandoned 30 years ago, because it failed to control both tremor and akinetic symptoms. These clinical findings suggest that the main mechanism of PVP does not involve the pallidothalamic pathway, which most investigators have considered as the main anatomical substrate of involuntary movements [1, 2, 6] (fig. 4).

According to Harnois et al. [16] and others [18], most P.m. neurons in the monkey project efferent fibers to the thalamus and to the PPN through two collaterals of a parent axon. The PPN reticular motor system is known to exert locomotion- and posture-regulating functions [19–21], being connected directly with the magnocellular reticular formation in the medulla oblongata that projects their fibers to the spinal motoneurons [22]. PVP, therefore, seems to participate primarily in the descending pallido-reticulo-spinal pathway, rather than the ascending pallido-thalamo-cortical circuit.

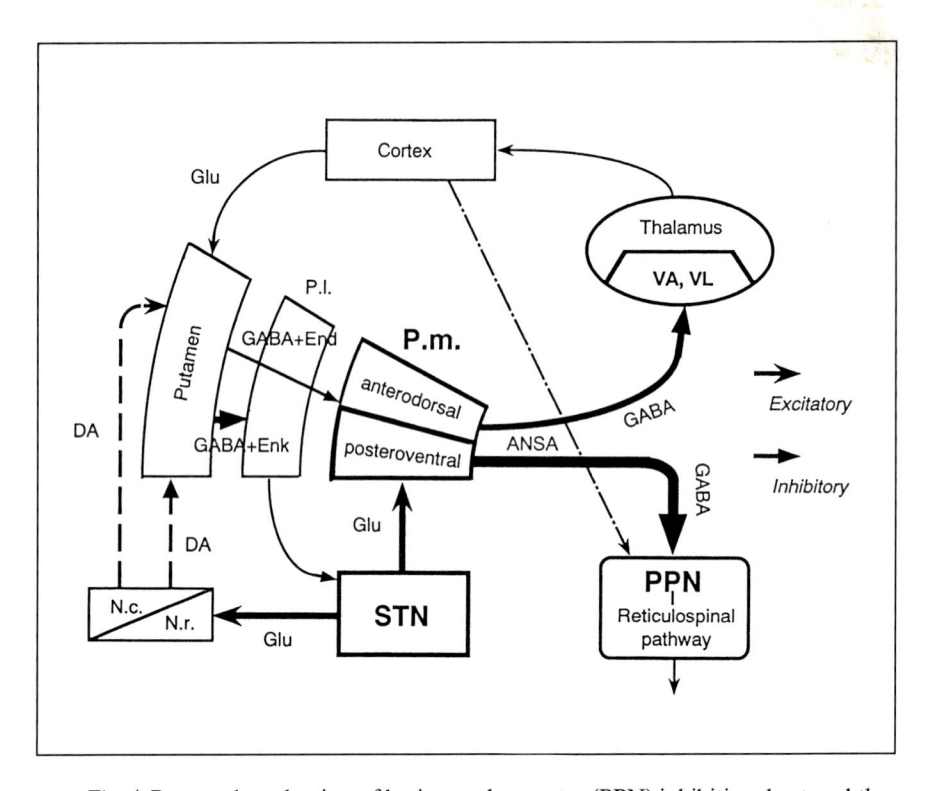

Fig. 4. Proposed mechanism of brainstem locomotor (PPN) inhibition due to subthalamic nucleus (STN) amplified pallidal outflow resulting in bradykinesia, gait freezing, and postural instability. Partial interruption of abnormal pallidal efferents by posteroventral pallidotomy (PVP) allows reversal of the brain stem inhibition, reversing akinetic symptoms. Concomitant effects on motor thalamus by PVP via the ansa lenticularis (ANSA) collaterals accounts for benefits of PVP on tremor and elimination of rigidity and, if present, dyskinesia.

The tonic overactivity of the P.m. in PD patients in the present study is compatible with the experimental findings in MPTP-induced disease in monkey models [11, 13, 29]. This is explained by the reciprocal alternation of the direct and indirect striopallidal pathway after the nigrostriatal dopaminergic deficiency [10, 13] (fig. 4). Decreased dopaminergic inhibition to the putamen allows excessive gabaergic inhibition of the P.l. within the indirect pathway, leading to disinhibition of the subthalamic nucleus, which in turn provides an excessive excitatory drive to the P.m. via glutaminergic outflow [30, 31]. This is reinforced by the reduced inhibitory input to the P.m. through the direct striopallidal pathway [10, 13]. Thus overactive gabaergic outflow of the P.m. excessively inhibits the thalamic nuclei and, via the pallidotegmental bundle, the PPN, underlying negative motor signs in PD.

The importance of the excessive inhibitory outflow through the pallidotegmental bundle of the ansa colaterals to the PPN in the mechanism of akinesia in PD [11, 26] has not been recognized. The dramatic therapeutic effects of PVP reveal the significance of pallidal inhibition of the mesencephalic locomotion reticulospinal pathway. This effect may differ from the reversal of akinesia by lesions of the subthalamic nucleus in animals which can provoke dyskinesia. The dual outflow from the pallidum to the PPN and the thalamus through the collaterals of the ansa resolves the apparent paradox of the concomitant reversal of akinesia with elimination of the dyskinesia [32, 33].

Hypokinetic symptoms consist of impairment in movement initiation and rhythm (akinesia), and reduction in the amplitude and velocity of voluntary movements (bradykinesia) [13]. We distinguished the akinetic type of PD from the bradykinetic type based on the predominant motor signs. Microelectrode recording in the akinetic patients revealed a low amplitude of the background activity in the medial pallidum, whose therapeutic lesion often failed to improve gait freezing and postural problems. These findings suggest that the neurons in the pallido-PPN pathway deteriorate partially in patients with akinetic type PD.

We reported previously that 27 out of 43 patients (63%) with various kinds of dystonia of idiopathic etiology benefit from stereotactic VL thalamotomy, whereas the others did not [9]. In the present study, most patients with dystonia that did not respond or responded only temporarily to the previous thalamotomy were improved by PVP. Good results of PVP were obtained for the dystonic syndromes, both focal and bilateral, especially in the case of proximal rather than distal symptoms, as compared with thalamotomy. The clinical results indicate that the mechanism of dystonia implicates both the pallidothalamic and the pallidoreticular pathways, and in individual patients they may act interdependently.

Some pallidal neurons in the dystonia patients indicated abnormal discharge relating to voluntary or involuntary movements. Single neuron analysis of the basal ganglia seems to be of great value to investigate pathophysiology of dystonia. In order to be able to interpret quantitative changes of the background activities in the basal ganglia, we need to collect further data in dystonia as well as in other movement disorders.

References

1 Page RD: The use of thalamotomy in the treatment of levodopa-induced dyskinesia. Acta Neurochir 1992;114:77–117.
2 Hassler R, Riechert T: Indikationen und Lokalisationsmethode der gezielten Hirnoperationen. Nervenarzt 1954;25:441–447.

3   Narabayashi H, Okuma T: Procaine-oil blocking of the globus pallidus for the treatment of rigidity and tremor of parkinsonism. Proc Jpn Acad 1953;29:134–137.
4   Spiegel EA, Wycis HT: Thalamotomy and pallidotomy for treatment of choreic movements. Acta Neurochir 1952;2:417–422.
5   Wycis HT, Spiegel EA: Ansotomy in paralysis agitans. Conf Neurol 1952;12:245–246.
6   Laitinen LV, Bergenheim AT, Hariz MI: Leksell's posteroventral pallidotomy in the treatment of Parkinson's disease. J Neurosurg 1992;76:53–61.
7   Laitinen LV, Bergenheim AT, Hariz MI: Ventroposterolateral pallidotomy can abolish all parkinsonian symptoms. Stereotact Funct Neurosurg 1992;58:14–21.
8   Svennilson E, Torvik A, Lowe R, et al: Treatment of parkinsonism by stereotactic thermolesions in the pallidal region. Acta Psychiatr Neurol Scand 1960;35:358–377.
9   Shima F: Stereotactic thalamotomy in treatment of idiopathic dystonia syndrome. Neurosurgeons 1993;12:154–160.
10   Alexander GE, Crutcher MD: Functional architecture of basal ganglia circuits. Trends Neurosci 1990;13:266–271.
11   Crossman AR, Clarke CE, Boyce S, Robertson RG, Sambrook MA: MPTP-induced parkinsonism in the monkey: Neurochemical pathology, complications of treatment and pathophysiological mechanisms. Can J Neurol Sci 1987;14:428–435.
12   DeLong MR: Activity of pallidal neurons during movement. J Neurophysiol 1971;34:414–427.
13   DeLong MR: Primate models of movement disorders of basal ganglia origin. Trends Neurosci 1990; 13:281–285.
14   Iansek R, Porter R: The monkey globus pallidus: Neuronal discharge properties in relation to movement. J Physiol 1980;301:439–455.
15   Robertson RG, Clarke CA, Boyce S, Sambrook MA, Crossman AR: The role of striatopallidal neurons utilizing gammaaminobutyric acid of the pathophysiology of MPTP-induced parkinsonism in the primate: Evidence from [^{3}H]flunitrazepam autoradiography. Brain Res 1990;531:95–104.
16   Harnois C, Filion M: Pallidal neurons branching to the thalamus and to the midbrain in the monkey. Brain Res 1980;186:222–225.
17   Hazrati LN, Parent A: Contralateral pallidothalamic and pallidotegmental projections in primates: An anterograde and retrograde labeling study. Brain Res 1991;567:212–223.
18   Percheron G, Filion M, Tremblay L, Fénelon G, François C, Yelnik J: The role of the medial pallidum in the pathophysiology of akinesia in primates. Adv Neurol 1993;60:84–87.
19   Garcia RE: The basal ganglia and the locomotor regions. Brain Res 1986;396:47–63.
20   Milner KL, Mogenson GJ: Electrical and chemical activation of the mesencephalic and subthalamic locomotor regions in the freely moving rats. Brain Res 1988;452:273–285.
21   Skinner RD, Kinjo N, Henderson V, Garcia RE: Locomotor projections from the pedunculopontine nucleus to the spinal cord. Neuroreport 1990;1:183–186.
22   Nakamura Y, Kudo M, Tokuno H: Monosynaptic projection from the pedunculopontine tegmental nuclear region to the reticulospinal neurons of the medulla oblongata. An electron microscope study in the cat. Brain Res 1990;524:353–356.
23   Shima F: Posteroventral pallidotomy for Parkinson's disease. Renewal of pallidotomy. Neurol Surg 1994;22:103–110.
24   Shima F, Kato M, Fukui M, Iacono PR: Posteroventral pallidotomy in treatment of Parkinson's disease. Funct Neurosurg 1992;31:51–58.
25   Jellinger K: The pedunculopontine nucleus in Parkinson's disease, progressive supranuclear palsy and Alzheimer's disease. J Neurol Neurosurg Psychiatry 1988;51:540–543.
26   Mitchell IJ, Clarke CE, Boyce S, Robertson RG, Peggs D, Sambrook MA, Crossman AR: Neural mechanisms underlying parkinsonian symptoms based upon regional uptake of 3-deoxyglucose in monkeys exposed to 1-methyl-4-phenyl-1,2,3,6-tetrahydropyridine. Neuroscience 1989;32:213–226.
27   Zweig RM, Jankel WR, Hedreen JC, Mayeux R, Price DL: The pedunculopontine nucleus in Parkinson's disease. Am Neurol Assoc 1989;26:41–46.
28   Narabayashi H, Kondo T, Yokochi F, Nagatsu T: Clinical effect of *L*-threo-3,4-dihydroxyphenylserine in cases of parkinsonism and pure akinesia. Adv Neurol 1986;45:593–602.

29 Miller WC, Delong MR: Altered tonic activity of neurons in the globus pallidus and subthalamic nucleus in the primate MPTP model of Parkinsonism; in Carpenter MB, Jayaraman A (eds): Advances in Behavioral Biology. New York, Plenum Press, 1987, pp 415–427.
30 Brotchie JM, Mitchell IJ, Sambrook AM, Crossman AR: Alleviation of parkinsonism by antagonism of excitatory amino acid transmission in the medial segment of the globus pallidus in rat and primate. Mov Disord 1991;2:133–138.
31 Hammond C, Rouzaire-Dubois B, Feger J, Jackson A, Crossman AR: Anatomical and electrophysiological studies on the reciprocal projections between the subthalamic nucleus and nucleus tegmenti pedunculopontinus in the rat. Neuroscience 1983;9:41–52.
32 Aziz TZ, Peggs D, Sambrook MA, Crossman AR: Lesion of the subthalamic nucleotomy alleviates parkinsonism in the 1-methyl-4-phenyl-1,2,3,6-tetrahydrophyridine (MPTP)-exposed. Br J Neurosurg 1991;6:575–582.
33 Bergman H, Wichmann T, Delong MR: Reversal of experimental parkinsonism by lesions of the subthalamic nucleus. Science 1990;249:1436–1438.

Fumio Shima, MD, PhD, Neurological Institute, Kyushu University, 3-1-1 Maidashi,
Higashi-ku, Fukuoka 812 (Japan)

Experiences from Human Stereotaxic Surgery

Hirotaro Narabayashi

Neurological Clinic, Tokyo, Japan

When medical treatment fails or proves unsatisfactory, stereotaxic surgery still remains as the most powerful tool for the treatment of patients with Parkinson's disease (PD) and other movement disorders originating from a disorder of the basal ganglia. Pharmacological therapy of PD, with *L*-dopa as its center, has made considerable progress in the last 20 years but, as is well recognized, there still exist many problems in its long-term clinical application. When rigidity and tremor do not satisfactorily respond to medicine or when dopa-induced dyskinesia (DID) limits the use of *L*-dopa, surgery often provides the most efficient solution, although it is obviously a symptomatic treatment [1]. Recently, it was also reported that the postural instability and difficulty of gait, such as festination or freezing of gait, seen in PD patients can be favorably influenced by surgery on the internal segment of the globus pallidus (GPi), which will be described later.

Rigidity and Tremor

Rigidity of the muscles of the extremities and trunk can totally be alleviated by producing a stereotaxic lesion either in the GPi or in the ventrolateral nucleus (VL) of the thalamus, where the pallidal efferent projections terminate, as previously described. On the contrary, tremor cannot be improved satisfactorily but only relatively or transiently by pallidal or VL surgery. Tremor in PD as well as other kinds of pathological tremor are totally alleviated by production of a lesion within the ventral intermediate nucleus (VIM), which lies in the ventral part of the thalamus bordering posteriorly to the VL and anteriorly to the ventral posterior nucleus (VP). The VIM receives no pallidal projection, but it receives projec-

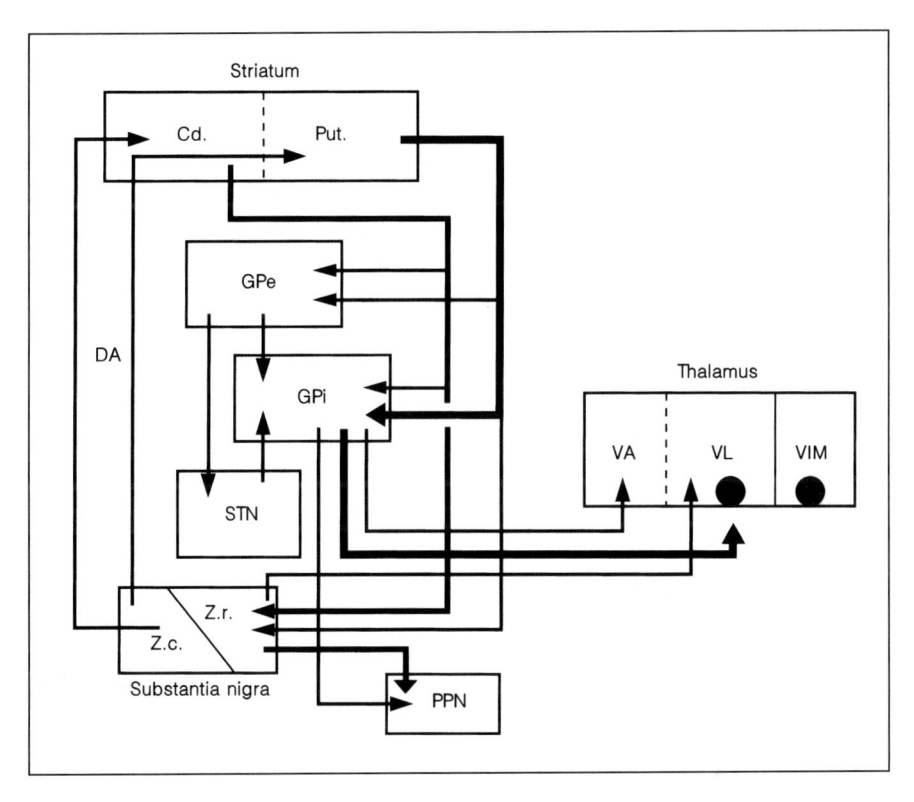

Fig. 1. Diagram of the outflow from the nigrostriatum proposed by the author. Cd. = Caudate nucleus; Put. = putamen; Gpe = external segment of globus pallidus; GPi = internal segment of globus pallidus; STN = subthalamic nucleus; Z.r. = reticular zone of the substantia nigra; Z.c. = compact zone of the substantia nigra; VA = ventral anterior nucleus; VL = ventrolateral nucleus; VIM = ventral intermediate nucleus; PPN = pedunculopontine nucleus.

tions from the deep cerebellar nuclei, the main vestibular nucleus, and also from the spinothalamic tract; the latter conducts the proprioceptive sensory input from the periphery. In cases with tremor, rhythmic neuronal discharges synchronous with peripheral tremor are recorded in the VIM due to this proprioceptive input from muscles and joints involved in tremor [2].

Localized high-frequency stimulation (40–60 Hz, 1 ms, 0.2–0.4 mA) of the GPi or of the VL thalamus often increases rigidity but influences tremor only secondarily. Tremor is more clearly and selectively suppressed by high-frequency stimulation of the VIM or by lesioning it. Therefore, two main positive motor symptoms, rigidity and tremor, are interpreted as different both from the anatomical and the physiological points of view (fig. 1).

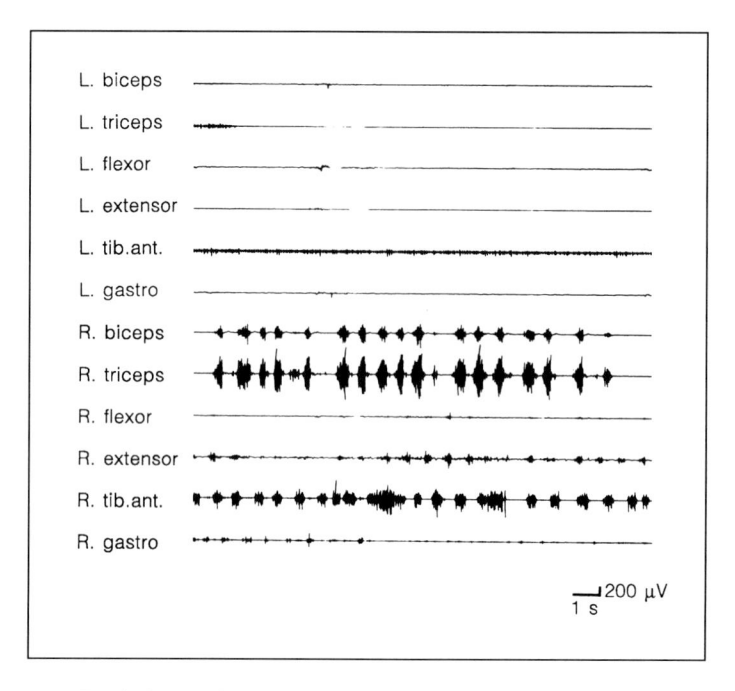

Fig. 2. Case K.S., 46 years old, female (after right thalamotomy). Dopa-induced dyskinesia does not appear on the operated side. The record was taken 1 week after right thalamotomy. Biceps = biceps brachii; triceps = triceps brachii; flexor = forearm flexor muscle; extensor = forearm extensor muscle; tib. ant. = tibialis anterior muscle; gastro. = gastrocnemius muscle.

Dopa-Induced Dyskinesia

Dopa-induced dyskinesia (DID) [3] is a frequently seen side effect of *L*-dopa treatment for PD, especially in cases of relatively younger onset. It is sometimes well influenced by coadministration of a dopamine (DA) agonist, but when it is of severe degree and not modified by such an agonist, it often prohibits the further increase of dose or even continuation of *L*-dopa treatment.

As already described, DID of the extremities can be totally abolished by lesioning the pallido-thalamic projection (fig. 2). In the thalamus, the VL lesion abolishes DID but lesioning the VIM does so only insufficiently. Therefore, the author concluded that both rigidity and DID are symptoms depending anatomically on the same pallido-thalamic projection but manifesting themselves differently depending on the pharmacological situation within the striatum. When DA is deficient, rigidity appears; and when *L*-dopa is overdosed or the striatal neurons are supersensitive, DID starts to appear.

Postural Instability and Difficulty of Gait such as Festination or Freezing of Gait

This is a relatively new topic for stereotaxic surgery. These symptoms are frequently seen in PD cases in the bilateral rigid state, and it has been demonstrated that either pallidotomy, Vo-thalamotomy or *L*-dopa can relieve these symptoms and ameliorate the rigidity. Therefore, these symptoms are interpreted as a difficulty secondary to the existence of muscle rigidity.

However, on the other hand, these difficulties in posturing and gait are also seen in cases of longstanding PD that have been under satisfactory *L*-dopa therapy. In these cases there is no sign of rigidity being relieved by long-term *L*-dopa therapy. Sometimes muscles are even a little hypotonic and a further increase in the *L*-dopa dose makes the difficulties and hypotonia worse. In view of the selective degeneration of the locus ceruleus and severely lowered dopamine-β-hydroxylase activity observed in elderly chronic parkinsonian cases, the author hypothesized that these difficulties in the chronic stage of PD without rigidity might be related to a norepinephrine (NE) deficiency. Administration of *L*-threo-DOPS, a synthetic precursor of NE, has been shown to improve these symptoms to various degrees; i.e., postural instability and the gait problem are alleviated in about half of the cases, as reported previously [4, 5].

However, the efficacy of this new compound on these problems tends gradually to decrease within several years, and afterwards there is no other treatment available.

As recently reported by Shima [6] and others and later confirmed by the author, stereotaxic posteroventral pallidotomy (PVP) is found to be effective on these difficulties in posturing and gait seen in the later period of the disease, for which thalamotomy is usually considered not to be indicated since there is no rigidity or tremor. PVP aims at intervening in the efferent pallidal outflow at the ansa lenticularis, which is located ventrally to the GPi. Also, even in the cases in which VL and VIM thalamotomy alleviated rigidity and tremor, these postural and gait disturbances are also observable during the long-term postoperative course, similar to the nonoperated longstanding cases.

These experiences may indicate that these symptoms or difficulties are not mediated by the pallidothalamic tract.

About the mechanism of the PVP effect, the still remaining efferent pallidal projection after thalamotomy is that going down to the pedunculopontine nucleus (PPN) in the upper pons. This projection down to the PPN is, therefore, assumed to be important in transmitting the neural impulses to control posture and gait. Strong inhibitory impulses from the pallidum to the PPN are suggested to cause disturbances of gait and posture; elimination of these impulses by PVP relieves such symptoms. Two cases of longstanding PD, under long-term *L*-dopa treatment, presented severe gait freezing and were well relieved by PVP. These cases will be reported in a forthcoming paper.

Lessons for Akinesia [7]

All these observations on the results of stereotaxic surgery provide a useful basis to answer the question: 'What is akinesia in PD?'

The term akinesia in PD is used relatively vaguely to describe a variety of difficulties of movement such as slowness and unskillfulness of movement, poverty or lack of movement and also difficulty in locomotion and gait. All these symptoms are observed mixed together in most of the cases of long-term illness, and therefore the mechanisms generating each of them are difficult to differentiate and to analyze separately. However, if we carefully observe the results of stereotaxic surgery on the different structures and the results of drug therapy, some understanding seems to be possible.

The author has classified akinesia into three subtypes depending on the responses to surgical and pharmacological treatment.

Only at the beginning or in the early period of the disease, difficulty of movement, such as slowness or unskillfulness of extremity movement, is interpreted as a secondary symptom caused by rigidity, because it is relieved immediately, and voluntary movement becomes normalized within 10 s with diminution of rigidity by surgery on the GPi or VL, or even by *L*-dopa (fig. 3). Therefore, this type of akinesia has been termed *secondary akinesia* by the author.

In contrast, *primary akinesia* is different from the former: it is not dependent on muscular rigidity and is not improved by surgical treatment of rigidity. The characteristics of such primary akinesia are considered to be poverty or lack of movement or loss of initiation of movement. It is alleviated by *L*-dopa.

The *third type of akinesia* is the difficulty of locomotion, i.e., disturbance of postural balance and difficulty of gait, especially freezing of gait. These difficulties are present throughout the course of the disease and have been attributed to rigidity. However, these difficulties are quite often observed in the later stage of the disease usually under sufficient *L*-dopa treatment, a stage in which rigidity has been relieved almost completely. Therefore, the difficulties are not related to rigidity and may be different from the two types of akinesia defined above.

As described in the previous chapter, two cases of PD under long-term *L*-dopa treatment, which presented severe gait freezing without rigidity, were favorably influenced, though moderately and transiently, by *L*-threo-DOPS and finally markedly relieved by PVP. This experience again shows that GPi surgery favors improvement in frozen gait even in the later stage of the illness and demonstrates its usefulness in alleviating rigidity and secondary akinesia in the earlier stage.

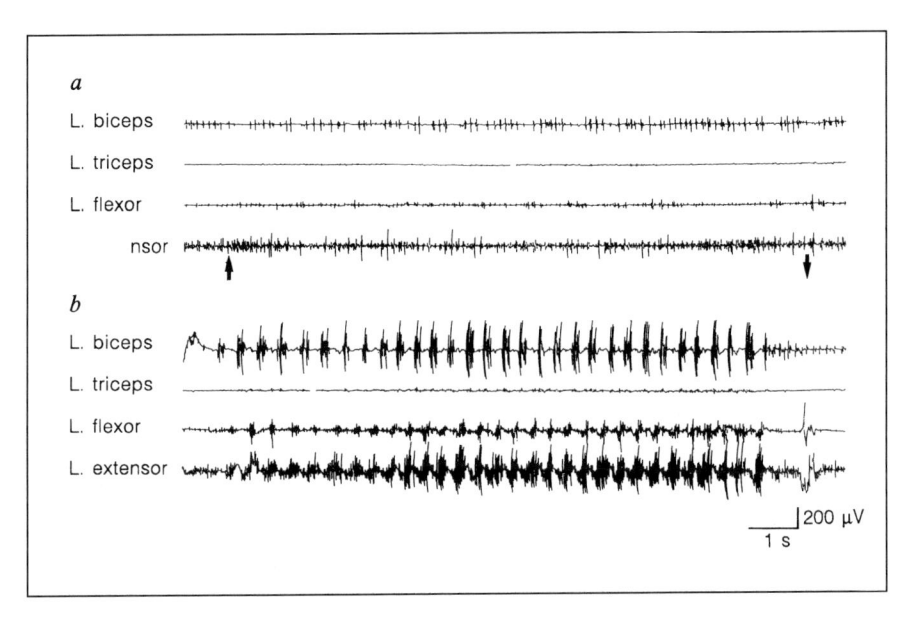

Fig. 3. Case T.K., 50 years old, male (left hemiparkinsonism, rigid type). Recovery of quick pronation and supination movement after alleviation of rigidity by right thalamotomy in a rigid parkinsonian patient. The movement was impossible to perform before surgery. Abbreviations as in figure 2. *a* Before surgery (pronation and supination movement). *b* After right VIM-VL thalamotomy.

Conclusion

The above-described observations and analysis indicate that the precise stereotaxic surgery using the microelectrode technique provides useful information for a better understanding of the neural mechanism responsible for generation of symptoms of PD. The neural network starting from the striatum operates via the GPi to produce a spectrum of parkinsonian symptoms, such as rigidity, secondary akinesia and postural and gait difficulties. GPi is the key structure. Its projection has two main terminations, the thalamus and the PPN. The data also suggest that the GPi may also be under the strong inhibitory influence of NE, which is a new challenging topic for future research. The axial symptom of PD, akinesia, can gradually be analyzed and explained through this therapeutic approach of stereotaxic surgery.

The generating mechanism of tremor is presumed to be relatively different from other parkinsonian symptoms and may relate to the cerebellar circuitry and the VIM, although it may also be indirectly linked to DA metabolism.

References

1 Narabayashi H: Surgical treatment in the levodopa era; in Stern G (ed): Parkinson's Disease. London, Chapman & Hall, 1990, pp 597–646.
2 Narabayashi H: Tremor: Its generating mechanism and treatment; in Vinken PJ, Bruyn GW, Klawans HL (eds): Handbook of Clinical Neurology. Amsterdam, Elsevier, 1986, vol 5(49), pp 597–607.
3 Narabayashi H: Levodopa-induced dyskinesia and thalamotomy. J Neurol Neurosurg Psychiatry 1984;47:831–839.
4 Narabayashi H, Kondo T, Hayashi A, Suzuki T, Nagatsu T: L-threo-3,4-dihydroxyphenylserine treatment for akinesia and freezing of parkinsonism. Proc Jpn Acad 1981;57(B):351–354.
5 Narabayashi H, Kondo T, Yokochi F, Nagatsu T: Clinical effects of L-threo-3,4-dihydroxyphenylserine in cases of parkinsonism and pure akinesia. Adv Neurol 1986;45:593–602.
6 Shima H: Posteroventral pallidotomy for Parkinson's disease: Renewal of pallidotomy. Neurol Surg 1994;22:103–110.
7 Narabayashi H: Three types of akinesia in the progressive course of Parkinson's disease. Adv Neurol 1993;60:18–24.

H. Narabayashi, MD, Neurological Clinic, 5-12-8 Nakameguro, Meguro-ku, Tokyo 153 (Japan)

Monoamine Neurons:
Gene and Gender Differentiation

Catecholaminergic Systems and the Sexual Differentiation of the Brain

Cellular Mechanisms and Clinical Implications[1]

Ingrid Reisert, Christof Pilgrim

Anatomie und Zellbiologie, Universität Ulm, BRD

Throughout the animal kingdom, nervous systems exhibit sexually dimorphic[2] traits. This concerns predominantly, but not exclusively, neural structures involved in the control of reproduction [1–3]. Classical examples of sexually dimorphic brain functions are reproductive behavior, i.e. courtship, copulation, and parental care, and the neuroendocrine feedback loops regulating secretion of adenohypophyseal hormones [4–6]. The concept underlying research on gender-specific differentiation of the brain is that these sexually dimorphic patterns of behavior and neuroendocrine control are based on gender-specific brain circuitry which, in turn, is thought to be represented by sex differences in neuronal morphology and function. Catecholamine (CA) systems are of particular interest in this regard because they are known to influence sexual behavior as well as pituitary functions [for reviews, see ref. 7–10]. There are numerous reports on sex differences in morphology and function of dopamine (DA) and noradrenaline (NA) systems of the brain. Most of these results have been acquired in the rat and concern distribution of transmitter-specific nerve cells and fibers, levels and turnover of CA, activities of synthesizing and metabolizing enzymes, receptors as well as CA-related behaviors. For a survey of sexual dimorphisms of adult CA systems, the reader is referred to recent review articles [11, 12].

Current dogma holds that sexual dimorphisms in the vertebrate brain are generated by the epigenetic action of gonadal hormones [13–15]. The classical

---

[1] Work conducted in the authors' laboratory was supported by the Deutsche Forschungsgemeinschaft.

[2] The term 'sexual dimorphism' refers to the existence of two distinct forms of behaviors, brain structures, and properties of nerve cells.

'organizational' hypothesis is that androgens organize male-type brain circuitry irrespective of the genetic sex. Androgen, after entering the brain, is aromatized to 17β-estradiol, the steroid thought to be responsible for the establishment of a male brain [3]. The organizing effect is irreversible and occurs during a 'critical period' or time-limited 'window' in development. Gonadal steroids secreted in adulthood are required to 'activate' the sex-specific circuitry [13]. Essentially two arguments speak for the assumption that the above concept may also apply to the sexual differentiation of CA systems. Experimental evidence has been reviewed in detail elsewhere [12]. First, sex differences in organization of CA systems and in functional indices of CA transmission begin to appear during the perinatal critical period. Second, experimental manipulation of the hormonal environment, such as gonadectomy and/or administration of sex steroids, during development leads to persisting changes in the sex-specific properties of CA neurons. Androgen or estrogen treatment of females during the critical period may result in masculinization of CA neuron numbers, CA levels or metabolism as well as receptor densities. Conversely, gonadectomy of male pups may prevent masculinization of CA neuron numbers as well as DA levels or metabolism. Not all experimental results seem to fit the androgen hypothesis: sex differences in NA levels, e.g., do not always change after gonadectomy or androgen treatment or may even be augmented by gonadectomy of female pups. Such exceptions to the rule may be taken as hints that the view of the process of sexual differentiation outlined above is too simplistic and that other factors and mechanisms must be invoked to fully explain this phenomenon.

In order to distinguish effects of gonadal hormones from other sex-linked factors, a new experimental approach was adopted by our group. Instead of raising cultures from pooled male and female rat embryos, we prepared gender-specific dissociated cell cultures after having separated the genders by inspection of the gonads. The cells were obtained from the main subdivisions of the embryonic day 14 rat brain stem. These cultures contain considerable numbers of tyrosine hydroxylase-immunoreactive (TH-IR) neurons which develop DA or NA properties according to the location in the brain stem from which they were taken. Surprisingly, region-specific morphological and functional sex differences of CA neurons were observed in control cultures raised in the absence of sex steroids [reviewed in ref. 11,12]. Female mesencephalic cultures were found to contain about 25% higher numbers of DA cells than male cultures. This was different with diencephalic DA neurons and rhombencephalic NA neurons where no statistically significant sex differences were observed. Morphometric measurements of soma sizes yielded a hormone-independent sex difference in diencephalic but not in mesencephalic cultures. Male DA neurons were about 30% larger than female neurons. A conclusive functional parameter to assess the maturation of neurons is their capacity for transmitter (re)uptake. Uptake of [^{3}H]DA was twice as high in female than in male diencephalon. The reverse was true for the mesencephalon where male cells took up 1.5 times more [^{3}H]DA than female cells. A sex difference in uptake of [^{3}H]NA was seen in rhombencephalic cultures cultivated for 6 days in vitro (DIV) but not yet in 3-DIV cultures. Investigations of developmental profiles of endogenous DA and metabolites, too, yielded hormone-independent sex differences. Higher levels of DA were measured in female than in male control cultures of both diencephalon and mesencephalon. However, when DA levels were related to numbers of TH-IR cells, the sex differences disappeared in mesencephalic but not in diencephalic cultures. Dihydroxyphenylacetic acid (DOPAC) levels and vesicular storage capacity did not exhibit sex differences. Sex differences in NA levels per TH-IR neuron were present in 3-DIV but no longer in 6-DIV rhombencephalic cultures. TH activity was assessed by measuring accumulation of dihydroxyphenylalanine (DOPA) in the presence of an inhibitor of aromatic amino acid decarboxylase. Diencephalic cultures raised from females produced markedly more DOPA than those raised from males. A temporary sex difference in DOPA synthesis (per CA neuron) was also seen in rhombencephalic cultures but not in mesencephalic cultures.

The salient point of the in vitro investigations on CA neurons developing in gender-specific cultures of embryonic rat brain is that sexual dimorphisms of CA neurons may develop independently of the action of gonadal steroids. This is in contradiction to the generally accepted theory which holds that sexual differentiation of the brain is caused by the organizational effect of androgen or its metabolite 17β-estradiol. The brain tissue used to raise the above cultures was removed at embryonic day 14, i.e., well before sex differences in hormonal environment of the rat embryo are known to develop [16, 17]. In order to cope with the remote possibility that sex steroid-dependent determinative events occur in utero before the brain tissue was taken into culture, additional cultures were prepared from embryos whose mothers had been treated with the estrogen antagonist tamoxifen or the androgen antagonist cyproterone acetate. In spite of this pretreatment, DOPA synthesis rates were again higher in diencephalic cultures raised from females than from males [18]. In conclusion, epigenetic control by the hormonal environment cannot be the only mechanism responsible for the generation of the sexual dimorphisms described above.

Although there is ample evidence to show that sex steroids influence cell death, neuronal growth, and synaptogenesis, data about sex differences in vertebrate brains are not consistent enough to support the notion that all sex differences depend on a differentiating hormone secreted by the gonads of the heterogametic sex. It has therefore been advocated that the role of gonadal steroids in brain differentiation be redefined [19]. First, the mircoenvironment of the developing neuron is not only determined by circulating hormones but also by cell-to-cell interactions. Steroid effects may be transmitted transsynaptically from steroid-sensitive to steroid-insensitive neurons [20]. This could take place both anterogradely and retrogradely. Steroids may exert these indirect effects by altering firing rates and/or neurotransmitter release of afferent neurons or by modulating synthesis and release of neurotrophic molecules from neurons or glia [15, 21]. A variety of environmental cues will eventually converge to direct the differentiation of a steroid-sensitive neuron. Intracellular signal transduction pathways activated by neurotransmitters or growth factors can interfere with sex steroid receptors and, through regulation of the steroid sensitivity of a neuron, create individual time windows for effects of sex steroids. It is, for instance, possible that a membrane-bound dopamine receptor activates an intracellular steroid receptor through a kind of 'cross-talk' involving adenylate cyclase and changes in phosphorylation of the steroid receptor [22]. Thus, signal transduction from a membrane-bound receptor could influence the binding of a steroid receptor to a hormone-responsive element in promoter regions of neuronal genes. In principle, such cross-coupling mechanisms can result in synergistic transcriptional activation as well as in down-regulation of the activity of a gene [23, 24]. Second, the possibility of hormone-independent realization of a sex-specific genetic program needs to be considered. The above considerations proceed from the assumption that gender-specific differentiation of the brain is an epigenetic phenomenon brought about by cell-extrinsic cues in the environment of the developing neuron. However, our results obtained with sex-specific dissociated cell cultures of embryonic rat brain have shown that CA neurons develop morphological and functional sex differences in the absence of sex steroids [11]. This suggests cell-intrinsic realization of genetic sex, which does indeed occur in insect neurons [25, 26]. In mammalian development, there is circumstantial evidence from behavioral research for direct, steroid-independent effects of the Y chromosome on neural development [27]. It remains to be seen whether identified genes located in the testis-determining region of the Y chromosome are active in developing neurons and are capable of influencing their differentiation.

Morphological, endocrinological and psychophysical data indicate that also the human brain is sexually dimorph. The mean weight of the brain, both absolutely and in relation to body weight, is higher in men than in women [28]. More specifically, morphological and functional data point to a sex-specific circuitry not only in the human hypothalamus but also in other brain regions. Similar to other mammals, neuroendocrine regulation of the pituitary is sex specific.

Gonadotropin-releasing hormone secretion is cyclic in the reproductive female but tonic in the male [29]. The response of growth hormone (GH) to GH-releasing hormone is more pronounced in females than in males [30] and it is the male whose GH secretion can be stimulated by cholinergic drugs [31]. This functional dimorphisms are paralleled by morphological observations. The interstitial nuclei 2 and 3 of the anterior hypothalamus (INAH) are twice as large in males as

Table 1. Sex-specific prevalence of disorders with a putative neurodevelopmental etiology

Disease	Ratio men:women	References
Attention-deficit hyperactivity disorder	8:1	60
Minimal brain dysfunction	4–8:1	61
Tourette syndrome	3–9:1	62
Stuttering	2–10:1	63
Autism	2.5–9:1	64, 65
Dyslexia	3.3:1	66
Severe mental retardation	1.9:1	67
Febrile convulsions	1.7:1	68

in females [32, 33]. Sex differences in size and cell numbers have been described in another nucleus of the human hypothalamus [34]. This nucleus was named 'sexually dimorphic nucleus' but seems to be identical to INAH 1 [33]. Aside from the hypothalamus, the human limbic system [35] as well as the neocortex contain sexually dimorphic components. The orbital area (A11 of Brodman) is characterized by a higher neuronal density in females than in males [36]. A left-right asymmetry in the shape of the planum temporale, the posterior region of the superior temporal gyrus, has been described, that shows up rarely in females but frequently in males [37]. Sex differences in functional asymmetry of the cortex [reviewed in detail in ref. 38] are believed to be the substrate for sex differences in certain behaviors, such as cognitive abilities. Boys and men outperform girls and women in mathematical reasoning and spatial processing [39–41]. Specialization of the right hemisphere for spatial processing develops earlier in boys [42]. Women are superior in their ability to discriminate pain [43]. Sex differences in cortical asymmetries might be related to sexual dimorphisms of commissures, i.e. anterior commissure [44] and corpus callosum [reviewed in ref. 45]. It may be of particular interest that the isthmus of the corpus callosum, which connects the temporoparietal cortical regions, is sexually dimorph. A correlation between hand preference and isthmal size is found in men but not in women [38].

Clinical observations indicate that sexual differentiation of the human brain is controlled by mechanisms similar to those found in other mammalian species. In particular, the developing human brain is sensitive to sex steroids and related drugs. Prenatal treatment with progestins, which mimick androgen effects, enhances masculine behavior, such as physical aggression, both in men and women [46]. Fetal exposure to diethylstilbestrol increases the rate of depression and anxiety in both sexes [47] and reverses sexual orientation in women [48]. Masculine behavior develops in females suffering from an overproduction of androgens in the course of congenital adrenal hyperplasia although female gender identity may be restored if the subjects undergo early treatment of the primary disease [49]. The importance of testosterone for the development of male gender identity can also be derived from observations on XY pseudohermaphrodites with primary 5α-reductase deficits or with an androgen insensitivity syndrome (testicular feminization) [50].

That sexual differentiation of the brain is controlled by additional mechanisms, not only in experimental rodents (cf. above) but also in man, is suggested by certain diseases likely to have a neurodevelopmental etiology. Table 1 lists a number of such disease that occur more often in the developing male than in the female. It is conceivable that the prevalence of these disorders in male children is caused by a male-specific vulnerability to temporary imbalances of neuroactive substances (e.g. neurotransmitters and hormones) or to the presence of infectious agents or environmental toxins in the environment of the brain. This would then result in a male-specific predisposition towards manifestation of the disease. On the other hand, there are neuropsychiatric disorders with a higher female prevalence. Chorea minor, tardive dyskinesia, migraine and depression are 2–3 times more frequent in women than in men [51–54]. Anorexia nervosa is found almost exclusively in young women [55]. Obviously, diseases with a male prevalence have an earlier onset than those with a female prevalence. The difference between the sex ratios of early and late onset diseases may relate to the concept that distinguishes 'organizational' and 'activational' effects of steroids on the nervous system (cf. above). There is ample evidence that estrogen has pronounced effects not only on developing but also on adult neurons, e.g. on transmitter metabolism and synaptic communication [56–58]. Female-specific disorders may thus be viewed as consequences of activational rather than organizational effects of estrogen.

Many of the above diseases can be related to defective CA transmission. Since CA neurons develop morphological and functional sexual dimorphisms under in vitro conditions, i.e. independently from the neural input from other brain regions, it is conceivable that CA neurons are primary targets of events that lead to a sexually differentiated brain and that they mediate sex-specific development of other neurons. By the same token, CA neurons could be responsible for the sex-specific prevalence of the disorders discussed above. This may hold for all stages of ontogenesis including degeneration and aging. The fact that Parkinson's disease is observed more often in men than in women [51, 59] suggests a male-specific component in the presumably multifactorial etiology of this neurodegenerative disorder. It remains to be seen whether male and female CA neurons do indeed differ in vulnerability to adverse conditions in their environment.

References

1. Beatty WW: Gonadal hormones and sex differences in nonreproductive behaviors in rodents: Organizational and activational influences. Horm Behav 1979;12:112–163.
2. Kelley DB: Sexually dimorphic behaviors. Annu Rev Neurosci 1988;11:225–251.
3. Hutchison JB: Hormonal control of behaviour: Steroid action in the brain. Curr Opin Neurobiol 1991;1:562–570.
4. Dyer RG: Sexual differentiation of the forebrain – Relationship to gonadotrophin secretion. Prog Brain Res 1984;61:223–236.
5. Jansson J-O, Edén S, Isaksson O: Sexual dimorphism in the control of growth hormone secretion. Endocr Rev 1985;6:128–150.
6. Arey BJ, Averill RLW, Freeman ME: A sex-specific endogenous stimulatory rhythm regulating prolactin secretion. Endocrinology 1989;124:119–123.
7. Numan M: Maternal behavior; in Knobil E, Neill J (eds): The physiology of reproduction. New York, Raven Press, 1988, vol 2, pp 1569–1645.
8. Everitt BJ: Cerebral monoamines and sexual behavior; in Money J, Musaph H (eds): Handbook of Sexology. Amsterdam, Elsevier/North Holland, 1977, pp 429–448.
9. Barclay SR, Cheng M-F: Role of catecholamines in the courtship behavior of male ring doves. Pharmacol Biochem Behav 1992;41:739–747.
10. Kordon C: Hypothalamo-hypophyseal mechanisms involved in the regulation of hormones and behaviour; in Briley M, Fillion G (eds): New Concepts in Depression. Basingstoke, Macmillan Press, 1988, pp 235–246.
11. Reisert I, Pilgrim C: Sexual differentiation of monoaminergic neurons – Genetic or epigenetic? Trends Neurosci 1991;14:468–473.
12. Reisert I, Küppers E, Pilgrim C: Sexual differentiation of central catecholamine systems; in Smeets WJAJ, Reiner T (eds): Phylogeny and development of catecholamine systems. Cambridge, Cambridge University Press, in press.
13. Goy RW, McEwen BS: Sexual differentiation of the brain. Cambridge, MIT Press, 1980.
14. Arnold AP, Gorski RA: Gonadal steroid induction of structural sex differences in the central nervous system. Annu Rev Neurosci 1984;7:413–442.
15. Toran-Allerand CD: On the genesis of sexual differentiation of the central nervous system: Morphogenetic consequences of steroidal exposure and possible role of α-fetoprotein. Prog Brain Res 1984; 61:63–97.
16. Weisz J, Ward IL: Plasma testosterone and progesterone titers of pregnant rats, their male and female fetuses, and neonatal offspring. Endocrinology 1980;106:306–316.
17. Baum MJ, Woutersen PJA, Slob AK: Sex difference in whole-body androgen content in rats on fetal days 18 and 19 without evidence that androgen passes from males to females. Biol Reprod 1991; 44:747–751.
18. Beyer C, Eusterschulte B, Pilgrim C, Reisert I: Sex steroids do not alter sex differences in tyrosine hydroxylase activity of dopaminergic neurons in vitro. Cell Tissue Res 1992;270:547–552.
19. Pilgrim C, Hutchison JB: Developmental regulation of sex differences in the brain: Can the role of gonadal steroids be re-defined? Neuroscience 1994;60:843–855.
20. Beyer C, Feder HH: Sex steroids and afferent input: Their roles in brain sexual differentiation. Annu Rev Physiol 1987;49:349–364.
21. Breedlove SM: Sexual dimorphism in the vertebrate nervous system. J Neurosci 1992;12:4133–4142.
22. Power RF, Lydon JP, Conneely OM, O'Malley BW: Dopamine activation of an orphan of the steroid receptor superfamily. Science 1991;252:1546–1548.
23. Wahli W, Martinez E: Superfamily of steroid nuclear receptors: Positive and negative regulators of gene expression. FASEB J 1991;5:2243–2249.
24. Schüle R, Evans RM: Cross-coupling of signal transduction pathways: Zinc finger meets leucine zipper. Trends Genet 1991;7:377–381.
25. Baker BS: Sex in flies: the splice of life. Nature 1989;340:521–524.
26. Taylor BJ, Truman JW: Commitment of abdominal neuroblasts in Drosophila to a male or female fate is dependent on genes of the sex-determining hierarchy. Development 1992;114:625–642.

27 Maxson SC: Potential genetic models of aggression and violence in males; in Driscoll P (ed): Genetically defined animal models of neurobehavioral dysfunctions. Basel, Birkhäuser, 1992, pp 174–188.
28 Swaab DF, Hofman MA: Sexual differentiation of the human brain. A historical perspective. Prog Brain Res 1984;61:361–374.
29 Grumbach MM: The neuroendocrinology of puberty; in Krieger DT, Hughes JC (eds): Neuroendocrinology. Sunderland, Sinauer, 1980, pp 249–258.
30 Benito P, Avila L, Corpas MS, Jiménez JA, Cacicedo L, Franco FS: Sex differences in growth hormone response to growth hormone-releasing hormone. J Endocrinol Invest 1991;14:265–268.
31 Barbarino A, Corsello SM, Tofani A, Sciuto R, Della Casa S, Rota CA, Barini A: Sexual dimorphism of pyridostigmine potentiation of growth hormone (GH)-releasing hormone-induced GH release in humans. J Clin Endocrinol Metab 1991;73:75–78.
32 LeVay S: A difference in hypothalamic structure between heterosexual and homosexual men. Science 1991;253:1034–1037.
33 Allen LS, Hines M, Shryne JE, Gorski RA: Two sexually dimorphic cell groups in the human brain. J Neurosci 1989;9:497–506.
34 Swaab DF, Gooren LJG, Hofman MA: The human hypothalamus in relation to gender and sexual orientation. Prog Brain Res 1992;93:205–219.
35 Allen LS, Gorski RA: Sex difference in the bed nucleus of the stria terminalis of the human brain. J Comp Neurol 1990;302:697–706.
36 Haug H: Macroscopic and microscopic morphometry of the human brain and cortex. A survey in the light of new results. Brain Pathol 1984;1:123–149.
37 Wada JA, Clarke R, Hamm A: Cerebral hemispheric asymmetry in humans. Arch Neurol 1975;32:239–246.
38 Witelson SF: Neural sexual mosaicism: Sexual differentiation of the human temporo-parietal region for functional asymmetry. Psychoneuroendocrinology 1991;16:131–153.
39 Benbow CP: Sex differences in mathematical reasoning ability in intellectually talented preadolescents: Their nature, effects, and possible causes. Behav Brain Sci 1988;11:169–232.
40 Kerns KA, Berenbaum SA: Sex differences in spatial ability in children. Behav Genet 1991;21:383–396.
41 Nordberg L, Rydelius P-A, Zetterström R: Psychomotor and mental development from birth to age of four years: Sex differences and their relation to home environment. Acta Paediatr Scand Suppl 1991;80(suppl 378):1–25-.
42 Witelson SF: Sex and the single hemisphere: Specialisation of the right hemisphere for spatial processing. Science 1976;193:425–426.
43 Feine JS, Bushnell MC, Miron D, Duncan GH: Sex differences in the perception of noxious heat stimuli. Pain 1991;44:255–262.
44 Allen LS, Gorski RA: Sexual orientation and the size of the anterior commissure in the human brain. Proc Natl Acad Sci USA 1992;89:7199–7202.
45 Witelson SF: Cognitive neuroanatomy: A new era. Neurology 1992;42:709–713.
46 Reinisch JM, Sanders SA: Prenatal gonadal steroidal influences on gender-related behavior. Prog Brain Res 1984;61:407–416.
47 Meyer-Bahlburg HFL, Ehrhardt AA: A prenatal-hormone hypothesis for depression in adults with a history of fetal DES exposure; in Halbreich U (ed): Hormones and Depression. New York, Raven Press, 1987, pp 325–338.
48 Ehrhardt AA, Meyer-Bahlburg HFL, Rosen LR, Feldman JF, Veridiano NP, Zimmerman I, McEwen BS: Sexual orientation after prenatal exposure to exogenous estrogen. Arch Sex Behav 1985;14:57–77.
49 Kimura D: Are men's and women's brains really different? Can Psychol 1987;28:133–147.
50 Imperato-McGinley J, Peterson R, Gautier T, Sturla E: The impact of androgens on the evolution of male gender identity; in Kogan SJ, Hafez ES (eds): Pediatric Andrology. Hingham, Kluwer, 1981, pp 125–140.
51 Scheid W: Lehrbuch der Neurologie, ed 5. Stuttgart, Thieme, 1983, pp 662–685.
52 Kane JM, Smith JM: Tardive dyskinesia. Arch Gen Psychiatry 1982;39:473–481.

53 Ziegler DK: Epidemiology of migraine; in Rose FC (ed): Handbook of Clinical Neurology: Headache. Amsterdam, Elsevier, 1986, pp 13–22.
54 Young MA, Fogg LF, Scheftner WA, Keller MB, Fawcett JA: Sex differences in the lifetime prevalence of depression: Does varying the diagnostic criteria reduce the female/male ratio? J Affective Disord 1990;18:187–192.
55 Tölle R: Psychiatrie, ed 8. Berlin, Springer, 1988.
56 McEwen BS: Gonadal steroids: Humoral modulators of nerve-cell function. Mol Cell Endocrinol 1980;18:151–164.
57 Maggi A, Perez J: Role of gonadal hormones in the CNS: Clinical and experimental aspects. Life Sci 1985;37:893–906.
58 Blaustein JD, Olster DH: Gonadal steroid hormone receptors and social behaviors; in Balthazart J (ed): Advances in Comparative and Environmental Physiology. Berlin, Springer, 1989, vol 3, pp 31–104.
59 Lilienfeld DE, Sekkor D, Simpson S, Perl DP, Ehland J, Marsh G, Chan E, Godbold JH, Landrigan PJ: Parkinsonism death rates by race, sex and geography: A 1980s update. Neuroepidemiology 1990;9:243–247.
60 Minde K: Hyperaktives Syndrom; in Remschmidt H, Schmidt MH (eds): Kinder- und Jugendpsychiatrie in Klinik und Praxis: Alterstypische, reaktive und neurotische Störungen. Stuttgart, Thieme, 1985, vol 3, pp 1–18.
61 Werner EE: Environmental interaction in minimal brain dysfunction; in Rie NE, Rie EO (eds): Handbook of Minimal Brain Dysfunction. New York, Wiley, 1980, pp 210–231.
62 Remschmidt H, Hebebrand J: Das Tourette-Syndrom: Eine zu selten diagnostizierte Tic-Störung? (abstract). Dtsch Ärzteblatt 1993;90:B-1287–B-1291.
63 Andrews G, Harris M: The syndrome of stuttering. London, Heinemann Medical Books, 1964.
64 Bryson SE, Clark BS, Smith IM: First report of a Canadian epidemiological study of autistic syndromes. J Child Psychol Psychiat 1988;29:433–445.
65 Wing L: Asperger's syndrome: A clinical account. Psychol Med 1981;11:115–129.
66 Rutter M, Tizard J: Intellectual and educational retardation: Prevalence and cognitive characteristics; in Rutter M, Tizard J, Whitmore K (eds): Education, Health and Behaviour. New York, Wiley, 1970, pp 39–53.
67 Broman S, Nichols PL, Shaughnessy P, Kennedy W: Retardation in young children. Hillsdale, Erlbaum, 1987.
68 Forsgren L, Sidenvall R, Blomquist HK, Heijbel J: A prospective incidence study of febrile convulsions. Acta Paediatr Scand 1990;79:550–557.

Prof. Dr. med. Ingrid Reisert, Anatomie und Zellbiologie, Universität Ulm, D–89069 Ulm (FRG)

Embryonic Striatal Grafting: Progress and Future Directions for Therapeutic Approaches to Neurodegenerative Diseases of the Basal Ganglia

Fu-Chin Liu[a], Stephen B. Dunnett[b], Ann M. Graybiel[a, 1]

[a] Department of Brain and Cognitive Sciences, Massachusetts Institute of Technology, Cambridge, Mass., USA;
[b] Department of Experimental Psychology, University of Cambridge, UK

One of the most devastating neurodegenerative diseases affecting the basal ganglia is Huntington's disease (HD), a late-onset genetic disease which is inherited through an autosomal dominant gene [1]. The major symptoms of HD are involuntary choreiform movements and decline of mental function. The most prominent pathology is degeneration of neurons in the striatum, a major station in basal ganglia circuits which are involved in movement, emotion and cognition. After a decade of effort, the mutated gene of HD, an expanded trinucleotide $(CAG)_n$ repeat in human chromosome 4, has recently been cloned [2]. However, the mechanism by which the product of HD gene causes neuronal death remains to be elucidated. In fact, the HD gene appears to be expressed throughout the central nervous system in both normal and affected individuals [31, 32]. As the primary degeneration involves particular striatal neurons expressing particular neurotransmitters, neuromodulators and receptors, it is intriguing to try to design treatments with pharmacological agents that could target the vulnerable striatal neurons while sparing those with different neurochemical phenotypes.

Alternatively, replacing the degenerating neurons with neural transplants should provide a promising therapeutic approach to HD therapy. Intracerebral neural grafting has been developed over the past two decades with the ultimate

---

[1] Our work was supported by NSF BNS-8720475, NATO grant RG. 85/0180 for International Collaboration in Research, NIH 1 R01 HD28341, the Seaver Institute, and the Whitaker Health Sciences Fund.

goal of replacing neurons lost in the wake of neurodegenerative progress [3–5]. Clinical trials of intracranial grafting for treating Parkinson's disease have achieved some degree of success [5]. With the same rationale, animal study of embryonic striatal grafting in experimental models of human neurodegenerative disease states has become a focus of intensive study [5, 6].

Modularity in Embryonic Striatal Grafts

In most of the grafting studies aimed at providing animal models of HD, cell suspensions derived from embryonic striatal primordia have been transplanted into the striatum of adult hosts in which prior excitotoxin lesions were made in the striatum. Such lesions produced a profile of degeneration similar to that in HD. Neurons in embryonic striata can survive grafting into the host striatum of adult rats, can form large grafts and can express a number of neurotransmitter-associated substances typical of those found in normal striatum. The most prominent feature of most embryonic striatal grafts is that they develop a tissue modularity in which neurochemically specialized patches (we have called P regions) are embedded in surrounding tissue (NP regions) lacking strong expression of these neurochemicals (fig. 1). The neurochemical substances highly expressed in the P regions include acetylcholinesterase (AChE), choline acetyltransferase (ChAT), tyrosine hydroxylase (TH), dopamine- and adenosine 3':5'-monophosphate-regulated phosphoprotein (DARPP-32), calbindin-D_{28kD} (calbindin), enkephalin-like peptide and substance P-like peptide [7–10]. All of these substances are neurochemical constituents of the normal adult striatum. The situation in the striatal grafts is thus one in which striatum-like P zones are embedded in surrounding zones lacking such striatal characteristics. Understanding this phenomenon is clearly crucial to designing appropriate grafting strategies for practical therapeutic use in HD.

Compartmentation in the Mammalian Striatum

The first possibility suggested was that the modularity in such grafts might reflect the neurochemical compartments of the normal striatum, striosomes and extrastriosomal matrix [11, 12]. These two compartments can be distinguished on the basis of at least four criteria: (1) neurons in the different compartments express different levels of neurochemical compounds; (2) the compartments have different patterns of connectivity with other brain regions; (3) differential regulation of gene expression can be induced in the two compartments, and (4) the neurons in the two compartments have different peak birthdates and times of matu-

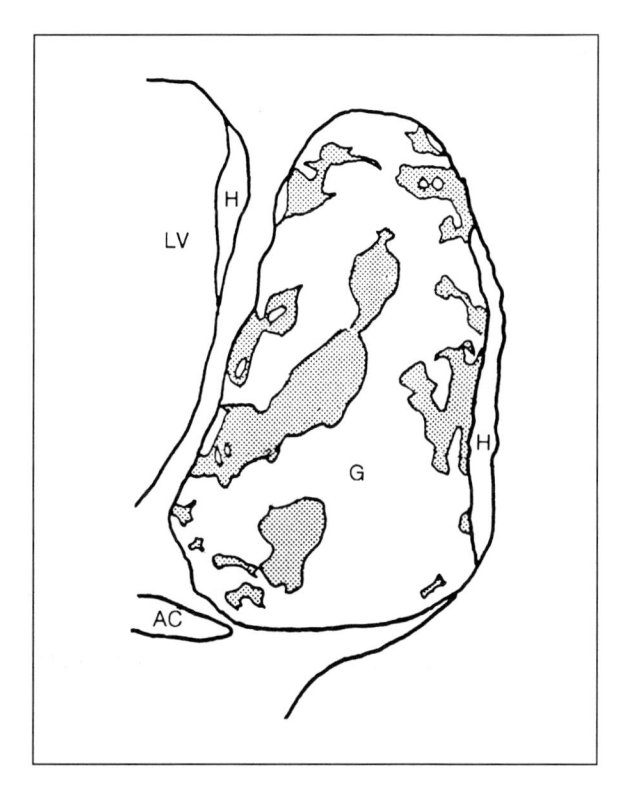

Fig. 1. Modularity in embryonic striatal grafts. ▨ = AChE-rich P regions; ☐ = AChE-poor NP regions. AC = Anterior commissure; H = host striatum; G = graft; LV = lateral ventricle.

ration, with striosomal cells leading those of the matrix compartment [for references, see ref. 3].

Before maturation is complete, future striosomes are characterized by high levels of AChE activity relative to the immature matrix, and are the sites first innervated by a high density of dopamine-containing nigrostriatal fibers in the developing striatum (the 'dopamine islands') [for references, see ref. 4]. These two characteristics hold also for the P regions of embryonic striatal grafts: they are AChE-rich and TH-rich regions in a surround that is AChE and TH poor. This finding led to the hypothesis that the modularity in embryonic striatal grafts represents a developmentally arrested compartmentation in which the AChE-rich P regions represent immature striosomes and the AChE-poor NP regions represent immature matrix [7]. This would represent an advantage of the grafts for reconstituting the cellular and neurochemical phenotypes of the striatum, but also would suggest the need for identifying and introducing factors that would lead to maturation of the graft tissue.

Modularity in Embryonic Striatal Grafts Reflects an Admixture of Striatal Tissue and Nonstriatal Tissue and/or Immature Striatal Tissue

Further work, however, led us to the viewpoint that the P regions may be the only striatum-like tissue in the grafts (fig. 1). We first carried out staining of such grafts for calbindin-D_{28kD}-, somatostatin- and met-enkephalin-like immunoreactivity. The phenotypes of neurons in the AChE-rich P regions were strikingly different from those in the AChE-poor P regions. Many neurons resembling those in the normal host striatum were present in the P regions, including medium-sized calbindin-positive neurons (fig. 2a), enkephalin-positive neurons and a few somatostatin-positive neurons. By contrast, many neurons with features atypical of striatal neurons appeared in the NP regions, such as very large calbindin- and somatostatin-positive neurons with well-stained multipolar dendrites (fig. 2b) and very small enkephalin-positive cells [8]. These neurons in the NP regions did not resemble neurons in the host striatum (fig. 2c), nor neurons in the striatum of normal adult rats. In good accord with these findings was the report by Wictorin et al. [9] that DARPP-32, a phosphoprotein enriched in striatum, is expressed by neurons in the P regions but rarely by neurons in the NP regions and that many neurons in the P regions, but not in the NP regions, extend their axons into the pallidum of the host brain.

Possible Origins of the Neurons with Nonstriatal Phenotype in the NP Regions of Embryonic Striatal Grafts

These different lines of evidence suggest that the modularity in embryonic striatal grafts may represent an admixture of striatal tissue and nonstriatal tissue and/or immature striatal tissue [8]. One possibility raised by this suggestion is that the neurons of the striatum-like P regions may actually be host striatal neurons that migrated into the graft following transplantation. If so, the grafts would represent islands of host tissue in a sea of donor nonstriatal tissue. We tested this possibility and found that very few striatal host neurons prelabeled in situ by fetal exposure of the future hosts to [^{3}H]thymidine migrated to the graft zones [15]. Thus the neurons in both the P and NP regions of grafts develop from the striatal primordia of the donors.

What could be the origin of the many apparently nonstriatal cells in the NP regions? There are at least four different possibilities. First, the multipolar calbindin-positive neurons in the NP regions resemble neurons normally present in regions adjacent to the striatum such as the ventrolateral cortex, pallidum and basolateral nucleus of amygdala. It is possible, then, that the donor tissue may not only contain striatal primordia, but also these nonstriatal primordia. Interesting-

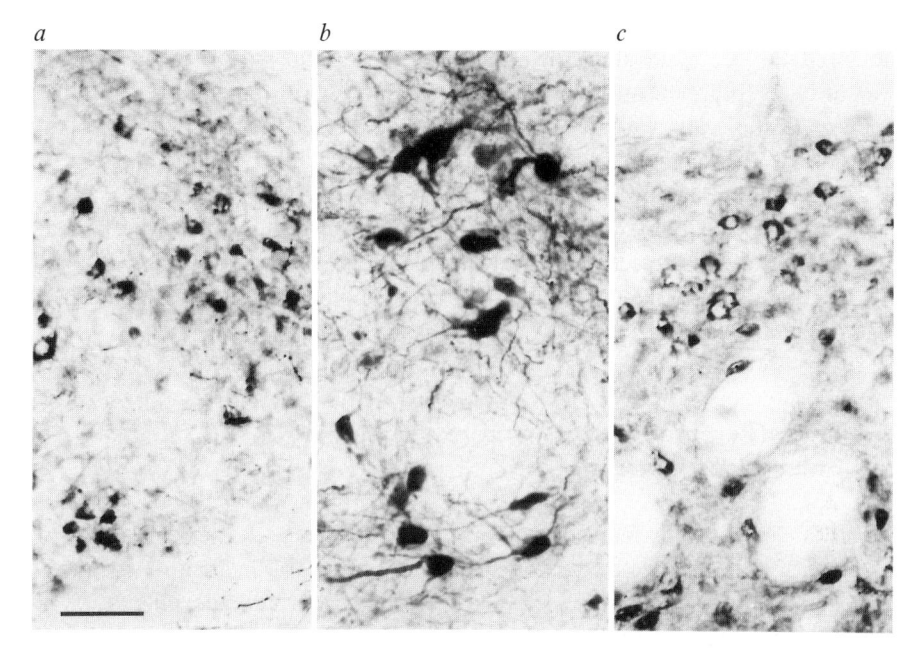

Fig. 2. Calbindin-positive neurons in the P regions, but not in the NP regions, resemble those in the host striatum. Scale bar indicates 50 μm. Modified from [8]. *a* Calbindin-positive neurons in the P regions of striatal grafts. *b* Large calbindin-positive neurons with multipolar well-stained dendrites in the NP regions of striatal grafts. *c* Calbindin-positive neurons in the host striatum.

ly, a population of multipolar calbindin-positive neurons similar to those in the NP regions is transiently present at the junction between the striatal anlagen and cortical anlagen of embryonic day (E) 14–15 rat forebrain [16]. As the donor tissues prepared for grafting by ourselves and others usually derive from E15 embryos, it is likely that the dissected donor tissues would include this transient population of multipolar calbindin-positive neurons. A second possibility is that the striatal primordium itself contains precursor cells of nonstriatal cells that, once generated in the ventricular zone of the developing striatal primordium, migrate through the developing striatum to their final nonstriatal destinations during early development. In fact, by immunostaining with Rat.401 antibody, a marker for radial glial cells [17], we have found in E16 rat forebrain that Rat.401-positive radial glial fibers stretch from the rostral ventricular zone across the striatal anlage to the lateral and ventral cortical plates, and that at caudal levels, Rat.401-positive radial glial fibers arch through the striatal anlage toward the basal forebrain regions [Liu and Graybiel, unpubl. observations]. As radial glial cells serve to guide the migrating neurons during development, it is possible that some nonstriatal cells migrate along the radial glial fibers in the striatal anlage on their way to other forebrain during early development. A third possibility is that the neurons in the grafts with nonstriatal phenotypes may be transformed striatal neurons. In a novel graft environment, a subpopulation of striatal precursor cells may differentiate into neurons with nonstriatal phenotype. Phenotypic transformation has been suggested for neuronal cell lines placed in a novel environment [18]. Finally, the dissected tissues may include different embryonic subdivisions, e.g. corresponding to the medial and lateral ganglionic eminences distinguished by some homeodomain genes [33].

Limited Reconstruction of Compartmentation in Embryonic Striatal Grafts

It is clear that the P regions of embryonic striatal grafts are comprised of tissue with many characteristics of mature striatum. To reconstitute striatum, however, it would be necessary to develop striosome/matrix compartmentation in the P regions. To test for this possibility, we assayed for calbindin, which is expressed by medium-sized matrix neurons, and for [^{3}H]naloxone binding, which is high in the striosomes. Our experiments with E15 striatal grafts indicated that, although many medium-sized calbindin-positive neurons are present in the P regions of the grafts, only a few patches of strong [^{3}H]naloxone binding sites occur in the grafts [8]. These findings suggest that the P regions comprise primarily striatal matrix tissue and there is limited reconstruction of striatal compartmentation in E15 striatal grafts. It would be of great interest to identify factors that could increase the expression of striosomal phenotype in the P regions.

Reconstitution of Cholinergic and Dopaminergic Systems in Embryonic Striatal Grafts

Pharmacological and clinical studies have long indicated that a balance between acetylcholine and dopamine in the striatum is important for executing normal movement [19]. As these neurochemical systems are crucial in motor control, it was important to determine how well they can develop in striatal grafts. We and others found a remarkably complete representation of the presynaptic and postsynaptic constituents of the striatal cholinergic and dopaminergic systems in the P regions (but not in the NP regions) of striatal grafts. AChE, ChAT, high-affinity choline uptake sites, muscarinic M1 receptor binding sites, TH, high affinity dopamine uptake sites, D_1-like and D_2-like dopamine receptor binding sites and DARPP-32, all are preferentially concentrated in the AChE-rich P regions [10, and references therein]. These results suggest that there is a reconstitution of the synthetic enzyme, degradative enzyme, uptake sites and receptor sites of cholinergic and dopaminergic systems in the P regions of embryonic striatal grafts. These findings provide an anatomical basis for reconstruction of functional interaction between cholinergic and dopaminergic systems in striatal grafts.

Functional Integration of Embryonic Striatal Grafts and Host Brain

Ultimately, the critical issue in evaluating striatal grafting strategy is whether the grafts are functionally integrated into the host brain. To address this question, we turned to the dynamic regulation of neuropeptides and transcription factors in the striatum.

Neuropeptides in the striatum are subject to regulation by the dopaminergic nigrostriatal system [for a review, see ref. 12]. Depletion of dopamine-containing inputs to the adult striatum, for example, results in up-regulation of enkephalin neuropeptide and proenkephalin mRNAs. In striatal grafts, the TH-positive fibers innervating the P zones are derived from nigrostriatal TH-containing afferents of the host [20–22]. To test whether enkephalin expression in the grafts could be influenced by such host dopaminergic innervation, we unilaterally destroyed the nigrostriatal innervation of hosts with 6-hydroxydopamine before grafting. We found that the dopaminergic denervation resulted in up-regulation of met-enkephalin-like immunoreactivity in the P regions of the grafts, as it did in the host striatum [22]. Similarly, depletion of dopamine-containing inputs to the striatum after grafting also up-regulates preproenkephalin mRNA levels in the P regions of such grafts [23]. These findings suggest that the neuropeptides expressed by grafted neurons are functionally regulated by the TH-containing afferents from the host that innervate the striatal parts of the grafts.

Immediate-early genes, including members of the *fos-jun* family of transcription factors, can be transiently induced in subpopulations of striatal neurons by dopamine agonists [24–26; see 27 for a review]. We used such stimulation protocols as a strategy to test whether embryonic striatal grafts are subject to host regulation at the signal transduction level. We found that nuclear Fos-like protein is induced in neurons of the P regions of grafts following challenge of the hosts with indirect dopamine agonists such as cocaine and amphetamine [28]. Mandel et al. [29] have made similar observations. These results demonstrate that neurons in striatal grafts are able to integrate sufficiently into neuronal circuits of the host brain to transduce extracellular signals from the host and to activate genetic programs in the grafts.

Other studies also indicate that striatal grafts are anatomically and functionally integrated into the host brain. The grafts establish partial synaptic connectivity with the host brain, are electrophysiologically responsive to stimulation of the cortex and thalamus of the host, and can reduce hypermetabolic activity and increase GABA release in the striatal target regions affected by the injections of excitotoxin into the host striatum before grafting [for references, see ref. 15]. This functional integration may underlie the partial behavioral recovery seen in the grafted animals.

Future Directions

Although the work here suggests that embryonic striatal grafting may have therapeutic potential for treating HD, there are still major issues remaining to be resolved. First, the main conclusion of our work is that large parts of striatal grafts do not have the properties of striatum. These large nonstriatal NP regions may accordingly limit the functional integration of the grafts into the basal ganglia circuitry of the host brain. A grafting procedure that would increase the proportion of striatal tissue (P regions) in the grafts clearly needs to be developed. This is an important future goal, as a similar P and NP region modularity has been found in human embryonic forebrain/striatal grafts implanted into the rat striatum [30]. A second serious concern is that striatal grafts may not survive in a host striatal environment in which neurodegeneration continues to occur. The progression of the degenerative process of HD may eventually eliminate the grafted cells. Third, a grafting technique capable of reconstituting striosome/matrix compartmentation, and ultimately, the matriosomal compartments typical of primates, needs to be developed. Much behavioral recovery might be possible without full reconstruction of the striatal neuronal network, as appears to be the case for grafting in Parkinson's disease. Ultimately, however, therapy for treating neurodegenerative diseases in the basal ganglia should involve reconstitution of at least the main elements of the neuronal tissue.

Note Added in Proof

Since this paper was submitted, Isacson and his colleagues [34] have reported that striatal grafts derived from the lateral ganglionic eminence are enriched in AChE-rich tissue, whereas striatal grafts derived from the medial ganglionic eminence are enriched in AChE-poor tissue.

References

1 Wexler NS, Rose EA, Housman DE: Molecular approaches to hereditary diseases of the nervous system: Huntington's disease as a paradigm. Annu Rev Neurosci 1991;14:503–529.
2 The Huntington's Disease Collaborative Research Group: A novel gene containing a trinucleotide repeat that is expanded and unstable on Huntington's disease chromosomes. Cell 1993;72:971–983.
3 Björklund A: Neutral transplantation – An experimental tool with clinical possibilities. Trends Neurosci 1991;14:391–322.
4 Björklund A: Dopaminergic transplants in experimental parkinsonism: Cellular mechanisms of graft-induced functional recovery. Curr Opin Neurobiol 1992;2:683–689.
5 Lindvall O: Prospects of transplantation in human neurodegenerative diseases. Trends Neurosci 1991;14:376–384.
6 Björklund A, Lindvall O, Isacson O, Brundin P, Wictorin K, Strecker R, Clarke DJ, Dunnett SB: Mechanisms of action of intracerebral neural implants: Studies on nigral and striatal grafts to the lesioned striatum. Trends Neurosci 1987;10:509–516.
7 Isacson O, Dawbarn D, Brundin P, Gage FH, Emson PC, Björklund A: Neural grafting in a rat model of Huntington's disease: Striosomal-like organization of striatal grafts as revealed by immunohistochemistry and receptor autoradiography. Neuroscience 1987;22:481–497.
8 Graybiel AM, Liu F-C, Dunnett SB: Intrastriatal grafts derived from fetal striatal primordia. I. Phenotypy and modular organization. J Neurosci 1989;9:3250–3271.
9 Wictorin K, Quimet CC, Björklund A: Intrinsic organization and connectivity of intrastriatal striatal transplants in rats as revealed by DARPP-32 immunohistochemistry: Specificity of connections with the lesioned host brain. Eur J Neurosci 1989;1:690–701.
10 Liu F-C, Graybiel AM, Dunnett SB, Baughman RW: Intrastriatal grafts derived from fetal striatal primordia. II. Reconstitution of cholinergic and dopaminergic systems. J Comp Neurol 1990;295:1–14.
11 Graybiel AM, Ragsdale CW, Jr: Histochemically distinct compartments in the striatum of human, monkey, and cat demonstrated by acetylthiocholinesterase staining. Proc Natl Acad Sci USA 1978;75:5723–5726.
12 Graybiel AM: Neurotransmitters and neuromodulators in the basal ganglia. Trends Neurosci 1990;13:244–254.
13 Liu F-C, Graybiel AM: Heterogeneous development of calbindin-D_{28K} expression in the striatal matrix. J Comp Neurol 1992;320:304–322.
14 Graybiel AM: Correspondence between the dopamine islands and striosomes of the mammalian striatum. Neuroscience 1984;13:1157–1187.
15 Liu F-C, Dunnett SB, Graybiel AM: Intrastriatal grafts derived from fetal striatal primordia. IV. Host and donor neurons are not intermixed. Neuroscience 1993;55:363–372.
16 Liu F-C, Graybiel AM: Transient calbindin-D_{28K}-positive systems in the telencephalon: Ganglionic eminence, developing striatum and cerebral cortex. J Neurosci 1992;12:674–690.
17 Hockfield S, McKay RDG: Identification of major cell classes in the developing mammalian nervous system. J Neurosci 1985;5:3310–3328.
18 Renfranz PJ, Cunningham MG, McKay RDG: Region-specific differentiation of the hippocampal stem cell line HiB5 upon implantation into the developing mammalian brain. Cell 1991;66:713–729.
19 Lehmann J, Langer SZ: The striatal cholinergic interneuron: Synaptic target of dopaminergic terminals? Neuroscience 1983;10:1105–1120.
20 Pritzel M, Isacson O, Brundin P, Wiklund L, Björklund A: Afferent and efferent connections of striatal grafts implanted into the ibotenic acid lesioned neostriatum in adult rats. Exp Brain Res 1986;65:112–126.
21 Wictorin K, Isacson O, Fischer W, Nothias F, Peschanski M, Björklund A: Connectivity of striatal grafts implanted into the ibotenic acid-lesioned striatum. I. Subcortical afferents. Neuroscience 1988;27:547–562.
22 Liu F-C, Dunnett SB, Graybiel AM: Influence of mesostriatal afferents on the development and transmitter regulation of intrastriatal grafts derived from embryonic striatal primordia. J Neurosci 1992;12:4281–4297.

23 Campbell K, Wictorin K, Björklund A: Differential regulation of neuropeptide mRNA expression in intrastriatal striatal transplants by host dopaminergic afferents. Proc Natl Acad Sci USA 1992; 89:10489–10493.
24 Robertson HA, Peterson MR, Murphy K, Robertson GS: D_1-dopamine receptor agonists selectively activate striatal c-*fos* independent of rotational behavior. Brain Res 1989;503:346–349.
25 Graybiel AM, Moratalla R, Robertson HA: Amphetamine and cocaine induce drug-specific activation of the c-*fos* gene in striosome-matrix and limbic subdivisions of the striatum. Proc Natl Acad Sci USA 1990;87:6912–6916.
26 Young ST, Porrino LJ, Iadarola MJ: Cocaine induces striatal c-Fos-immunoreactive proteins via dopaminergic D_1 receptors. Proc Natl Acad Sci USA 1991;88:1291–1295.
27 Graybiel AM: Acute effects of psychomotor stimulant drugs on gene expression in the striatum. NIDA Res Monogr 1993;125:72–81.
28 Liu F-C, Dunnett SB, Robertson HA, Graybiel AM: Intrastriatal grafts derived from fetal striatal primordia. III. Induction of modular patterns of Fos-like immunoreactivity by cocaine. Exp Brain Res 1991;85:501–506.
29 Mandel RJ, Wictorin K, Cenci MA, Björklund A: *fos* expression in intrastriatal striatal grafts: Regulation by host dopaminergic afferents. Brain Res 1992;583:207–215.
30 Wictorin K, Brundin P, Gustavii B, Lindvall O, Björklund A: Reformation of long axon pathways in adult rat central nervous system by human forebrain neuroblasts. Nature 1990;347:556–558.
31 Li S-H, Schilling G, Young WS III, Li X-J, Margolis RL, Stine OC, Wagster MV, Abbott MH, Franz ML, Ranen NG, Folstein SE, Hedreen JC, Ross CA: Huntington's disease gene (IT15) is widely expressed in human and rat tissues. Neuron 1993;11:985–993.
32 Strong TV, Tagle DA, Valdes JM, Elmer LW, Boehm K, Swaroop M, Kaatz KW, Collins FS, Albin RL: Widespread expression of the human and rat Huntington's disease gene in brain and nonneural tissues. Nat Genet 1993;5:259–265.
33 Bulfone A, Puelles L, Porteus MH, Frohman MA, Martin GR, Rubenstein JLR: Spatially restricted expression of Dlx-1, Dlx-2 (Tes-1), Gbx-2, and Wnt-3 in the embryonic day 12.5 mouse forebrain defines potential transverse and longitudinal segmental boundaries. J Neurosci 1993;13:3155–3172.
34 Pakzaban P, Deacon TW, Burns LH, Isacson O: Increased proportion of acetylcholinesterase-rich zones and improved morphological integration in host striatum of fetal grafts derived from the lateral but not the medial ganglionic eminence. Exp Brain Res 1993;97:13–22.

Fu-Chin Liu, PhD, Massachusetts Institute of Technology,
Department of Brain and Cognitive Sciences, E25-618,
45 Carleton Street, Cambridge, MA 02139 (USA)

Basic Fibroblast Growth Factor in the Substantia nigra in Parkinson's Disease and Normal Aging

Patrick L. McGeer[a], Ikuo Tooyama[b], Hiroshi Kimura[b], Edith G. McGeer[a]

[a] Kinsmen Laboratory of Neurological Research and the Neurodegenerative Disorders Centre, The University of British Columbia, Vancouver, Canada;
[b] Institute of Molecular Neurobiology, Shiga University of Medical Science, Otsu, Shiga, Japan

The dopamine (DA) neurons of the brain constitute the neurotransmitter system which has been perhaps the most intensively studied over the past 25 years. One reason is the early development of methods for their histochemical localization and for the quantitative assay of DA and its metabolites. An even more important reason was the early recognition, made possible by the use of these methods, that loss of DA neurons probably plays a key role in the pathology of Parkinson's disease (PD), while their functional overactivity seems a basic problem in schizophrenia. Another interesting aspect of the literature is the rather large loss of striatal DA with advanced age which has been generally found in primates, including humans [1]. The reasons for the vulnerability of these neurons with aging and PD are not known, but a number of hypotheses have been advanced. Thus, it has been suggested that these neurons may be poisoned due to the accumulation of melanin or to the action of an environmental or endogenously formed neurotoxin resembling 6-hydroxy-DA (6OHDA) or 1-methyl-4-phenyl-1,2,3,6-tetrahydropyridine (MPTP). They might be destroyed by iron-catalyzed or other free-radical-generating reactions. Or they might die due to loss of some neurotrophic factor essential to their survival. Neurotrophic factors are presently receiving a great deal of attention and a number of them have been reported to promote the survival of DA neurons in culture and/or to protect neurons of the substantia nigra (SN) against the action of 6OHDA or MPTP (table 1). Immunohistochemical studies have shown that isoforms of one of these factors, basic fibroblast growth factor (bFGF), appears in DA neurons of the SN in rats

Table 1. Brief summary of some of the literature on the effects of various neurotrophic factors on dopaminergic neurons or dopaminergic parameters

Factor	References reporting effects (or absence of effects) on dopaminergic neurons	
	in culture	in vivo
Platelet-derived growth factor	2, (3)	
Epidermal growth factor	4, 5[1], 6[1]	5[1]
Basic fibroblast growth factor	3, 6[1], 7, 8	9[1], 10[2]
Acidic fibroblast growth factor	3	
Insulin-like growth factors I and II	11	
Nerve growth factor	(12, 13)	14[1]
Brain-derived growth factor	12, 15, 16[1], 17[1]	(18, 19[3], 20[2])
Glial-derived neurotrophic factor	21	
Ciliary neurotrophic factor		22[3]
Neurotrophin-3	(13)	
Interleukin-1	(3)	

Negative reports are in parentheses.
[1] Study involved a neurotoxin.
[2] Study involved transplantation.
[3] Study involved axotomy.

[23–25], monkeys [23] and humans [23, 26]. We have reported [26], and summarize here, immunohistochemical studies suggesting that loss of bFGF precedes death of DA neurons in PD. On the other hand, there does not appear to be preferential loss of bFGF from DA neurons with normal aging.

Methods

Tissue Preparation
The brains used were from 6 cases of PD (71 ± 2.9 years), 6 age-matched neurologically normal controls (73 ± 3.8 years), plus 4 younger controls; all brains were obtained within 7–12 h after death. The clinical diagnosis in each PD case was confirmed at autopsy.

In all cases, the left half of each midbrain was dissected out, fixed for 2 days at 4°C with 4% paraformaldehyde in 0.1 M phosphate-buffered saline (PBS), pH 7.4, and then transferred to a solution of 15% sucrose in PBS where it was maintained in the cold until used. Serial sections of 30 μm thickness were cut from the level of the oculomotor nerve to the upper pons from each brain, and stained immunohistochemically for bFGF or tyrosine hydroxylase (TH), or by the Klüver-Barrera and HE methods for anatomical verification.

Immunohistochemistry

Sections were treated as previously described in detail [26]: (1) with 0.5% H_2O_2 to destroy endogenous peroxidase; (2) with 5% skim milk to block nonspecific binding of immunoglobulins; (3) with the bFGF monoclonal antibody (MAB78, 3.8 µg/ml) or with rabbit polyclonal anti-TH (Eugene Biochemicals, 1:5,000); (4) with the appropriate secondary antibody (Vector Laboratory), and (5) with the avidin-biotin-horseradish peroxidase (ABC, Vectastain; 1:2,000). A dark purple color was developed using a solution containing 0.02% diaminobenzidine tetrahydrochloride, 0.0045% H_2O_2, 0.05% imidazole and 0.6% nickel ammonium sulfate.

Characterization of the bFGF antibody has been previously described [25, 27]; it recognizes an epitope in the amino terminal residue 1–9 region. On Western blot examination of cytosolic extracts of control SN, the bFGF antibody recognized three bands with molecular weights of 18, 27 and 29 kD. These correspond to the 18-kD classical bFGF, and two higher molecular weight forms of bFGF [28]. Further proof of specificity was that no positive staining was obtained in control or PD brain using the antibody preincubated with human recombinant bFGF.

Image Analysis

The numbers of neurons per 30-µm section which were pigmented or bFGF-positive were counted at a single level by a computer analyzer using the MCID system (Imaging Research Inc., Canada). The level chosen for comparative counting was that having the highest density of dopaminergic cells. It corresponds to plate XXXVI in the Olszewski and Baxter atlas [29], in the mid portion of the SN where the oculomotor nerve emerges. Structures with areas less than 134 µm² (diameter in spheres <13 µm) were automatically erased to avoid counting positive glial cells and extracellular melanin deposits. The number counted by the computer consisted of all pigmented and immunopositive neurons. Since it was difficult to distinguish melanin and immunopositive reactions by computer for technical reasons, melanin-positive, immunonegative cells and melanin-negative, immunopositive cells were picked up on the same screen by eye and counted. Further details on the counting procedure have been published [26].

Statistical analyses between PD and age-matched control groups were performed using Anova followed by Dunnett's t test. The significance levels for correlations of cell counts with age in the controls were determined by regression analysis.

Results

A comparison of SN sections from PD patients and the age-matched controls showed the expected severe loss in the PDs of the pigmented and/or TH-positive neurons. It was evident, however, that the loss of bFGF-like immunoreactivity was even more severe than that of the DA neurons themselves. In the controls about 94% of the pigmented neurons were positive with the bFGF antibody. In contrast, only about 8% of the remaining DA neurons in the PD sections showed any reaction with the bFGF antibody (fig. 1), although there was a significant proportion of such neurons which stained more intensely than the neurons in controls [26].

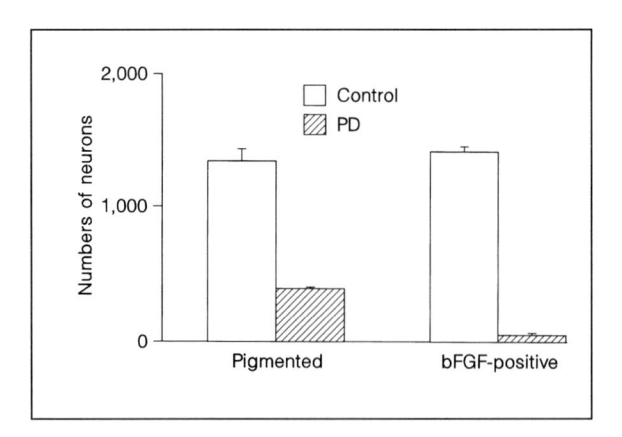

Fig. 1. Numbers (± SEM) of neurons containing melanin or bFGF in 30-mm sections of the SN at the level of the oculomotor nerve in Parkinsonian patients and age-matched controls. n = 6 in each group. In each case, the PDs differ from controls at p < 0.0001.

An analysis of the number of pigmented cells in sections at the oculomotor level from controls revealed a significant decrease with age (fig. 2a). The percent of such cells which showed bFGF-like immunoreactivity did not, however, change significantly with age (fig. 2b).

Discussion

Our finding that most pigmented neurons of the SN are immunopositive for bFGF in normal control cases is in accord with previous reports [23, 24]. In PD, on the other hand, only 8.2% of the remaining pigmented cells were immunopositive for bFGF. The fact that a significant proportion of the remaining neurons stained intensely for bFGF may indicate up-regulation as a survival mechanism. None of the control cases, including the youngest (15 years old) and the oldest (82 years old), showed neurons with such strong bFGF immunoreactivity. This alteration, therefore, would seem to be related to PD pathology and not to age.

Several reports have shown that bFGF has a trophic effect on mesencephalic dopaminergic neurons (table 1). The presence of FGF receptor mRNA in SN neurons has been shown by in situ hybridization studies [30]. This suggests that bFGF may act on these neurons in either an autocrine or a paracrine fashion, being taken up in the latter case by a receptor-mediated mechanism. Such a mechanism is consistent with the report [31] that injection of radioactive bFGF into the rat SN results in the appearance within a few hours of radioactivity in the

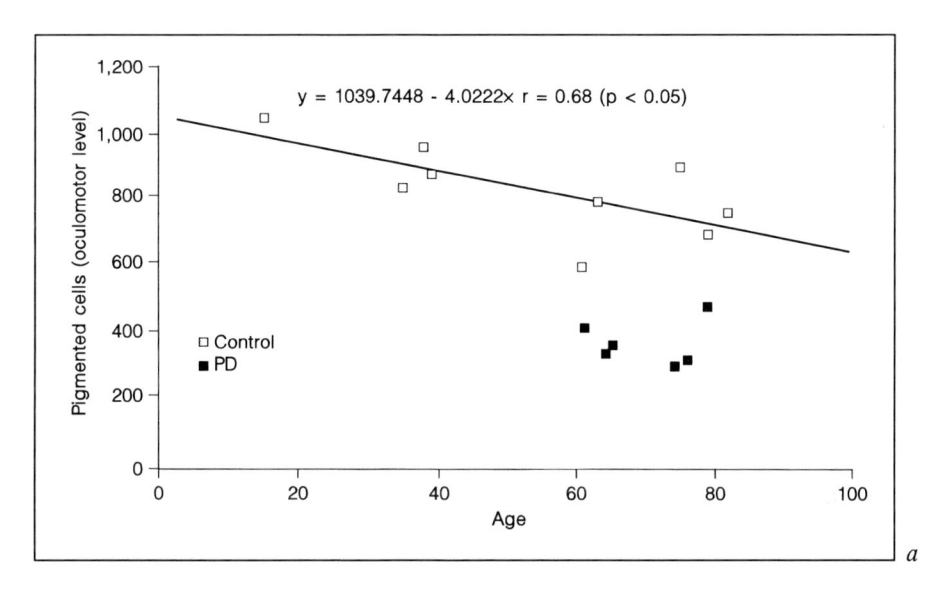

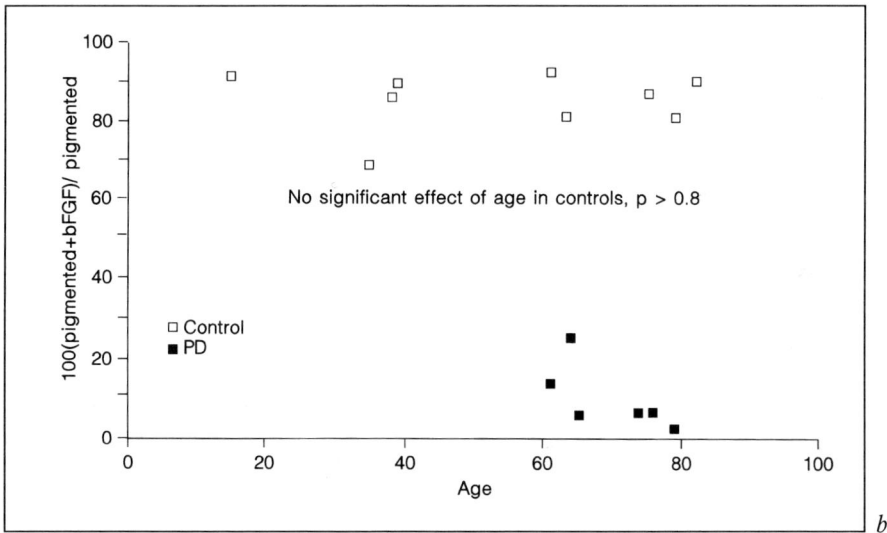

Fig. 2. *a* Numbers of pigmented cells in a section of the SN at the level of the oculomotor nerve plotted against age for 9 neurologically normal controls and 6 cases of PD. The line of correlation shown is calculated for the control group. *b* Age plotted against the percent of the pigmented cells in a section of the SN at the level of the oculomotor nerve which also contain bFGF.

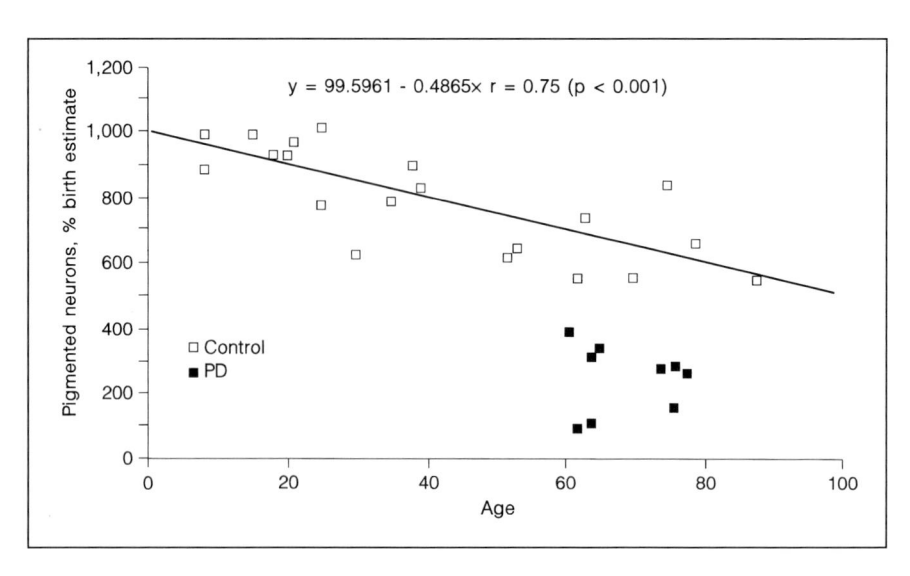

Fig. 3. Plot of relative pigmented cell numbers against age for 20 controls and 9 cases of PD. Data are combined from the counts obtained at the level of the oculomotor nerve (compare fig. 2a) and counts previously obtained [32] in every tenth section through the SN. In order to allow the data to be combined, the counts in each subject were expressed as a percent of the estimated number at birth as indicated by the line of correlation for controls against age. The line of correlation and equation shown here are for the combined control data.

striatum of intact, but not 6OHDA-lesioned animals. Further insight into the mechanism could be obtained by in situ hybridization studies on the SN for the mRNA of bFGF itself.

The loss of bFGF in PD SN neurons might be interpreted as part of a common down-regulation of growth factors and other proteins as a preterminal event. However, it seems clear from a comparison of sections stained for TH with those stained for bFGF that the loss in bFGF precedes any observable loss in neurotransmitter synthetic capacity [26]. Moreover it is difficult to explain the intensification of bFGF staining seen in some neurons in the PD cases on the basis of terminal down-regulation. Some dysfunction must exist which results in a down-regulation in most DA neurons, but an upregulation in some. If a paracrine mechanism is involved in the action of bFGF on DA neurons, the loss of staining might be due to a deficiency in the supply of either bFGF or in the receptors which mediate its transport. The latter mechanism would be more consistent with the observed extremes of bFGF staining. If an autocrine mechanism is involved, either explanation is possible. An examination of the expression of bFGF receptors in SN neurons of PD and control cases would be of great interest. A normal or enhanced receptor activity on surviving neurons would suggest that a deficiency of bFGF may be linked to the disease process.

The significant decrease with age in controls found in this study of the numbers of pigmented neurons at the oculomotor level (fig. 2a) is consistent with some previous reports where other methods of counting were used. In particular, our initial report of a decrease in DA neurons in the SN with age in controls [32] was based upon counts of every tenth section throughout the SN. The absolute numbers obtained in that study were therefore very different from those plotted in figure 2a, but, if the number in each case is expressed as a percent of the estimated number at birth, the two sets of data can be combined to give a highly significant line of regression (fig. 3). It has previously been suggested that DA cell counts at the level of the oculomotor nerve give an accurate reflection of the percentage of SN neuronal loss in PD [33]; it would appear that this is also true for the much lesser loss seen in normal aging.

Despite this similarity, it would appear that the process of cell loss in PD at death is quite different from that in normal aging since the former appears to involve prior loss of bFGF while the latter does not (compare fig. 1, 2b). It has been suggested [34] that PD may be due to a process that causes localized neuronal damage at an early age but does not produce symptoms until several decades of age-related neuronal attrition has taken place. We have previously reported that the appearance of reactive microglia [35] and complement proteins [36] in the SN in PD argues against this hypothesis and suggests a much more active pathological process occurring at the time of death. The contrast found here between the effects of PD and normal aging on the appearance of bFGF in DA neurons would also favor such a more active pathology.

Acknowledgments

We thank Dr. Donald Calne for clinical details of the parkinsonian cases and Ms. Joane Sunahara for technical assistance. This research was supported by grants from the Japan Foundation for Aging and Health, the MRC, the Parkinson Foundation of Canada and the Alzheimer Society of British Columbia, as well as donations from individual British Columbians.

References

1 Beal MF: Neurochemical aspects of aging in primates. Neurobiol Aging 1993;14:707–709.
2 Nikkhah G, Odin P, Smits A, Tingstrom A, Othberg A, Brundin P, Funa K, Lindvall O: Platelet-derived growth factor promotes survival of rat and human mesencephalic dopaminergic neurons in culture. Exp Brain Res 1993;92:516–523.

3 Engele J, Bohn MC: The neurotrophic effects of fibroblast growth factors on dopaminergic neurons in vitro are mediated by mesencephalic glia. J Neurosci 1991;11:3070–3078.
4 Casper D, Mytilineou C, Blum M: EGF enhances the survival of dopamine neurons in rat embryonic mesencephalon primary cell culture. J Neurosci Res 1991;30:372–381.
5 Hadjiconstantinou M, Fitkin JG, Dalia A, Neff NH: Epidermal growth factor enhances striatal dopaminergic parameters in the 1-methyl-4-phenyl-1,2,3,6-tetrahydropyridine-treated mouse. J Neurochem 1991;57:479–482.
6 Park TH, Mytilineou C: Protection from 1-methyl-4-phenylpyridinium (MPP+) toxicity and stimulation of regrowth of MPP(+)-damaged dopaminergic fibers by treatment of mesencephalic cultures with EGF and basic FGF. Brain Res 1992;599:83–97.
7 Ferrari G, Minozzi MC, Toffano G, Leon A, Skaper SD: Basic fibroblast growth factor promotes the survival and development of mesencephalic neurons in culture. Dev Biol 1989;133:140–147.
8 Knüsel B, Michel PP, Schwaber JS, Hefti F: Selective and nonselective stimulation of central cholinergic and dopaminergic development in vitro by nerve growth factor, basic fibroblast growth factor, epidermal growth factor, insulin and the insulin-like growth factors I and II. J Neurosci 1990; 10:558–570.
9 Otto D, Unsicker K: Basic FGF reverses chemical and morphological deficits in the nigrostriatal system of MPTP-treated mice. J Neurosci 1990;10:1912–1921.
10 Steinbusch HWM, Vermeulen RJ, Tonnaer JADM: Basic fibroblast growth factor enhances survival and sprouting of fetal dopaminergic cells implanted in the denervated rat caudate-putamen: Preliminary observations. Prog Brain Res 1990;82:81–86.
11 Knüsel B, Hefti F: Trophic actions of IGF-I, IGF-II and insulin on cholinergic and dopaminergic brain neurons. Adv Exp Med Biol 1991;293:351–360.
12 Knüsel B, Winslow JW, Rosenthal A, Burton LE, Seid DP, Nikolics K, Hefti F: Promotion of central cholinergic and dopaminergic neuron differentiation by brain-derived neurotrophic factor but not neurotrophin 3. Proc Natl Acad Sci USA 1991;88:961–965.
13 Snyder SH: Parkinson's disease. Fresh factors to consider. Nature 1991;350:195.
14 Garcia E, Rios C, Sotelo J: Ventricular injection of nerve growth factor increases dopamine content in the striata of MPTP-treated mice. Neurochem Res 1992;17:979–982.
15 Hyman C, Hofer M, Barde YA, Juhasz M, Yancopoulos GD, Squinto SP, Lindsay RM: BDNF is a neurotrophic factor for dopaminergic neurons of the substantia nigra. Nature 1991;350:230–232.
16 Skaper SD, Negro A, Facci L, Dal Toso R: Brain-derived neurotrophic factor selectively rescues mesencephalic dopaminergic neurons from 2,4,5,-trihydroxyphenylalanine-induced injury. J Neurosci Res 1993;34:478–487.
17 Spina MB, Squinto SP, Miller J, Lindsay RM, Hyman C: Brain-derived neurotrophic factor protects dopamine neurons against 6-hydroxydopamine and N-methyl-4-phenylpyridinium ion toxicity: Involvement of the glutathione system. J Neurochem 1992;59:99–106.
18 Altar CA, Boylan CB, Jackson C, Hershenson S, Miller J, Wiegand SJ, Lindsay RM, Hyman C: Brain-derived neurotrophic factor augments rotational behavior and nigrostriatal dopamine turnover in vivo. Proc Natl Acad Sci USA 1992;89:11347–11351.
19 Knüsel B, Beck KD, Winslow JW, Rosenthal A, Burton LE, Widmer HR, Nikolics K, Hefti F: Brain-derived neurotrophic factor administration protects basal forebrain cholinergic but not nigral dopaminergic neurons from degenerative changes after axotomy in the adult rat brain. J Neurosci 1992;12:4391–4402.
20 Lapchak PA, Beck KD, Araujo DM, Irwin I, Langston JW, Hefti F: Chronic intranigral administration of brain-derived neurotrophic factor produces striatal dopaminergic hypofunction in unlesioned adult rats and fails to attenuate the decline of striatal dopaminergic function following medial forebrain bundle transection. Neuroscience 1993;53:639–650.
21 Lin L-FH, Doherty DH, Lile JD, Bektesh S, Collins F: GDNF: A glial cell-derived neurotrophic factor for midbrain dopaminergic neurons. Science 1993;260:1130–1132.
22 Hagg T, Varon S: Ciliary neurotrophic factor prevents degeneration of adult rat substantia nigra dopaminergic neurons in vivo. Proc Natl Acad Sci USA 1993;90:6315–6319.
23 Bean AJ, Elde R, Cao YH, Oellig C, Tamminga C, Goldstein M, Pettersson RF, Hökfelt T: Expression of acidic and basic fibroblast growth factors in the substantia nigra of rat, monkey, and human. Proc Natl Acad Sci USA 1991;88:10237–10241.

24 Cintra A, Cao Y, Oellig C, Tinner B, Bortolotti F, Goldstein M, Pettersson RF, Fuxe K: Basic FGF is present in dopaminergic neurons of the ventral midbrain of the rat. NeuroReport 1991;2:597–600.
25 Tooyama I, Walker D, Yamada T, Hanai K, Kimura H, McGeer EG, McGeer PL: High molecular weight basic fibroblast growth factor-like protein is localized to a subpopulation of mesencephalic dopaminergic neurons in the rat brain. Brain Res 1992;593:274–280.
26 Tooyama I, Kawamata T, Walker D, Yamada T, Hanai K, Kimura H, Iwane M, Igarashi K, McGeer EG, McGeer PL: Loss of basic fibroblast growth factor in substantia nigra neurons in Parkinson's disease. Neurology 1993;43:372–376.
27 Seno M, Iwane M, Sasada R, Moriya N, Kurokawa T, Igarashi K: Monoclonal antibodies against human basic fibroblast growth factor. Hybridoma 1989;8:209–221.
28 Baird A, Böhlen P: Fibroblast growth factors; in Sporn MB, Roberts AB (eds): Peptide growth factors and their receptors. I. Berlin, Springer, 1990, pp 369–418.
29 Olszewski J, Baxter D: Cytoarchitecture of the human brainstem, ed 2. Basel, Karger, 1982;54.
30 Wanaka A, Johnson EM Jr, Milbrandt J: Localization of FGF receptor mRNA in the adult rat central nervous system by in situ hybridization. Neuron 1990;5:267–281.
31 McGeer EG, Singh EA, McGeer PL: Apparent anterograde transport of bFGF in the rat nigrostriatal dopamine system. Neurosci Lett 1992;148:31–33.
32 McGeer PL, McGeer EG, Suzuki JS: Aging and extrapyramidal function. Arch Neurol 1977;34:33–35.
33 Zarow C, Chui HC: A simple method for assessing neuronal number in the human substantia nigra [abstract]. Soc Neurosci Abstr 1991;17(pt2):1450.
34 Calne DB, Eisen A, McGeer EG, Spencer P: Alzheimer's disease, Parkinson's disease and motor neuron disease: A biotrophic interaction between aging and environment. Lancet 1986;i:1067–1070.
35 McGeer PL, Itagaki I, Akiyama H, McGeer EG: Rate of cell death in Parkinsonism indicates active neuropathological process. Ann Neurol 1988;24:574–576.
36 Yamada T, McGeer PL, McGeer EG: Lewy bodies in Parkinson's disease are recognized by antibodies to complement proteins. Acta Neuropathol (Berl) 1992;84:100–104.

Dr. Patrick L. McGeer, Kinsmen Laboratory of Neurological Research,
University of British Columbia, 2255 Wesbrook Mall, Vancouver, BC, V6T 1W5 (Canada)

Closing Remarks

Nobuo Yanagisawa

Department of Medicine (Neurology), Shinshu University School of Medicine, Matsumoto, Japan

The key words of the symposium on Age-Related Monoamine-Dependent Disorders and Their Modulation by Gene and Gender, held in Tokyo from November 12–13, 1993, were basal ganglia, catecholamine neurons, age and gender. The methodology used in research presented at the symposium included molecular biology, anatomy, chemistry, physiology, pharmacology, neuroimaging, clinical studies, stereotaxy and tissue transplantation. In addition to basic research on neural circuitries, transmitters and neuronal activities in the basal ganglia, studies on various diseases were presented, including hereditary progressive dystonia with marked diurnal fluctuation (HPD), dopa-responsive dystonia (DRD), juvenile parkinsonism (JP), Gilles de la Tourette syndrome, idiopathic torsion dystonia, Parkinson's disease (PD), and experimental models.

Among comprehensive and informative presentations on basal ganglia disorders of juvenile onset, the relationship of HPD to DRD was an intriguing issue. Since the proposal of a concept of dopa-responsive dystonia [1], relations between HPD and DRD have been a matter of debate at several meetings. The Segawa symposium of the year 1990 focused on this issue from various aspects [2], where D.B. Calne raised three aspects of research for discrimination of HPD from DRD: (1) genetic separation by molecular biology; (2) chemical and histological pathology, and (3) physiology including polysomnography and ocular movements.

Originally, HPD was proposed by Segawa et al. [3, 4] as a nosological entity. Criteria for diagnosis or characteristics of this condition have been detailed with an accumulation of data on the natural course of Japanese patients and reports on the same conditions from Europe and Asia [5, 6]. On the other hand, dopa-responsive dystonia was proposed as a more comprehensive condition of young-onset dystonia with obvious effects of *L*-dopa [1]. In the 1993 symposium, data

Table 1. Common features of hereditary progressive dystonia (HPD) and dopa-responsive dystonia (DRD)

1	Inheritance: autosomal dominant
2	Age of onset: childhood, most frequent around 6 years
3	Gender: female more predominant
4	Clinical signs Dystonia beginning from foot Muscular rigidity Hyperreflexia Tremor, but its nature is different between the two conditions
5	Favorable response to levodopa with little dyskinesia as a side-effect
6	PET findings: uptake of 18 fluorodopa in the striatum is normal

were presented to cast more light on the relations between the two conditions, which are summarized below.

Common findings of HPD and DRD are listed in table 1. The mode of inheritance, age of onset, gender preference, clinical signs, PET findings and favorable response to *L*-dopa are common to HPD and DRD. However, there are some dissimilarities between the two conditions.

The natural course is different. DRD develops parkinsonism whereas HPD does not. However, the terminology on parkinsonism is a matter of dispute. Clinical pictures in HPD change little up to 50–60 years of age. A more marked fluctuation of symptoms was reported in HPD, but this may become insignificant with time.

There were differences in the nature of tremor between the two conditions according to presentations at the symposium. In HPD, postural tremor at a rate of 8–10 Hz is observed, while DRD shows resting tremor. This difference may lead to a notion whether the condition develops parkinsonism or not. Both HPD and DRD show tremor at later stages of illness. The preference of the affection for the left side has been claimed in HPD since the early reports, but no side preference was mentioned in DRD.

The new data presented at the symposium concern gene detection, pathology of DRD and basal ganglia mechanisms.

Abnormal gene at 14q was detected both in DRD and HPD. Nygaard et al. [7], and Endo et al. [8, this volume] mapped the gene for DRD/HPD to chromosome 14q in a 22-cM region between D14S47 (14q11.2–14q22) and D14S63 (14q11–14q24.3) by linkage analysis on DRD/HPD families.

The first autopsy of DRD showed a depigmentation of nerve cells in the substantia nigra. The depigmentation of remaining nigral cells was a unique finding in a patient with autosomal recessive disease characterized by marked dystonia with rigidity of young onset [9]. Both siblings of this family with one autopsied case showed quite severe dystonia, rigidity with contracture at joints in the lower extremity with a marked forward-bending posture [10]. Takahashi et al. [11, this volume] reported that the fluorodopa uptake in the striatum of the younger brother, TsO [10] was reduced to an extent similar to young-onset idiopathic parkinsonism while the fluorodopa uptake in DRD was normal. The significance of depigmented nigral cells in dopa-responsive dystonia and juvenile parkinsonism will be an important issue in the future.

Regarding the basal ganglia mechanism of dystonia, Crossman and colleagues reported increased activity in putaminopallidal and pallidosubthalamic pathways, and decreased activity in the subthalamopallidal and pallidothalamic pathways in a primate model of dopamine agonist-induced dystonia [12].

Thus data have been accumulated that indicate HPD and the strictly defined DRD are the same disease. Other than HPD and DRD, young-onset diseases, including the Tourette syndrome and idiopathic dystonia, were discussed comprehensively, promoting the understanding of the age dependency of monoamine-mediated systems in the basal ganglia and their disorders. The aims of the organizers, Drs. Segawa and Nomura, have been successfully achieved, making issues for future study clear and concrete.

References

1. Nygaard TG, Marsden CD, Duvoisin RC: Dopa-responsive dystonia. Adv Neurol 1988;50:377–384.
2. Segawa M (ed): Hereditary Progressive Dystonia with Marked Diurnal Fluctuation. New York, Parthenon Publishing, 1993.
3. Segawa M, Ohmi K, Itoh S, Aoyama M, Hayakawa H: Childhood basal ganglia disease with remarkable response to L-dopa, 'hereditary basal ganglia disease with marked diurnal fluctuation'. Shinryo (Tokyo) 1971;24:667–672.
4. Segawa M, Hosaka A, Miyagawa F, Nomura Y, Imai H: Hereditary progressive dystonia with marked diurnal fluctuation. Adv Neurol 1976;14:215–233.
5. Ouvrier RA: Progressive dystonia with marked diurnal fluctuation. Ann Neurol 1978;4:412–417.
6. Aggarwal R, Bagga A, Kolra V: Progressive dystonia with marked diurnal variation. Indian J Paediatr 1984;51:747–749.
7. Nygaard TG, Wilhelmsen KC, Risch NJ, Brown DL, Trugman JM, Gilliam TC, Fahn S, Weeks DE: Linkage mapping of dopa-responsive dystonia (DRD) to chromosome 14q. Nature Genet 1993;5: 386–391.

8   Endo K, Tanaka H, Saito M, Tsuji S, Nygaard TG, Weeks DE, Nomura Y, Segawa M: The gene for hereditary progressive dystonia with marked diurnal fluctuation maps to chromosome 14q. Basel, Karger, 1995. Monogr Neural Sci, vol 14, pp 120–125.
9   Gibb WRG, Narabayashi H, Yokochi M, Iizuka R, Lees AJ: New pathological observations in juvenile onset parkinsonism with dystonia. Neurology 1991;41:820–822.
10 Yanagisawa N: Juvenile parkinsonism with pallidal posture and spastic paraplegia; in Segawa M (ed): Hereditary Progressive Dystonia with Marked Diurnal Fluctuation. New York, Parthenon Publishing, 1993, pp 205–214.
11 Takahashi H, Snow B, Nygaard T, Yokochi M, Calne D: Fluorodopa PET scans of juvenile parkinsonism with prominent dystonia in relation to dopa-responsive dystonia. Basel, Karger, 1995. Monogr Neural Sci, vol 14, pp 87–94.
12 Mitchell IJ, Luquin R, Boyce S, Clarke CE, Robertson RG, Sambrook MA, Crossman AR: Neural mechanisms of dystonia: Evidence from a 2-deoxyglucose uptake study in a primate model of dopamine agonist-induced dystonia. Mov Disord 1990;5:49–54.

Nobuo Yanagisawa, MD, Department of Medicine (Neurology), Shinshu University
School of Medicine, Asahi 3–1–1, Matsumoto 390 (Japan)

Subject Index

Acetylcholinesterase, expression in embryonic striatal grafts 226–228
Akinesia
  mechanism in Parkinson's disease 205, 212
  surgical correction 212
  types 212
2-Amino-5-phosphonopentanoic acid, effect on mice rotational behavior 161–165

Basal ganglia
  context-dependent oculomotor control 59–61, 67, 68
  descending fibers
    functions 142, 143
    labeling 142
    nigrotectal neurons 143
  disinhibition 60, 67, 68
  dopaminergic neurons 136
  feedback loops connecting the cerebral cortex 163, 164
  higher brain functions 59
  neural circuitry 134–137, 143, 163, 175
  nigrostriatal fiber function 135
  nuclei types 134, 135
  pathways 178, 179
  role
    indirect pathway in oculomotor control 178–186
    locomotion 175
  striatum
    cholinergic neurons 137
    compartments 137
    function 135
    structure in Tourrette's syndrome 42, 43
  thalamocortical circuitry
    cortical projections 142
    limbic striatum 141, 142
    loops 138, 139
    motor circuit 140, 141
    prefrontal areas 131
    subloop organization 141
Basic fibroblast growth factor
  aging effect on levels 241
  effect on dopamine neurons 235, 236, 238, 240
  imaging 237
  immunostaining 236, 237
  levels in Parkinson's disease 236–238, 241
  receptors 240, 241
Brain
  sectioning 148, 179, 189
  sexual differentiation 218–221
  staining 148, 189

Catecholaminergic systems
  effect of sex hormones 217–219, 221
  sexual dimorphism 216, 217, 221

Chronic tic syndrome, twitch movement in REM sleep 69, 70, 72–75
Clonidine, treatment of Tourrette's syndrome 44

Dihydroxyphenylalanine
  sex differences in metabolism 218
  therapy
    psychic response 28
    side-effects 50, 51, 95, 210
Dopa-induced dyskinesia onset 4, 210
  absence in hereditary progressive dystonia 16, 17, 21, 32
  correction by surgery 208, 210
L-Dopamine
  aging effect on neuronal content 241
  degradation 77
  fluorodopamine metabolism 88, 91
  loss in neurodegenerative brain disorders 101, 102, 241
  receptor
    antagonists 78
    expression in neurons 165, 166
    modulation 5
    PET tracers 77, 78
    signal transduction 219
    types 77
  reuptake 77
  rostrocaudal gradient 101, 103, 104, 106, 107, 148, 154
  synthesis 77
  uptake rates
    rodents and primates 147
    sexual differences 217, 218
Dopaminergic fibers
  bilaminar innervation pattern 154, 155
  cause of destruction 235
  cingulate cortex 156
  cortex 157, 158
  dopamine uptake 147, 154
  electron microscopy 151, 153, 156
  immunostaining 148–151, 156
  light microscopy 152, 153
  medial wall of cerebral hemisphere 149, 151, 156
  rostrocaudal gradient of dopamine 148, 154

Dopa-responsive dystonia, *see also* Hereditary progressive dystonia, Juvenile parkinsonism
  age of onset 1, 25–28, 37, 38, 111
  anticholinergic therapy 105
  chromosome linkage 2, 21
  clinical characteristics 10, 18, 26–28, 37, 38, 102–104, 110–112, 115, 245
  D_2 receptor
    density 98, 99
    raclopride binding 95–100
    regulation 4
  diagnosis 115, 245
  direct projection of the striatum 14, 15
  diurnal fluctuation 38, 56, 111, 116, 121
  dopa response 25, 27, 28, 37, 87, 95, 99, 103, 104, 110–113
  dopamine
    induction of dyskinesia 121
    levels 3, 104, 105
    rostrocaudal gradient 104, 106, 107
    uptake rate 18, 91–93, 95, 247
  gene locus 114, 115, 124, 247
  heredity 37, 38, 113
  incidence 110
  linkage analysis 114, 247
  motor disorders 56
  nosology 109
  pathophysiology 105–107, 114
  PET findings 79, 80, 88–93, 114
  serotonin metabolites in cerebrospinal fluid 114, 115
  sex differences 113
  tremor rate 246
  tyrosine hydroxylase activity 105, 116
  voluntary saccades 66
Dystonia-parkinsonism
  age of onset 2
  dopamine uptake rate 19
  heredity 2
  nosology 27, 79, 80
  PET findings 79–81
  sex distribution 2, 20

Early-onset dystonia
  age of onset 127
  clinical features 127

ethnicity 127, 129
gene
  cloning 130
  locus 126–129
  markers 129
  mutations 128, 129
  screening 130
heredity 126, 130, 131
treatment 128
Early-onset idiopathic parkinsonism, see Young onset Parkinson's disease
Electromyogram
  briskness of involuntary movement 54
  correspondence to clinical observation 53
  integrated relationship to isometric contraction tension 51
  patterns
    athetosis-dystonia 52
    ballism 52
    basal ganglia diseases 53
    chorea 52
    tremor 52
  surface, for involuntary movement disorders 50–53
Embryonic striatal grafts
  compartmentalization 226, 227
  functional integration with host brain 231, 232
  immunostaining 228, 229
  modularity of neurochemical expression 226, 227
  neuropeptide regulation 231
  origin of neurons with nonstriatal phenotype 228–230
  reconstruction
    cholinergic systems 230, 231
    compartmentalization 230
    dopaminergic systems 230, 231
    transcription factor induction 231
Estrogen, effect on brain development 217–221
External pallidum neurons
  directional cells 186
  electrode placement 179
  inhibition 178

latency of excitatory response 184
tonic response 183
visuo-oculomotor control 181–186

Gilles de la Tourette syndrome, see Tourrette's syndrome
Globus pallidus
  lesions 208
  projections 213
  role in Parkinson's disease 204, 205, 213
Glutamate, effect on motor functions 160, 163, 164
Gross movement, hereditary progressive dystonia 12, 14, 16
GTP cyclohydrolase I
  defects in hereditary progressive dystonia 6
  gene locus 6
  sex differences in activity 6

Hereditary progressive dystonia
  absence of action dystonia 16, 18
  age of onset 2, 17–19, 25–28, 120
  aging effects on tremor frequency 57
  chromosome linkage 2, 21
  clinical characteristics 10, 11, 18–20, 26–28, 56, 120, 246
  diagnosis 245
  direct projection of the striatum 14, 15
  diurnal fluctuation 10, 12, 17, 18, 21, 111, 120, 121
  dopa response 12, 17, 21, 25, 27, 28, 87, 120
  dopamine
    levels 14, 20, 21
    uptake rate 18, 91–93
  gene locus 6, 121, 123, 124, 247
  GTP cyclohydrolase I defects 6
  heredity 20, 122
  incidence 110
  linkage analysis 121, 122, 124, 247
  motor disorders 56
  movements in REM sleep
    gross 12, 14, 16
    twitch 12, 13, 70, 72–75

Subject Index   251

Hereditary progressive dystonia
(continued)
  nigrostriatal dopamine neuron
    activity 4
    lesion 12, 14
  nosology 245, 246
  pathophysiology 2, 3, 11, 121
  PET findings 88–93
  preservation
    D_2 receptor 17, 20
    pallidofugal pathway 17, 18
  serotonin metabolites in cerebrospinal
    fluid 121
  sex distribution 2, 19, 20, 27
  tremor rate 246
  tyrosine hydroxylase activity 14, 20, 21
  voluntary saccades 15, 16, 66
Huntington's disease
  age-related motor symptoms 50, 51
  dystonia development 54
  EMG 53, 54
  gene 225
  heredity 225
  pathology 225
  treatment
    embryonic striatal graft 232
    intracerebral neural grafting 225, 226

Juvenile parkinsonism
  age of
    death 31
    onset 2, 19, 25–28, 30–32, 88, 110,
      111, 113
  aging effects on tremor frequency 56, 57
  clinical characteristics 19, 26–28, 55,
    88, 111, 112, 115
  D_2 receptor regulation 4
  diagnosis 88, 115
  diurnal fluctuations 112, 116
  dopa response 25–27, 31, 89, 91, 113
  dopamine
    levels 3, 29
    uptake rate 91–93
  heredity 113
  linkage analysis 115, 116
  motor disorders 55
  pathology 29–31, 114

  PET findings 88–93, 114
  serotonin metabolites in cerebrospinal
    fluid 114
  sex distribution 2, 6, 19, 113
  types 3, 29–33
  tyrosine hydroxylase activity 29
  voluntary saccades 66

Lewy bodies, association with Parkinson's
  disease 33, 80, 88, 110
Locomotion
  context-dependent adaptation of move-
    ment 174–176
  generation of rhythm 168
  origin of drive signals 168
  role
    basal ganglia 175
    higher-order neurons 169, 170
    reticulospinal neurons 169, 170,
      174–176

1-Methyl-4-phenyl-1,2,3,6-tetrahydro-
  pyridine
  effect on response to cortical stimula-
    tion 195
  models of Parkinson's disease
    cat 195
    primate 62, 191, 195
  potency 62, 235
1-Methyl-4-phenylpyridium
  effect
    behavior in cats 190, 191
    brain histology 192
    corticostriatal signal transmission
      192, 194, 195
    inactivation of nigrostriatal system 188,
      189, 195
  infusion in cats 189
MK-801
  effect on motor stimulation 165
  NMDA antagonist 160
  target sites 160
Motor cortex, electrical stimulation 189

Nigrostriatal dopamine neurons
  aging effects 1, 5, 12
  circadian fluctuation in terminal 12
  destruction by MPP ion 188, 189
  effect
    dopamine depletion on corticostriatal
      projections 192–195
    unilateral destruction 188
  function
    chronic tic syndrome 69, 74–76
    dopa-responsive dystonia 88–93
    hereditary progressive dystonia
      88–93
    juvenile parkinsonism 88–93
    Parkinson's disease 88–93
  lesions 1, 2, 14
  measurement of activity 74, 88
  neuropeptide regulation 231
  PET imaging 88–93
  receptor supersensitivity 3, 16, 17, 75,
    76, 99
  tyrosine hydroxylase activity 92
Noradrenaline, sexual differences in
  uptake 218
Norepinephrine, deficiency in Parkinson's
  disease 211

Obsessive-compulsive disorder
  PET findings 81–83
  Tourrette's syndrome association 41,
    44, 47, 81
  treatment 44
Opioid receptor, PET tracers 78

Pallidoreticular pathway, see Posteroventral pallidotomy
Parkinson's disease
  age of onset 25–28, 110
  aging effects on tremor frequency 56, 57
  akinesia mechanism 205, 212
  basic fibroblast growth factor levels
    236–238, 241
  clinical characteristics 26–28, 79
  dopa response 27, 28, 37, 99, 110, 211
  dopamine
    age-related motor side effects 50, 51
    levels 37, 101, 102, 235

receptor density 79, 98, 99
    rostrocaudal gradient 101, 103,
      107
    uptake rate 18, 90, 92, 93
  dystonia onset 1, 2
  EMG 53, 56
  nosology 30–32
  PET findings 77, 79–81, 88–93
  raclopride binding to D_2 receptors
    95–100
  sex distribution 2, 5, 221
  surgical intervention 197–205, 208
  tonic overactivity in medial segment of
    the globus pallidus 204, 205
  treatment by intracerebral neural
    grafting 226
  voluntary saccades 16, 60–63, 66
Polysomnography
  hereditary progressive dystonia 12, 16
  parameters 70
Positron emission tomography
  dopa-responsive dystonia 79–81, 88–93,
    114
  dopamine uptake rate determination 88,
    91, 95
  dystonia-parkinsonism 79–81
  fluorine-18 detection 91, 95
  juvenile parkinsonism 88–93, 114
  Parkinson's disease 77, 79–81, 88–93
  Tourrette's syndrome 81–83
  tracers
    dopaminergic
      postsynaptic 78, 95, 96
      presynaptic 77, 78, 88
    opioid 78
Posteroventral pallidotomy
  correction of gait 211, 212
  effect
    descending pallidoreticulospinal
      pathway 203–205
    spontaneous neural activity 201,
      202
  indications 198
  mechanism of effects 197, 205, 211
  outcome
    dystonia 201, 202
    Parkinson's disease 200–203

Posteroventral pallidotomy (continued)
  surgical procedure
    anesthesia 199
    electrical stimulation of target area 199
    measurement of neuronal activity 199
    MRI 198–200
    radiofrequency thermocoagulation 200
Posture
  control system 174
  role of vermis 170

Quinpirole, effect on mice rotational behavior 162, 163

Raclopride
  binding to D_2 receptors
    dopa-responsive dystonia 95–100
    Parkinson's disease 95–100
  effect on mice rotational behavior 162, 163
  PET 95–100
  uptake
    effects of aging 97, 99
    index 96
Rapid eye movement, see also Twitch movement
  measurement 70
  modulation 75, 76
Reticulospinal neurons
  branching pattern 170, 171
  convergence
    corticoreticular fibers 172, 173
    fastigioreticular fibers 172, 173, 175
  feedback mechanisms 175
  higher-order control of reflex 174–176
  role in locomotion 169, 170
  termination mode 171, 172
Rigidity, mechanism 208, 209

Saccades, see Voluntary saccades
SCH 23390, effect on mice rotational behavior 162, 163
Segawa's syndrome, see Dopa-responsive dystonia

SKF 38393, effect on mice rotational behavior 162, 163, 165
Stereotaxic surgery, see also Posteroventral pallidotomy, Ventrolateral nucleus thalamotomy
  mice 161
  treatment
    akinesia 212
    difficulty of gait 211
    dystonia 201, 202
    Parkinson's disease 197, 200–203, 205, 208, 210, 211
    postural instability 211

Testosterone, effect on brain development 217–221
Torsion dystonia, see also Dopa-responsive dystonia
  aging effects on tremor frequency 57
  classification 36, 37, 55
  direct projection of the striatum 75
  EMG 55
  gene locus 39
  heterogeneity 39
  motor disorders 55
Tourette's syndrome
  age of onset 2, 41
  behavioral problems 41, 44, 47, 81
  clonidine therapy 44
  D_2 receptor regulation 4
  etiology 3, 41, 45, 81, 82
  heredity 2, 45, 46
  nigrostriatal dopamine neuron activity 4
  pathogenesis role
    basal ganglia structure 42, 43
    brain metabolism 42
    dopaminergic systems 43, 44, 82
    maternal stress 46, 47
    noradrenergic system 44
    opioid system 83
    prenatal factors 46
    testosterone 45
  PET findings 81–83
  risk factors 46, 47
  sex distribution 2, 20, 41, 45, 220
  twitch movement in REM sleep 70, 74, 75

Tremor, mechanism 208, 209
Twitch movement
  amplitude 74, 75
  defined 70, 74
  REM sleep events
    age dependence 72, 74
    chronic tic syndrome 69, 70, 72–75
    hereditary progressive dystonia 12, 13, 70, 72–75
    normal 71, 73
    Tourrette's syndrome 70, 74, 75
Tyrosine hydroxylase
  aging effect on activity 12, 74, 121
  circadian fluctuation 12, 20, 21
  levels
    dopa-responsive dystonia 105
    hereditary progressive dystonia 12, 20, 21
    juvenile parkinsonism 29
  nigrostriatal dopamine neuron distribution of activity 92
  tetrahydrobiopterin cofactor deficiency in disease 105, 114

U 991194A, stimulation of locomotor activity 164

Ventral intermediate nucleus
  effect of lesions 208, 209
  projections 208, 209
Ventrolateral nucleus thalamotomy, treatment of Parkinson's disease 197, 205, 208, 210, 211
Voluntary saccades
  abnormalities
    dopa-responsive dystonia 66
    hereditary progressive dystonia 15, 16, 66
  juvenile parkinsonism 66
  Parkinson's disease 16, 60–63, 66
    age variation 5
    aging effects 63–65
    amplitude 64, 66
    control by basal ganglia 59–61, 67, 68
    direction of response 182, 183, 186
    external pallidum neuron control 181–186
    initiation 61, 63
    latency 64, 66, 183, 184
    localization of visuo-oculomotor cells 185
    memory guidance 60–63
  monkeys
    dopamine-deficient model 62
    effect of muscimol injection 67
    latency 183, 184
    testing 60, 179–181
  peak velocity 65, 66

Wilson's disease, age-related motor symptoms 50, 51

X-linked dystonia with parkinsonism
  gene locus 126, 127
  heredity 126

Young onset Parkinson's disease
  age of onset 32, 87, 90
  diagnosis 88
  dopamine uptake rate 18, 91–93
  incidence 87
  nosology 87
  pathology 88